# gcse geography

## edexcel B

Series editor: Bob Digby

Cameron Dunn
Dave Holmes
Sue Warn
Dan Cowling
Russell Chapman

**OXFORD**
UNIVERSITY PRESS

**OXFORD**
UNIVERSITY PRESS

Great Clarendon Street, Oxford OX2 6DP

Oxford University Press is a department of the University of Oxford.
It furthers the University's objective of excellence in research,
scholarship, and education by publishing worldwide in

Oxford   New York

Auckland  Cape Town  Dar es Salaam  Hong Kong  Karachi
Kuala Lumpur  Madrid  Melbourne  Mexico City  Nairobi
New Delhi  Shanghai  Taipei  Toronto

With offices in
Argentina  Austria  Brazil  Chile  Czech Republic  France  Greece
Guatemala  Hungary  Italy  Japan  Poland  Portugal  Singapore
South Korea  Switzerland  Thailand  Turkey  Ukraine  Vietnam

Oxford is a registered trade mark of Oxford University Press
in the UK and in certain other countries

© Oxford University Press 2009

Authors: Bob Digby, Cameron Dunn, Dave Holmes, Sue Warn, Dan
Cowling, Russell Chapman

The moral rights of the authors have been asserted

Database right Oxford University Press (maker)

First published 2009

British Library Cataloguing in Publication Data

Data available

ISBN 978-0-19-913490-8

10 9 8 7 6 5

Printed in Singapore by KHL Printing Co. Pte Ltd

Paper used in the production of this book is a natural, recyclable product made
from wood grown in sustainable forests. The manufacturing process conforms to
the environmental regulation s of the country of origin.

**Acknowledgements**
The publisher and authors would like to thank the following for permission to
use photographs and other copyright material:
p8 Jean-Claude Revy, Ism/Science Photo Library; p8 (inset) Photolibrary; p9 IODP/
JAMSTEC; p11 Winfried Stricker/Fotolia; p17 Roger Ressmeyer/Corbis; p19 Sayyid
Azim/Associated Press; p21 Koichi Kamoshida/Staff/Getty Images News/ Getty
Images AsiaPac/ Getty Images; p23 Andy Wong/Associated Press; p24 Bryan &
Cherry Alexander Photography; p26t Dave Harlow/U.S. Geological Survey; p26b
Courtesy of SOHO/[instrument] consortium. SOHO is a project of international
cooperation between ESA and NASA; p28 Wolfgang Kaehler/Corbis; p29 The
Print Collector/Alamy; p34 Original photograph taken in 1928 of the Upsala
Glacier: ©Archivo Museo Salesiano/De Agostini Recent comparison image taken
in 2004. ©Greenpeace/Daniel Beltrá; p35 Warren Faidley/Corbis; p36 Skyscan/
Corbis; p37 Robert Dear/Associated Press; p38 Jacques Descloitres, MODIS
Land Science Team/NASA; p43t Juan Carlos Munoz/Photolibrary; P43m Len Jr/
Photolibrary; P43b Pavel Filatov/Alamy; p46bl Bill Bachman/Alamy; p46br Per-
Anders Pettersson/Contributor/Getty Images News/Getty Images AsiaPac/Getty
Images; p48tl John Mosseso; p48tr Dave Watts/Alamy; p48ml Denis Scott/Corbis;
p48mr Tony Camacho/Science Photo Library; p48b Gerard Lacz/Photolibrary;
p50t Photolibrary; p50ml OUP Picture Bank; p50mr AlaskaStock/Photolibrary;
p50b Ho New/Reuters; p51 Mike Jubb/Photographers Direct; p55 Kevin Schafer/
Corbis; p56t NASA Images/Alamy; p56m Fallsview/Dreamstime; p56b Paul Hardy/
Corbis; p57t Craig Tuttle/Corbis; p57b Stephanie Cabrera/Zefa/Corbis; p65
Margaret Courtney-Clarke/Photolibrary; p66 Balkanpix.com/Rex Features; p70
Pascal Deloche/Godong/Corbis; p72tl Ashley Cooper/Alamy; p72tr Graeme Ewens/
Science Photo Library; p74 Cameron Dunn; p75 Flasshary; p79t John Farmar/
Ecoscene; p79m Simmons aerofilms; p81 Mark Lynas/Still Pictures; p83 Graeme
Ewens/Science Photo Library; p84 John Boyes/World of Stock; p85 Bob Digby;
p88 Bob Digby; p90t Bob Digby; p90b Bob Digby; p92 Bob Digby; p94t photo:

webbaviation.co.uk; p94b Neil Holmes Freelance Digital/Alamy; p95 Andrew Holt/
Alamy; p96 Kelly Dorset: www.flickr.com/delanthear/; p97 Wendy North; p100
Dave Thompson/Associated Press; p101t Stringer UK/Reuters; p101b Rimmington
CE/Photographers Direct; p104l Rotherham Investment and Development
Office; p104r Geogphotos/Alamy; p107 Paweł Borowka/Dreamstime; p111 Andre
Maslennikov/Dinodia; p116 Ross Frid/Alamy; p117 Mark Webster/Photolibrary;
p121 Ho New/Reuters; p122 Bob Digby; p124 Aldo Pavan/Corbis; p125t Martin
Harvey/Corbis; p125m Martin B Withers/FLPA; p125b Eric and David Hosking/
Corbis; p126 Paul Mayall/Photographers Direct; p127 Bob Digby; p128m Richard
McDowell/Alamy; p128b Winfred Wisniewski; Frank Lane Picture Agency/Corbis;
p129t Susanna Bennett/Alamy; p129m Bob Digby; p130bl Joern/Shutterstock;
p130br Susanna Bennett/Alamy; p135 Jeremy Hartley/Panos Pictures; p140
Damir Sagolj/Reuters; p141 Elena Yakusheva/Shutterstock; p146 Yuriko Nakao/
Reuters; p147 John Warburton-Lee Photography/Alamy; p148 Keith Dannemiller/
Corbis; p150 Stringer Shanghai/Reuters; p154 Danny Lehman/Corbis; p157
Ariana Cubillos/Associated Press; p158 Hulton Archive/Stringer/Getty Images;
p160t Tim Graham/The Image Bank/Getty Images; p160b Justin Sullivan/Staff/
Getty Images News/Getty Images North America/Getty Images; p161l Martin
Wright/Still Pictures; p161b Ace Stock Limited/Alamy; p162t worldmapper.org;
p162m worldmapper.org; p162b Ben Smith/Shutterstock; p164 Antoine Gyori/
AGP/Corbis; p166l Rune Hellestad/Corbis; p166r Content Mine International/
Alamy; p167 Jeff Haynes/AFP; p168 Helene Rogers/Alamy; p169 Diaphor La
Phototheque/Photolibrary; p172 10radio.org; p173 David; p177t Mark Edwards/
Still Pictures; p177m Mark Edwards/Still Pictures; p179 Tim Ayers/Alamy; p186
John MacNeill; p187 Julian Makey/Rex Features; p195 Verityjohnson/Shutterstock;
p195m Mick Rock/Cephas Picture Library/Photolibrary; p196 newcastlephotos.
blogspot.com/; p199 Studio18; p200 Robert Brook/Alamy; p201l Andrew Fox/
Alamy; p201r Conrad Elias/Alamy; p203 Marion Kaplan/Alamy; p204l Jon Arnold
Images Ltd/Alamy; p204r Kees Metselaar/Alamy; p205 Mike Hughes/Alamy; p209l
Grant Smith/Photolibrary; p209r Dapei Xin/EPA; p210 Robert Harding Picture
Library Ltd/Alamy; p212t VIEW Pictures Ltd/Alamy; p212b David Forster/Alamy;
p213 Martin Jones/Ecoscene; p214t Peter Jordan/Alamy; p214m David Goddard/
Contributor/Getty Images News/Getty Images Europe/Getty Images; p215ml Sam
Toren/Alamy; p215mr Ashley Cooper/Alamy; p216 www.CartoonStock.com; p217
Nick Turner/Alamy; p217 (inset) Ray Roberts/Alamy; p218m Geogphotos/Alamy;
p218b Lorraine Brooks, Dedham Vale Society; p220 Mark Boulton/Alamy; p222
Nigel Pavitt/John Warburton-Lee Photography/Photolibrary; p223 Jorgen Schytte/
Still Pictures; p224t Penny Tweedie/Alamy; p224b F1online digitale Bildagentur
GmbH/Alamy; p225 Peter Davey/FLPA; p226t Arjen van de Merwe/IFMSLP; p226b
Nigel Pavitt/John Warburton-Lee Photography; p227 Dave Hoisington/CIMMYT;
p229 Bob Digby; p230 World Pictures/Alamy; p231 Ashley Cooper/Alamy; p233l
John Malley; p233r John Malley; p235 Davlan/Alamy; p236 Age Fotostock/
Photolibrary; p240 IndiaPicture; p241 Prakash Singh/AFP; p243 Shamik Mehta;
p244t Anand S; p244b Kamal Kishore/Reuters; p246t Martin Wright/Still Pictures;
p246b Robert Nickelsberg/Contributor/Time & Life Pictures/Getty Images; p247
Jayanta Shaw/Reuters; p248t John Henry Claude Wilson/Photolibrary; p248b Jiri
Rezac/Alamy; p250 Mark Goebel/Photographers Direct; p255 Vincent Du/Reuters;
p259 Beatrix Stampfli/Associated Press; p262 Jeff Koterba/King Features; p263
Stringer India/Reuters; p264 Associated Press; p265 Alex Segre/Alamy; p267m
John Warburton-Lee/John Warburton-Lee Photography; p267b Chad Ehlers/Jupiter
Images; p269t Kkaplin/Shutterstock; p269b Sergio Pitamitz/Corbis; p270 Bob
Digby; p272t Bob Digby; p272b Bob Digby; p273 Bob Digby; p275tl Bob Digby;
p275tr Bob Digby; p275ml Bob Digby; p275mr Bob Digby; p275bl Bob Digby;
p275br Bob Digby; p276t Bob Digby; p276b Bob Digby; p277 Bob Digby;  p284t
James Jagger/Photographers' direct; p274tm David hills/Istockphoto; p284mb
Nigel Francis Ltd/Robert Harding; p284b David Hughes/Robert Harding; p287 t & c
Bob Digby; p290b Bob Digby; p296 samc / Alamy; p297 Bob Digby;p301 Bob Digby

The graph on page 261 was drawn using data published by A.T. Kearney.

# Contents

*continued ...*

# Unit 1 Dynamic planet

➕ In this section you will learn about the theme of unit 1 – how our dynamic planet links together into one physical system which supports us, but can also bite back.

> **+ Dynamic** means constantly changing.

## Supporting life

As far as we know, ours is the only planet that supports life. The Earth is a survivor; it is 4.6 billion years old! The planet has slowly evolved into a complex life-support system that keeps humans, plants and animals alive. Four spheres – which really means layers – provide the vital services.

| Sphere | What is it? | Why do we need it? |
|---|---|---|
| Atmosphere | The layer of gases which make up the air around us. | • Oxygen, needed for animals to breathe.<br>• Carbon dioxide for plant growth.<br>• The greenhouse effect to keep the planet warm.<br>• Weather and climate. |
| Hydrosphere | The layer of water – seas, rivers, lakes, groundwater and ice – on the Earth's crust. | • Water for animal and plant life.<br>• Water cycle which moves water around the planet. |
| Biosphere | The very thin layer of living things – plants, animals and humans – living on the crust. | • Plants and animals provide us with food.<br>• Many living things can also be used to make medicines and as fuel. |
| Geosphere | The rocks of the Earth's crust, and deeper into the Earth towards the core. | • The Earth's core makes a magnetic field protecting us from space radiation.<br>• We use rocks, minerals and fossil fuels as resources for fuel, building, and much more. |

The four spheres are all linked together. The living layer, the biosphere, is the most important. It is also unique, because other planets do not have a biosphere. The biosphere interacts with the other spheres in a number of ways:

- Plants and animals turn rock and sediment (the geosphere) into soil.
- Plants are part of the water cycle (the hydrosphere), because of transpiration.
- Plants take in carbon dioxide, and give out oxygen, keeping the air (the atmosphere) breathable.

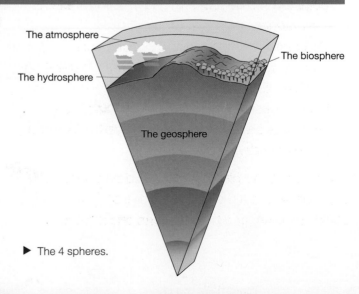

▶ The 4 spheres.

## Making life harder

The Earth does a very good job of supporting life, but sometimes it can harm humans. One part of the Earth is very restless – the geosphere. It can produce earthquakes, volcanoes and tsunami, which cause disasters when humans get in the way. The hydrosphere can also turn nasty too, when flooding threatens lives and homes. The atmosphere produces hurricanes, storms and tornadoes, which can cause havoc. The 'spheres' make our lives possible, but they can also threaten us.

**On your planet**

+ There may be between 2 and 80 million species in the world, but some scientists think half of them will be extinct in 100 years.

## Planet in peril?

Humans are changing and damaging all four of our life-support spheres. This is surprising, because we depend on them so much. Damaging the spheres places the planet in danger, and might even threaten the future of the human race.

- Humans have polluted the atmosphere, increasing the amount of carbon dioxide in it by 40%.
- We have used up most of the fossil fuels buried in the geosphere, and will soon need to look for new energy sources.
- Deforestation, farming and pollution are causing extinctions in the biosphere faster than ever before.
- Human use of water means that many rivers and lakes are polluted. Water supplies in some places are drying up.

The rest of this unit explores the four spheres and explains what they are like. It looks at how they affect humans, and how humans affect them.

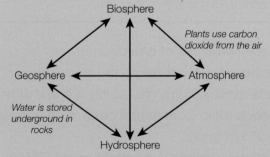

▲ Damaging the spheres…

## your questions

1 Write down a definition of each of the four spheres.
2 How have you used the four spheres today? Write down one way you have used each sphere.
3 Draw a large version of this diagram. It will need to be a whole page.

Biosphere

*Plants use carbon dioxide from the air*

Geosphere — Atmosphere

*Water is stored underground in rocks*

Hydrosphere

See if you can describe the links between the four spheres. Two have been done for you.
4 State four ways in which humans might be harmed by the spheres.
5 How could your life be made worse, if humans keep damaging the spheres?

➕ **In this section you will learn about the interior of the Earth.**

## Journey to the centre

No one has ever seen the inside of the Earth. The deepest we have managed to get is 3.5 km in the Witwatersrand gold mine in South Africa. Miners would have to drill another 6365 km to reach the centre of the Earth.

What we know about the Earth's interior comes from direct and indirect evidence. We can get direct evidence from the Earth's surface. Indirect evidence, like earthquakes and material from space, also helps us to understand the Earth.

## Examining the crust

The diagram on the right shows the structure of the Earth. The crust is the surface of the Earth. It is a rock layer forming the upper part of the **lithosphere**. The lithosphere is split into **tectonic plates**. These plates move very slowly, at 2-5 cm per year, on a layer called the **asthenosphere**.

> ➕ The **lithosphere** is the uppermost layer of the Earth. It is cool and brittle. It includes the very top of the mantle and, above this, the crust.

There are two types of crust:
- **Continental crust** forms the land. This is made mostly of granite, which is a low density igneous rock. Continental crust is on average 30-50 km thick.
- Under the oceans is **oceanic crust**. This is much thinner, usually 6-8 km thick. It is also denser and made of an igneous rock called basalt.

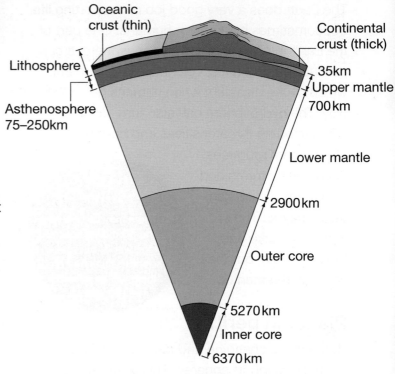

Oceanic crust (thin)
Continental crust (thick)
Lithosphere
Asthenosphere 75–250km
35km
Upper mantle
700km
Lower mantle
2900km
Outer core
5270km
Inner core
6370km

▲ **Geologists** (scientists who study the Earth and its structure) think the Earth has a layered structure, like an onion.

▲ Geologists collect samples of the crust and compare the two types: polished granite from continental crust (left) and basalt from oceanic crust (right).

## The asthenosphere and mantle

The movement of the tectonic plates is evidence that there is a 'lubricating' layer underneath the lithosphere. This is the asthenosphere. You might think that this layer would be a liquid. But if it was a liquid, the heavy tectonic plates would sink into it. Geologists think the asthenosphere is partly molten rock and partly solid rock. It may be like very thick, dense, hot porridge!

The asthenosphere is in the top layer of the **mantle**. The mantle is the largest of the Earth's layers by volume, and is mostly solid rock. We know this because sometimes you can see the top of the mantle attached to an overturned piece of crust.

Earthquake waves tell us about the physical state of the Earth. They speed up, change direction or stop when they meet a new layer in the Earth. Some waves travel easily through the crust, mantle and inner core, but not through the outer core. This suggests that the outer core has a different physical state and may be liquid, not solid.

▲ The Japanese have built a 57 000-tonne international scientific drilling ship, the *Chikyu*, which will drill 7 km through the oceanic crust to reach the mantle – further than anyone has ever gone before.

## Clues from space

At a depth of 2900 km, we can't sample the Earth's core. Geologists think that it is metal – mostly nickel and iron. Evidence for this comes from **meteorites**, which are fragments of rock and metal that fall to Earth from space. Most come from the asteroid belt between Mars and Jupiter. Meteorites come in several types:

- Stony meteorites, with a similar composition to basalt.
- Stony-iron meteorites, containing a lot of the mineral olivine.
- Iron meteorites, which are solid lumps of iron and nickel.

These meteorites may be fragments of the lithosphere, mantle and core of a shattered planet. Iron meteorites may show that the Earth's core is made up of iron and nickel.

| Layer | | Density (grams / cm³) | Physical state | Composition | Temp (°C) |
|---|---|---|---|---|---|
| Lithosphere | Continental crust | 2.7 | Solid | Granite | Air temp - 900 |
| | Oceanic crust | 3.3 | Solid | Basalt | Air temp - 900 |
| Mantle | Asthenosphere | 3.4-4.4 | Partially molten | Peridotite | 900-1600 |
| | Lower mantle | 4.4-5.6 | Solid | | 1600-4000 |
| Core | Outer core | 9.9-12.2 | Liquid | Iron and nickel | 4000-5000 |

### your questions

1. Draw a cross-section of the Earth. You need a large circle divided into layers. Label the layers with details of density, temperature and physical state.
2. What are the main differences between the lithosphere and asthenosphere?
3. Go through all of the evidence we have on the Earth's structure. Draw a table and classify the evidence into 'direct' and 'indirect'.

➕ In this section you will learn how the Earth's core drives the process of plate tectonics.

### Hot rocks

Inside the Earth it is hot. We know this because of:

- molten lava spewing from active volcanoes
- hot springs and geysers.

Heat from inside the Earth is called **geothermal** ('Earth-heat'). The heat is produced by the **radioactive decay** of elements such as uranium and thorium in the core and mantle. This raises the core's temperature to over 5000 °C.

> ➕ Some elements are naturally unstable and radioactive. Atoms of these elements release particles from their nuclei and give off heat. This is called **radioactive decay**.

### On your planet

➕ In a few billion years the core and mantle will stop convecting, because the radioactive heat will run out. This will shut down our magnetic field, and life on Earth will be destroyed by radiation from space.

The inner core is so deep and is under such huge pressure that it stays solid. The outer core is liquid because it is under lower pressure. As heat rises from the core, it creates **convection currents** in the liquid outer core and mantle (see below). These vast mantle convection currents are strong enough to move the tectonic plates on the Earth's surface. The convection currents move about as fast as your fingernails grow. Radioactivity in the core and mantle is the engine of plate tectonics.

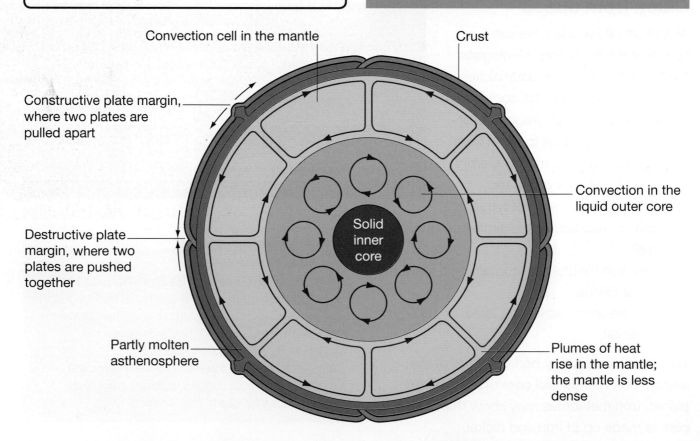

Convection cell in the mantle

Crust

Constructive plate margin, where two plates are pulled apart

Convection in the liquid outer core

Destructive plate margin, where two plates are pushed together

Solid inner core

Partly molten asthenosphere

Plumes of heat rise in the mantle; the mantle is less dense

## Plumes

The parts of convection cells where heat moves towards the surface are called **plumes**. These are concentrated zones of heat. In a plume, the mantle is less dense. Plumes bring **magma** (molten rock) to the surface. If magma breaks through the crust, it erupts as **lava** in a volcano.

- Some plumes rise like long sheets of heat. These form **constructive plate boundaries** at the surface.
- Other plumes are like columns of heat. These form **hot spots**. Hot spots can be in the middle of a tectonic plate, like Hawaii and Yellowstone in the USA.

**On your planet**

✦ Did you know that the Earth's magnetic field sometimes 'flips', so north becomes south and south becomes north?

## Magnetic field

The Earth is surrounded by a huge invisible magnetic field called the **magnetosphere**. This is a force field. It protects the Earth from harmful radiation from space and the sun (see right).

The Earth's magnetic field is made by the outer core. As liquid iron in the outer core flows, it works like an electrical dynamo. This produces the magnetic field.

▶ Sometimes you can see the magnetosphere. The northern lights (aurora borealis) form when radiation from space hits the magnetosphere and lights up the sky.

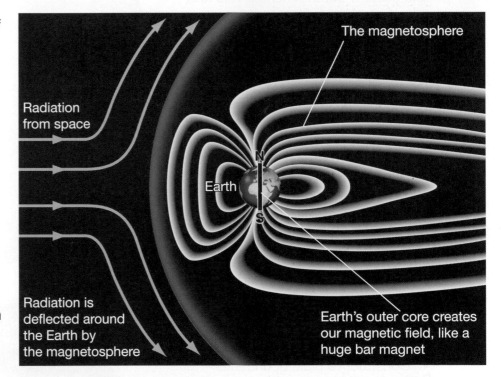

The magnetosphere

Radiation from space

Earth

Radiation is deflected around the Earth by the magnetosphere

Earth's outer core creates our magnetic field, like a huge bar magnet

---

### your questions

1 Why is the centre of the Earth at over 5000 °C?
2 Look at the diagram of the Earth's convection currents. What happens to the crust at the top of one of the convection currents in the mantle?
3 Why do tectonic plates move?
4 Find out why Mars is a 'dead' planet with no plate tectonics. You might like to visit this website: http://mars.jpl.nasa.gov/classroom/students.html

5 Each set of words below has an odd one out. For each:
  a say which is the odd one out
  b explain your choice.
    - inner core, outer core, mantle, crust
    - convection, northern lights, plume, cell, current
    - lava, uranium, magma, geyser

In this section you will find out how the Earth's tectonic plates have moved in the past, and about plate boundaries.

## Pangea, the supercontinent

Scientists know that the continents were once all joined together. They formed a supercontinent called **Pangea**. The diagram on the right shows the position of the continents 250 million years ago. Identical rocks and fossils dating from this time have been found in West Africa and eastern South America. This tells us that Africa and South America were once joined. Pangea started to split apart about 200 million years ago. Since then, plate tectonics has moved the continents to the positions they are in today.

## Moving plates

Today, the Earth's lithosphere is split into 15 large **tectonic plates** and over 20 small ones. These are like the patches that make up a football. The plates move very slowly on the asthenosphere. Where two plates meet, there is a **plate boundary**. There are three types of plate boundary, as shown on the map opposite:

- Constructive plate boundaries – formed when two plates move apart.
- Destructive plate boundaries – formed when two plates collide.
- Conservative plate boundaries – formed when two plates slide past each other.

Plate boundaries are where the 'action' is. Most earthquakes and volcanoes are found on plate boundaries.

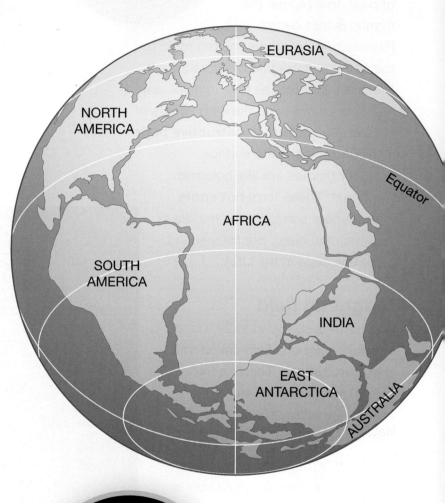

*On your planet*

+ Every year, the distance between the UK and the USA grows by about 2 cm. This is because the mid-Atlantic constructive plate boundary creates new oceanic crust.

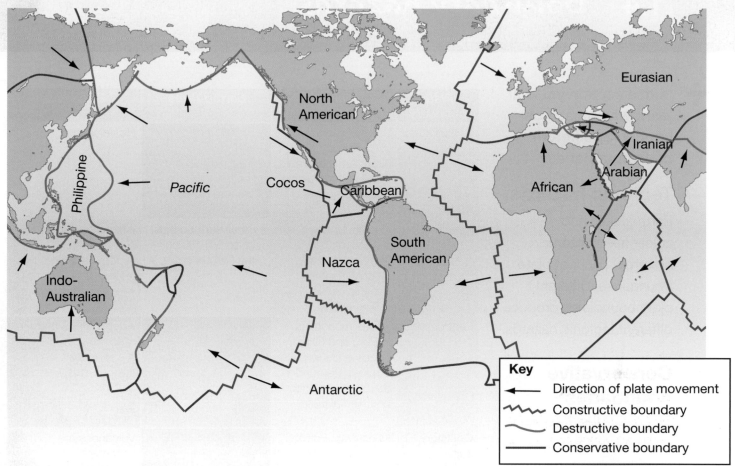

**Key**

| | |
|---|---|
| ←— | Direction of plate movement |
| ∿∿ | Constructive boundary |
| ～ | Destructive boundary |
| —— | Conservative boundary |

## Crust: old and new

Most continental crust is 3-4 billion years old. The oldest oceanic crust is only 180 million years old. Why the age difference?

New oceanic crust forms constantly at constructive plate boundaries:

- Convection currents bring magma up from the mantle.
- The magma is injected between the separating plates.
- As the magma cools, it forms new oceanic crust.
- The plates continue to move apart, allowing more magma to be injected.

Old oceanic crust is destroyed by **subduction** at destructive plate boundaries – it is 'recycled' by the Earth. Continental crust was formed billions of years ago, and has not formed since. It is less dense than oceanic crust, so can't be subducted and destroyed.

**+ Subduction**
describes oceanic crust sinking into the mantle at a destructive plate boundary. As the crust subducts, it melts back into the mantle.

*On your planet*

✛ Look at the map of plates and plate boundaries above. The circle of plate boundaries around the Pacific Ocean is called the 'Pacific ring of fire, because it has many active volcanoes.

### your questions

1 Compare the map of Pangea with a modern world map from an atlas. Describe how India has moved since the time of Pangea.
2 Look at the map of tectonic plates.
   a Which plate is the UK on?
   b Name a country which is split by two plates.
   c Name two plates that are moving apart.
   d Name two plates that are colliding.
3 **Exam-style question** Explain how tectonic plates move. (4 marks)

**+** In this section you will learn which hazards happen at different plate boundaries.

## Tectonic hazards

Earthquakes and volcanoes (**tectonic hazards**) occur at plate boundaries. Different plate boundaries produce different tectonic hazards.

## Conservative boundaries

As plates slide past each other, friction between them causes earthquakes. These are rare but very destructive, because they are shallow (close to the surface). The San Andreas fault is shown below.

| Plate boundary | Example | Earthquakes | Volcanoes |
|---|---|---|---|
| Conservative | San Andreas fault in California, USA. North American and Pacific plates sliding past each other. | | No volcanoes. |
| Constructive | Iceland, on the mid-Atlantic ridge. The Eurasian and North American oceanic plates pulling apart. | • Small earthquakes up to 5.0-6.0 on the Richter scale. | • Not very explosive or dangerous.<br>• Occur in fissures (cracks in the crust).<br>• Erupt basalt lava at 1200 °C. |
| Destructive | Andes mountains in Peru and Chile. Nazca oceanic plate is subducted under the South American continental plate. | | |
| Collision zone | Himalayas. Formed as the Indian and Eurasian continental plates push into each other. | | Volcanoes are very rare. |

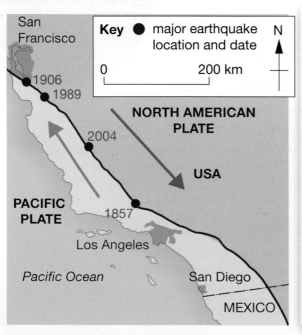

**+ Collision zones** are a type of destructive boundary. They form mountain ranges like the Himalayas (see right). Two continental plates of low-density granite collide, pushing up mountains. Earthquakes happen on faults (huge cracks in the crust) in collision zones.

## Constructive boundaries

As plates move apart, magma rises up through the gap, as the diagram below shows. The magma is **basalt** and is very hot and runny. It forms **lava flows** and shallow sided volcanoes.

Earthquakes are caused by **friction** as the plates tear apart. These earthquakes are small. They don't cause much damage.

*On your planet*
+ Every year there are about 100 000 earthquakes strong enough to be felt. The largest earthquake recorded was a magnitude 9.5 in Chile in 1960.

+ **Tectonic hazards** are natural events that effect people and property.

## Destructive boundaries

As the plates push together (see the diagram below), oceanic plate is **subducted**. As it sinks, it melts and makes magma called **andesite** (after the Andes). Sea water is dragged down with the oceanic plate. This makes the magma less dense so it rises through the continental crust. The water erupts as steam making volcanoes very explosive.

Sinking oceanic plate can stick to the continental plate. Pressure builds up against the friction. When the plates finally snap apart, a lot of energy is released as an earthquake. These earthquakes can be devastating, especially if they are shallow.

**Constructive plate boundary**

**Destructive plate boundary**

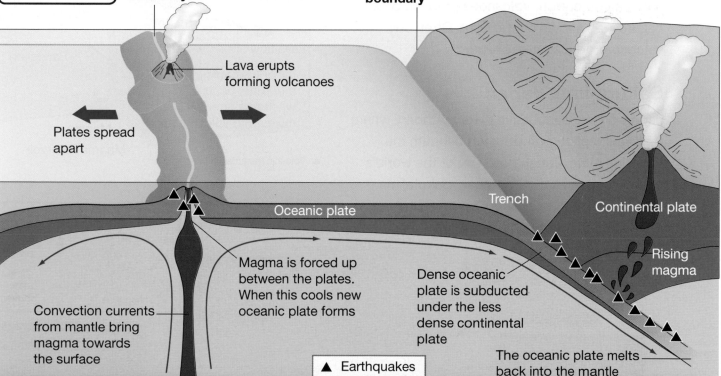

Lava erupts forming volcanoes

Plates spread apart

Convection currents from mantle bring magma towards the surface

Magma is forced up between the plates. When this cools new oceanic plate forms

Oceanic plate

Trench

Continental plate

Dense oceanic plate is subducted under the less dense continental plate

Rising magma

The oceanic plate melts back into the mantle

▲ Earthquakes

### your questions

1 Which type of plate boundary is most dangerous for humans to live on?
2 Match the words below into pairs:
constructive  fault  collision zone
fissures  landslides  explosive
destructive  conservative

Write a brief explanation of your pairs.
3 Look at the diagram of the San Andreas fault. The two plates can 'lock', stopping them from sliding. Why do you think this is?
4 Which part of the San Andreas fault might be due for a big earthquake?

✚ In this section you will examine the impact of volcanoes on developed countries.

**On your planet**

✚ There are about 100 active volcanoes in the world. On average, 80 erupt every year.

## Destructive power

The most devastating volcanoes are the most explosive ones. The Volcanic Explosivity Index (VEI) measures destructive power on a scale from 1 to 8. Mount St Helens, which erupted in May 1980, measured 4. Modern humans have never experienced an eruption measuring 8.

Volcanoes often produce **lava flows**. These destroy property, but rarely kill people because they can be outrun. Volcanoes produce other hazards, as shown on the right. Often one volcano produces many hazards.

## Sakurajima, Japan

Japan is on a destructive plate boundary where the Pacific plate is subducted beneath the Eurasian plate. Japan has 10% of the world's active volcanoes. One of these, Sakurajima (see below), has been erupting since the 1950s. It can erupt 200 times in one year.

Prevailing wind

Ash and gas column

Eruption cloud

Acid rain

Ash fall builds up on roofs, causing buildings to collapse

Lava bombs can kill people close to the crater

Pyroclastic flow

Pyroclastic flows are deadly clouds of hot ash and gas that sweep along at 200 km/h

Lava flow

Lahar (volcanic mudslide) occurs when rain or snow mixes with volcanic ash

Landslide

Magma

Magma reservoir

▲ Volcanic hazards

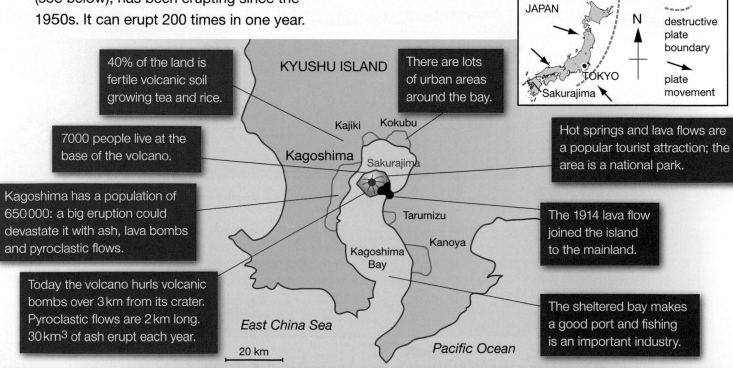

40% of the land is fertile volcanic soil growing tea and rice.

KYUSHU ISLAND

There are lots of urban areas around the bay.

7000 people live at the base of the volcano.

Kajiki

Kokubu

Kagoshima

Sakurajima

Hot springs and lava flows are a popular tourist attraction; the area is a national park.

Kagoshima has a population of 650000: a big eruption could devastate it with ash, lava bombs and pyroclastic flows.

Tarumizu

Kanoya

The 1914 lava flow joined the island to the mainland.

Today the volcano hurls volcanic bombs over 3km from its crater. Pyroclastic flows are 2km long. 30km$^3$ of ash erupt each year.

Kagoshima Bay

East China Sea

Pacific Ocean

20 km

The sheltered bay makes a good port and fishing is an important industry.

JAPAN

N

TOKYO

Sakurajima

- - - destructive plate boundary

→ plate movement

Sakurajima is a **stratovolcano** (or composite cone volcano) over 1000 metres high. Stratovolcanoes are dangerous and explosive (see right). They erupt andesite lava.

## Living with the threat

Volcanic eruptions can be **predicted** – scientists can say when a volcano will erupt. They can then warn people to take shelter or **evacuate**. The diagram below shows how Sakurajima is monitored and also the evacuation procedure there.

Japan is a developed country. It can afford to spend money on monitoring, protection and evacuation. When Sakurajima does erupt, it will probably not cause many deaths. Homes, crops and industries will be destroyed, but most people have insurance and the Government will help to repair the damage. In developed countries, tectonic hazards damage property (economic costs) but cause less harm to people (social costs).

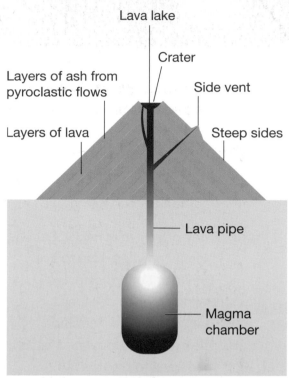

Lava lake

Crater

Layers of ash from pyroclastic flows

Side vent

Layers of lava

Steep sides

Lava pipe

Magma chamber

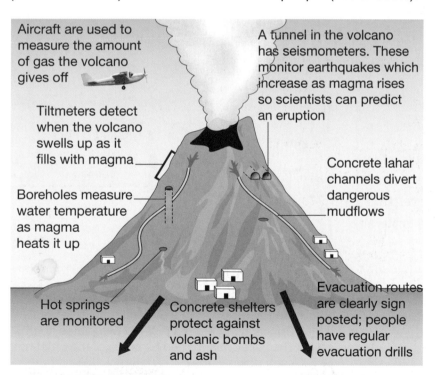

Aircraft are used to measure the amount of gas the volcano gives off

A tunnel in the volcano has seismometers. These monitor earthquakes which increase as magma rises so scientists can predict an eruption

Tiltmeters detect when the volcano swells up as it fills with magma

Boreholes measure water temperature as magma heats it up

Concrete lahar channels divert dangerous mudflows

Hot springs are monitored

Concrete shelters protect against volcanic bombs and ash

Evacuation routes are clearly sign posted; people have regular evacuation drills

▲ Two men take cover in a lava bomb shelter during an eruption of Sakurajima (in the background).

## your questions

1 Use Google to search for 'VEI scale'. Research the VEI of famous eruptions such as Krakatoa in 1883 and Mount Pinatubo in 1991.
2 Living close to Sakurajima is risky. Draw a spider diagram to show factors attracting people to this area.
3 Make a table like the one on the right:

| Protection | Prediction |
|---|---|
|  |  |

Use it to list methods used to protect people from Sakurajima, and how scientists predict an eruption.

4 **Exam-style question** Using examples, explain how volcanic eruptions can be predicted. (4 marks)

✚ In this section you will learn how volcanoes can have devastating consequences for people in the developing world.

## At risk

In the developing world, people are at greater risk from tectonic hazards than those in developed countries:

- They often live in risky locations, because there is nowhere else to live.
- They can't afford safe, well-built houses. Buildings often collapse.
- They don't have insurance.
- Their governments don't have the money and resources to provide **aid**.
- Communications are poor, so warning and evacuation may not happen.

Most volcanic eruptions with high death tolls are in the developing world (see the table).

| Volcano | Year | Number of deaths | Cause of deaths |
|---|---|---|---|
| Mount Pinatubo, Philippines | 1991 | 700 deaths | Pyroclastic flows and lahars |
| Nevado del Ruiz, Columbia | 1985 | 23 000 deaths | Lahars |
| El Chichon, Mexico | 1982 | 3500 deaths | Pyroclastic flows |

## Mount Nyiragongo

Mount Nyiragongo lies in the Democratic Republic of Congo in Africa. This volcano sits in the African Rift Valley — a constructive plate boundary where the continent of Africa is literally being pulled apart (see the diagram).

In January 2002, very hot and runny basalt lava poured out of Mount Nyiragongo. A river of lava, 1000 metres wide, flowed 20 km into the city of Goma (see photo). 14 villages were destroyed.

- 100 people died, mostly from poisonous gas and getting trapped in lava.
- As lava gushed out, it triggered earthquakes and buildings collapsed.
- 12 500 homes were destroyed.
- The eruption was predicted. 400 000 people were evacuated.
- Many became **refugees**.

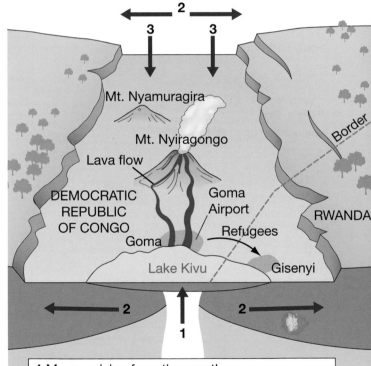

1 Magma rising from the mantle
2 Convection currents pull the two plates apart
3 The valley has formed as some of the crust has sunk downwards

✚ **Refugees** are people who are forced to move due to natural hazards or war.

✚ **Aid** is help. It can be short-term such as food given in emergency, or long-term such as training in health care.

## Relief effort

Within days, people began returning to Goma. Over 120 000 were homeless. This was a crisis and people needed help quickly. With little clean water, food and shelter, diseases like cholera could spread. The United Nations and Oxfam began a **relief effort** to help.

- The United Nations sent in 260 tonnes of food in the first week. Families got 26 kg of rations.
- In the UK, a TV appeal asked people to give money to help.
- Governments around the world gave $35 million to get aid to the refugees.
- Emergency measles vaccinations were carried out by the World Health Organisation.

In a developing country, the main problem is poverty. Most people fled from the lava with nothing. It was months before many could start building new homes. By June 2002, however, some roads had been cleared of lava and the water supply repaired.

## Future threats

Mount Nyiragongo was active again in 2005. It could erupt at any time. There is also volcanic activity under Lake Kivu. Gases like carbon dioxide and sulfur dioxide rise through the Earth into the lake. They get trapped in mud on the lake bottom. An earthquake could shake these gases free. In 1986 this happened to volcanic Lake Nyos in Cameroon, also in Africa. 1700 people suffocated from breathing in too much carbon dioxide.

▲ Lava destroyed 40% of Goma, covered half the airport and destroyed 45 schools. Water and electricity supplies were also cut off by the lava.

+ A **relief effort** is like aid. It is help given by organisations or countries to help those facing an emergency.

**On your planet**
+ No lava flows faster than Mount Nyiragongo's. It flows at 100 km per hour. A cheetah could just out run it - but you couldn't!

### your questions

1 Explain in your own words what we mean by aid and relief effort.
2 The airport at Goma was covered by lava. Why was this a problem?
3 How successful was the relief effort in helping people affected by Mount Nyiragongo's eruption?
4 Why do you think people still live around Mount Nyiragongo and Lake Kivu?
5 In pairs, discuss ideas for giving long-term help to people near Mount Nyiragongo.

+ In this section you will learn how earthquakes are measured and about their awesome power.

## Why is the ground shaking?

Earthquakes can't be predicted. They start without warning and can be catastrophic. An earthquake is a sudden release of **energy**. It's a bit like bending a pencil until it suddenly snaps. Underground, tectonic plates try to push past each other – building up pressure. The pressure is suddenly released along **faults** (cracks in the crust), sending out a huge pulse of energy. This travels out in all directions as earthquake waves (see the diagram on the far right).

An earthquake starts at the **focus**. The **epicentre** is the point on the Earth's surface above the focus, and is the first place to shake.

## Magnitude

The power of an earthquake – how much the ground shakes – is its **magnitude**. A **seismometer** measures this using the **Richter scale** (see near right).

## Niigata, Japan

In July 2007, a magnitude 6.8 earthquake struck Niigata. This is the same magnitude as the 1995 Kobe earthquake, also in Japan.

How can two earthquakes of the same magnitude have such different effects (see the table)?

The shaking is worse on the surface if the focus is shallow

Earthquake waves travel out in all directions

Epicentre

Fault

Focus

Plates overcome friction and give way, releasing energy

Plates try to move past each other building up pressure

| Key | \ Plate movement |

Huge devastation

Lots of devastation and deaths

Damage begins, but deaths are rare

Extent to which ground shakes

Great

Major

Strong

Moderate

Small

Minor

Not felt

Magnitude

*On your planet*

+ A magnitude 7.0 earthquake releases the same energy as a large nuclear bomb.

◀ The Richter scale is a logarithmic scale. A magnitude 6.0 earthquake is 10 times more powerful than a magnitude 5.0.

| Niigata, 2007 | Kobe, 1995 |
|---|---|
| • Kashiwaziki, the city affected, has 90 000 people. 11 died, 1000 were injured. | • Kobe is a city of 1.5 million. |
| • Other areas affected were farms and villages with a low population density. Only 350 buildings were destroyed. | • Population density is very high. 5000 died, 26 000 were injured. |
| • A tsunami warning was issued, but it was a false alarm. | • Many fires started, and rescuers could not reach them due to collapsed buildings. The damage was $200 billion. |
| • The epicentre was offshore, so there was less shaking on land. | • The epicentre was close to Kobe. |
| • It happened at 6:00pm. People were alert and remembered their earthquake drill. | • Soft ground made the shaking worse. |
| | • It happened at 6:00am. People were asleep and became confused in the dark. |

## Long-term planning

The secret of survival is long-term planning. Japan is a developed country, so it can afford to do this. There is a 70% **probability** (chance) of a magnitude 7.2 earthquake hitting Toyko in the next 30 years. It could kill 7000 and injure 160 000. There is no way of **predicting** when it might happen.

- Every year Japan has earthquake drills.
- Emergency services practise rescuing people.
- People keep emergency kits at home containing water, food, a torch and radio.

▲ The result of the Niigata earthquake.

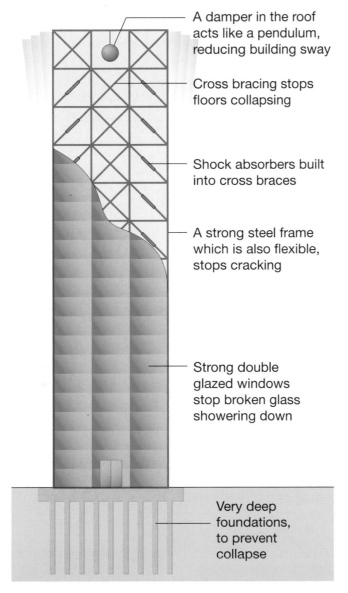

A damper in the roof acts like a pendulum, reducing building sway

Cross bracing stops floors collapsing

Shock absorbers built into cross braces

A strong steel frame which is also flexible, stops cracking

Strong double glazed windows stop broken glass showering down

Very deep foundations, to prevent collapse

▲ Many buildings in Japan are earthquake proof like this one. They can withstand a major earthquake. Gas supplies automatically shut off, reducing the risk of fire.

*On your planet*

+ If you feel an earthquake, just remember: 'duck, cover and hold'. Duck down onto the ground, cover yourself under a table (hide), and hold onto its legs!

### your questions

1 Earthquake damage is reduced as you move away from the epicentre. Why is this?
2 a Draw a large version of the Richter scale diagram
  b Use a website, like Wikipedia, to find a list of earthquakes since 2000. Mark their magnitudes and death tolls on your diagram.
  c Is there a link between magnitude and death toll?
3 Japan is on a destructive plate boundary. Why does it have frequent earthquakes?
4 a Classify earthquake impacts in Japan into social and economic.
  b Which are greater? Explain your answer.

In this section you will find out about the impacts of earthquakes on developing countries, and how people respond to them.

+ A **tsunami** is a series of waves produced by an undersea earthquake. Tsunami waves travel across the ocean at over 400 km/h. They can be 10 metres high when they hit the coast.

## Death and destruction

Earthquakes in the developing world often have very high death tolls compared to volcanoes. Destructive earthquakes happen regularly, as the table shows.

| Location | Year | Deaths | Magnitude | Key facts |
|----------|------|--------|-----------|-----------|
| Java, Indonesia | 2006 | 5800 | 6.3 | 135000 homes destroyed. 1.5 million homeless |
| Kashmir, Pakistan | 2005 | 86000 | 7.6 | One third of deaths were due to landslides. Many children died in collapsed schools. |
| Aceh, Indonesia | 2004 | 290000 | 9.3 | The second largest death toll in history, mostly due to the **tsunami** it created. |
| Bam, Iran | 2003 | 30000 | 6.6 | Many people were trapped when their mud-brick homes collapsed. |

## Earthquake in Sichuan

Sichuan is a province in central China. On 12 May 2008 it was hit by a magnitude 8.0 earthquake. There was no warning. The **primary effects** of the earthquake were huge:

- 70000 people died
- 400000 were injured
- 5 million were made homeless
- Up to $75 billion in damage

There were up to 200 **aftershocks**. On 27 May, a magnitude 6.0 aftershock caused 420000 buildings to collapse.

+ **Primary effects** – caused instantly by the earthquake. Roads crack, walls collapse and landslides happen.

+ **Secondary effects** – in the hours, days and weeks after the earthquake. Fires break out, disease spreads and food and water run short.

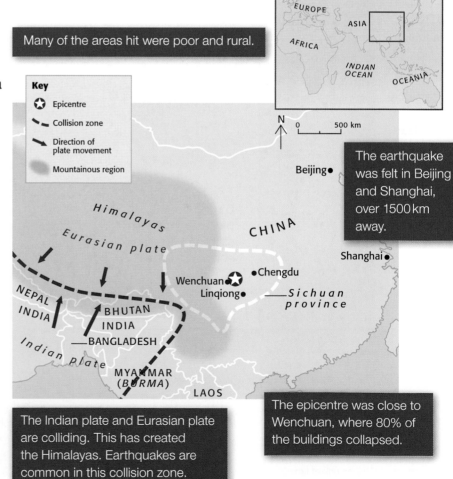

Many of the areas hit were poor and rural.

**Key**
- ✪ Epicentre
- Collision zone
- Direction of plate movement
- Mountainous region

The earthquake was felt in Beijing and Shanghai, over 1500 km away.

The Indian plate and Eurasian plate are colliding. This has created the Himalayas. Earthquakes are common in this collision zone.

The epicentre was close to Wenchuan, where 80% of the buildings collapsed.

## Local responses

Heavy rain, landslides and aftershocks made the rescue effort difficult:

- The Prime Minister, Wen Jiabao, flew to the area very soon after the earthquake.
- 50 000 soldiers were sent to help dig for survivors.
- Helicopters were used to reach the most isolated areas.
- Chinese people donated $1.5 billion in aid.

## International responses

China quickly asked the rest of the world to help:

- Some countries sent money. The UK gave $2 million.
- Finland sent 8000 six-person tents, and Indonesia sent 8 tonnes of medicines.
- Rescue teams flew in from Russia, Hong Kong, South Korea and Singapore.

## Building for the future?

The Sichuan earthquake caused over 700 schools to collapse. China has strict building rules, so schools should have withstood the shaking. Even in a poorer developing country, buildings can be made cheaply to withstand earthquakes. The diagram on the right shows how.

**+ Aftershocks** often occur as the fault 'settles' into its new position. They can injure or kill rescuers. In the developing world, aftershocks often destroy buildings that were weakened by the first earthquake.

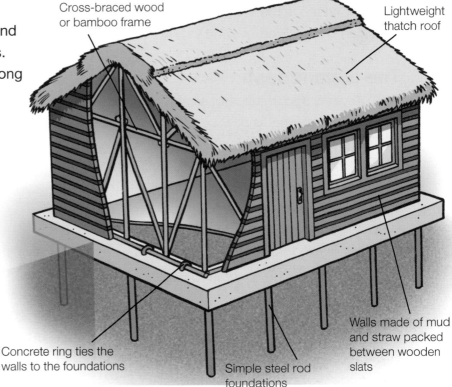

Cross-braced wood or bamboo frame

Lightweight thatch roof

Walls made of mud and straw packed between wooden slats

Simple steel rod foundations

Concrete ring ties the walls to the foundations

### your questions

1 Explain why the Sichuan earthquake happened. Remember to include plate names and boundary types in your answer.
2 a Make a list of all the effects of the Sichuan earthquake.
  b Use two colours to circle social effects (effects on people) and economic effects (effects to do with money).

3 Look back on the effects of the two volcanoes and two earthquakes you have studied.
  a Compare impacts each had on people and property.
  b Explain the differences.
4 **Exam-style question** Explain why earthquakes happen on destructive plate margins. You may draw a diagram to help with your answer. (4 marks)

✚ In this section you will learn how climate was very different from that of today in both the recent, and distant, past.

## What is climate?

People say climate is what you expect, but weather is what you get! If you plan a holiday to Majorca in August, you expect it to be hot and sunny (climate). If it rains when you get there, you've got weather!

Weather is short-term day-to-day changes in things like temperature, wind, cloud cover and rainfall. Climate is the *average* of these weather conditions, measured over 30 years.

## The distant past

120 000 years ago, rhinoceroses and elephants roamed around what is now London. At other times in the past, huge ice sheets stretched from the North Pole as far south as London. Scientists know that climate was different in the past. They use physical evidence such as:

- fossilised animals, plants and pollen that no longer live in the UK
- landforms, like the U-shaped valleys left by retreating glaciers
- samples from the ice sheets of Greenland (see the photo) and Antarctica.

Ice sheets are like a time capsule. They contain layers of ice, oldest at the bottom, youngest at the top. Each layer is one year of snowfall. Trapped in the ice layers are air bubbles. These preserve air from the time the snow fell. Locked in the air bubble is carbon dioxide. **Climatologists** can reconstruct past temperatures (shown in the graph on the right) by drilling a core through the ice and measuring the amount of trapped carbon dioxide in ice layers.

▲ A lump of ice the size of a house falls off the Greenland ice sheet into the sea.

✚ A **climatologist** is a scientist who is an expert in climate and climate change.

*On your planet*

✚ The level of carbon dioxide in the atmosphere is higher today than at any time in the past 800 000 years.

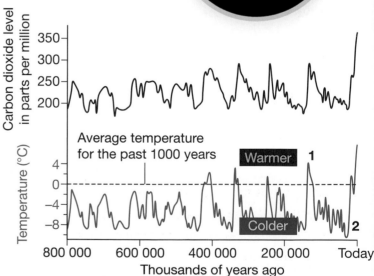

**Key**

**1** Warm periods, called **interglacials.**

**2** Cold periods, called **glacials.** Some glacial periods became Ice Ages.

During the **Quaternary** (the last 2.6 million years of geological time), warm periods (**interglacials**) lasted for between 10 000 and 15 000 years. Cold periods (**glacials**) lasted about 80 000–100 000 years. During some glacial periods, it became so cold that the Earth plunged into an ice age. Huge ice sheets extended over the continents in the northern hemisphere. There were also vast areas of floating sea ice. The last time this happened was between 30 000 and 10 000 years ago, in the last ice age (see right). The ice sheets were 400–3000 metres thick, and so heavy that they made the Earth's crust sag. So much water was locked up in the ice sheets that sea levels fell by over 100 metres.

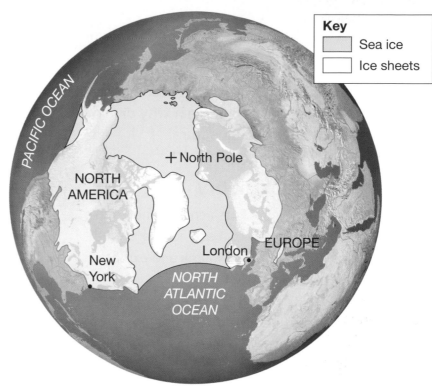

**Key**
Sea ice
Ice sheets

▲ Ice cover in the last ice age.

## The recent past

There is also evidence for climate change in more recent times. Evidence comes from:

- old photographs, drawings and paintings of the landscape
- written records, such as diaries, books and newspapers
- the recorded dates of regular events, such as harvests, the arrival of migrating birds and tree blossom.

These sources are often not very accurate, because they were not intended to record climate. However, they can still give us some idea of overall climate trends in the recent past. This type of evidence suggests that climate changes regularly – every few hundred years. Average temperatures over the past 2000 years or so have probably varied between 1–1.5 °C colder or warmer than average temperatures today (see the graph on the right).

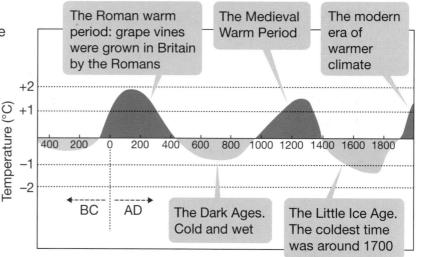

The Roman warm period: grape vines were grown in Britain by the Romans

The Medieval Warm Period

The modern era of warmer climate

The Dark Ages. Cold and wet

The Little Ice Age. The coldest time was around 1700

**your questions**

1 What is a climatologist?
2 Look at the graph of carbon dioxide levels and temperature over the last 800 000 years.
   a Describe the variations in the temperature graph.
   b Does the Earth warm up faster, or cool down faster?
   c How closely are temperature and carbon dioxide levels linked?
3 Look at the map of ice sheets. Which parts of the land were covered by ice 20 000 years ago, and are now ice-free?
4 **Exam-style question** Describe some of the evidence that tells us climate was different in the past. ( 4 marks)

In this section you will learn that there are three main theories that climatologists use to explain why climate has changed in the past. These changes are all natural.

▼ Mount Pinatubo erupting in 1991.

### The eruption theory

Big volcanic eruptions can change the Earth's climate. Small eruptions have no effect – the eruption needs to be very large and explosive. Volcanic eruptions produce:

- ash
- sulphur dioxide gas.

If the ash and gas rise high enough, they will be spread around the Earth in the **stratosphere** by high-level winds. The blanket of ash and gas will stop some sunlight reaching the Earth's surface. Instead, the sunlight is reflected off the ash and gas, back into space. This cools the planet and lowers the average temperature.

> + The **stratosphere** is the layer of air 10–50 km above the Earth's surface. It is above the cloudy layer we live in, called the troposphere.

In 1991, Mount Pinatubo in the Philippines erupted, releasing 17 million tonnes of sulphur dioxide (see the top photo). This was enough to reduce global sunlight by 10%, cooling the planet by 0.5 °C for about a year.

Mount Pinatubo was very small-scale compared to the 1815 eruption of Tambora in Indonesia. This was the biggest eruption in human history. In 1816, temperatures around the world were so cold that it was called 'the year without a summer', and up to 200 000 people died in Europe as harvests failed. The effects lasted for four to five years. In general, volcanoes only affect climate for a few years.

2001/03/29 09:36 UT

▲ Sunspots on the surface of the sun in March 2001.

**On your planet**

+ In 1947, astronomers observed the Great Sunspot, the largest ever recorded. It was about 30 times bigger than Earth.

## The sunspot theory

Over 2000 years ago Chinese astronomers started to record sunspots. These are black areas on the surface of the sun (see the bottom photo opposite). Sometimes the sun has lots of these spots. At other times they disappear. Even though the spots are dark, they tell us that the sun is more active than usual. Lots of spots mean more solar energy being fired out from the sun towards Earth.

Cooler periods, such as the Little Ice Age, and warmer periods, such as the Medieval Warm Period, may have been caused by changes in sunspot activity. Some people think that, on average, there were more volcanic eruptions during the Little Ice Age, and that this added to the cooling. However, climate change on timescales of a few hundred years, and 1–2 °C, cannot be explained by volcanoes – but it might be explained by sunspot cycles (see the top right diagram).

## The orbital theory

Over very long timescales, there have been big changes in climate. Cold glacial periods and ice ages were 5–6 °C colder than today. Some interglacials were 2–3 °C warmer than today. Such big changes need a big cause. Scientists think they know what this is – changes in the way the Earth orbits the sun.

You might think that the Earth's orbit does not change, but over very long periods it does, as the diagrams on the right show.

- The Earth's orbit is sometimes circular, and sometimes more of an ellipse (oval).
- The Earth's axis tilts. Sometimes it is more upright, and sometimes more on its side.
- The Earth's axis wobbles, like a spinning top about to fall over.

These three changes alter the amount of sunlight the Earth receives. They also affect where sunlight falls on the Earth's surface. On timescales of thousands of years, the changes would be enough to start an ice age, or end one. These changes are called **Milankovitch Cycles**.

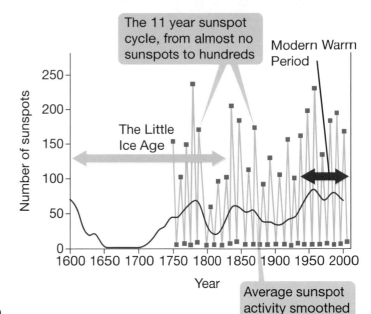

The 11 year sunspot cycle, from almost no sunspots to hundreds

Modern Warm Period

The Little Ice Age

Average sunspot activity smoothed out over time

It takes 100 000 years for the Earth's orbit to change from being more circular, to an ellipse, and back again.

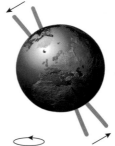

It takes 41 000 years for the Earth's axis to tilt, straighten up, and tilt back again.

It takes 26 000 years for the Earth's axis to wobble, straighten up, and wobble again.

### your questions

1 Explain how big volcanic eruptions might change our climate.
2 Draw and complete a table to compare the three theories of climate change.
3 Write down the names of the three orbital changes and the time each change takes.

➕ In this section you will examine some of the impacts of past natural climate change. Do past events offer lessons that might help us cope with global warming?

## Viking Greenland

The Viking saga of 'Erik the Red' tells us that Erik was a Norse Viking, whose parents fled to Iceland. In 982, Erik was banished to Greenland for murder. In western Greenland, he and about 500 other Vikings found a land that was largely free of ice. This was the start of the Medieval Warm Period, when Greenland was very different from today. The warmer climate meant that by 1100 Greenland had:

- over 200 farms – keeping goats, sheep and cows, and growing hay (grass) to feed the animals
- a population of 3000–4000 Vikings
- trade links with Iceland and Norway
- summer hunting expeditions north of the Arctic Circle, for seals and whales.

Life on Greenland was very hard. The Greenland Vikings survived there for over four centuries, but soon after 1410 they died out (see below). What happened? We don't know exactly what killed off the last few Greenland Vikings, but we do know why they declined – the Little Ice Age.

## Learning from the Vikings

The Little Ice Age made life impossible for the Greenland Vikings. They ran out of food and died out. However, they did not help their own survival:

- Deforestation and soil erosion meant that the Viking farms probably produced very little food. The Vikings had damaged the land they depended on.
- They also depended too much on trade with Iceland and Norway. When this stopped, because of sea ice, they had no one else to turn to.
- The Greenland Vikings were not very adaptable. They tried to live as they did in Norway and Iceland.

When the climate became colder, the Greenland Vikings were isolated in a damaged environment with no real way of changing their lifestyle to cope.

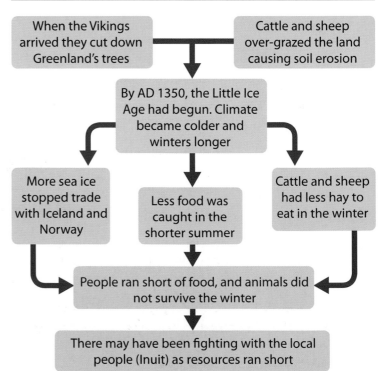

When the Vikings arrived they cut down Greenland's trees

Cattle and sheep over-grazed the land causing soil erosion

By AD 1350, the Little Ice Age had begun. Climate became colder and winters longer

More sea ice stopped trade with Iceland and Norway

Less food was caught in the shorter summer

Cattle and sheep had less hay to eat in the winter

People ran short of food, and animals did not survive the winter

There may have been fighting with the local people (Inuit) as resources ran short

▲ Greenland today.

# The Little Ice Age

As the world cooled, the Little Ice Age began to have wider impacts. In England, 'Frost Fairs' were held on the River Thames when it froze over in the winter (see right). The first Frost Fair was in 1608 and the last in 1814.

During the Medieval Warm Period, Europe's population grew. New land was given over to farming, often on hillsides and areas that had not been farmed before. The warmer climate made this possible. This growth crashed to a halt in 1315, when the Little Ice Age took hold.

- Cold and rain lashed Europe in the spring and summer of 1315.
- Wheat and oats did not ripen and the harvest failed.
- The cool wet weather continued in 1316 and 1317.
- By 1317 the 'Great Famine' had begun. It lasted until 1325.
- In some areas, 10–20% of the peasant farmers may have died of hunger.

Things did not improve much for Europe. In 1349, it was struck by the Black Death (bubonic plague). This was to kill far more than the Great Famine, but was probably made worse by the colder climate and more difficult farming conditions.

In the Alps, many valley glaciers grew in the colder climate. In the 1820s and 1850s they advanced down valleys, destroying villages and farmland. Many farmers stopped growing wheat, because it needs warm summers. The crop of choice became the cold and wet loving potato, imported from South America in about 1530.

The Little Ice Age caused many problems in Europe. However, people did adapt. They learned to farm new crops, abandoned farms high on hillsides and learned to enjoy fairs on frozen rivers. People can adapt to climate change, but it takes time.

▲ A Frost Fair on the River Thames in London in 1683.

## your questions

1 How did the climate at the time affect Erik the Red's prospects when he was banished to Greenland?

2 Imagine you were one of the last farmers in Greenland. Write a brief letter to a relative in Iceland explaining how difficult life has become.

3 Greenland's Vikings did not get on with the local Inuit people. How might befriending them have helped the Vikings to survive?

4 How did people in Europe adapt to the colder climate of the Little Ice Age?

✚ **In this section you will discover that, as well as humans, plants and animals are vulnerable to climate change.**

## Ecosystems

Plants and animals live together in ecosystems. They depend on each other, and are linked together in **food chains**. They also depend on the physical environment around them, which includes the climate. Together, plants, animals and the physical environment make up ecosystems like the one in the diagram. Ecosystems can be small, such as a pond, or large, such as the tropical rainforest.

If one part of an ecosystem changes, the other parts will also change. When climate has changed in the past, it has spelled disaster for some plants and animals.

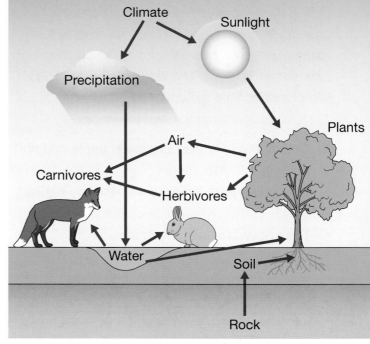

▲ The parts of an ecosystem.

## The dinosaurs

65 million years ago many dinosaurs suddenly became **extinct**. Two possible causes are:
- a strike by a massive asteroid in Mexico
- a huge volcanic eruption in Deccan, India, lasting up to 1 million years.

Both of these events are known to have happened at that time. They may have happened together. Both would have thrown up huge amounts of dust, ash and gas into the air, blocking out the sun. Plants would have struggled to grow as the climate cooled. Ecosystems would have broken down as food chains collapsed, sealing the dinosaurs' fate.

✚ **Extinction** means a species of plant or animal dying out completely, so none survive.

✚ In a **food chain**, plants provide food for plant-eating animals (herbivores). Herbivores provide food for meat-eating animals (carnivores). Plants and animals are linked together, and depend on each other.

# Ice age megafauna extinction

If we fast forward in time from the dinosaurs, we come to a more recent 'mass extinction', when a number of species died out together. The animals were the Quaternary **megafauna**, and the time was 10 000–15 000 years ago.

> + **Megafauna** means 'big animals'. Most weighed over 40 kg and included the woolly mammoth, giant elk, ground sloth, sabre-tooth cat, giant beaver and glyptodon.

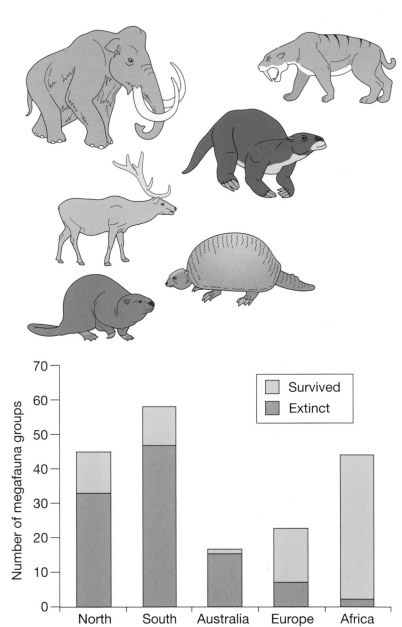

What happened to the megafauna? It seems that they were the victims of two new stresses at once – humans, and climate change. As the last major ice age ended, and the climate warmed up by about 6 °C in only 1000 years, many animals had to move.

- They migrated and tried to find new areas to live in, where the climate suited them.
- However, finding the right plants to eat in the new areas would have been difficult.
- This would have disrupted food chains, leaving some animals short of food.
- Climate change stress may also have made the megafauna weaker than normal.

As the climate warmed, humans also migrated into new areas. They hunted some of the megafauna, meaning less prey for carnivores. Some herbivores may have been hunted to extinction, leaving carnivores with nothing to prey on. Humans and climate change seem to have acted together to cause the extinction shown in the graph.

The extinct ice age megafauna provide a lesson for us today. There are many humans on the planet and our climate seems to be changing. Is it surprising that some scientists think that up to 30 000 species are becoming extinct every year?

## your questions

1 What is an ecosystem?
2 Use the words below to draw a simple food chain:
   Rabbit   Sun   Grass   Wolf   Bacteria
3 Look at the graph of megafauna extinctions.
   a Which regions lost:
      i) the largest number of groups of megafauna?
      ii) the least number of groups of megafauna?
   b Work out the percentage of groups of megafauna that became extinct for each region.
4 **Exam-style question** Using examples, describe how ecosystems were affected by climate change in the past. (4 marks)

+ In this section you will learn how our **atmosphere** is being changed by human activity.

## The greenhouse effect

Earth's atmosphere is vital to life. The gases which make up the atmosphere are important:

- Nitrogen (78.1%) is an important nutrient for plant growth.
- Oxygen (20.9%) is breathed in by animals, which breath out carbon dioxide.
- Carbon dioxide (0.03%) is breathed in by plants, which breath out oxygen.
- Water vapour (about 1%) forms clouds, which are a key part of the water cycle.

Carbon dioxide is a very important gas, even though it makes up only a tiny fraction of the atmosphere. This is because it helps to regulate the temperature on Earth – it is a **greenhouse gas**. Greenhouse gases make the planet warmer by about 16 °C. This keeps the Earth comfortably warm. Without greenhouse gases, most of the planet would be a frozen wasteland. It is important to understand that the **greenhouse effect** is natural (see the diagram).

## Greenhouse gases

Carbon dioxide is the most common greenhouse gas, but there are others (see the table).

+ The **atmosphere** is a layer of gases above the Earth's surface.

+ The **greenhouse effect** is the way that gases in the atmosphere trap heat from the sun. The gases act like the glass in a greenhouse. They let heat in, but prevent most of it from getting out.

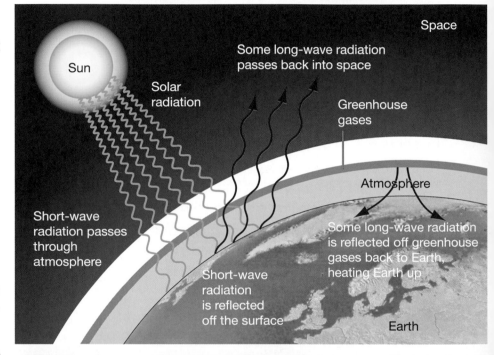

| Greenhouse gas | % of greenhouse gases produced | Sources | Warming power compared to carbon dioxide | % increase since 1850 |
|---|---|---|---|---|
| Carbon dioxide | 89% | Burning fossil fuels (coal, oil and gas), deforestation which releases carbon dioxide. | 1 | +30% |
| Methane | 7% | Gas pipeline leaks, farming rice in paddy fields, cattle farming. | 21 times more powerful | +250% |
| Nitrous oxide | 3% | Jet aircraft engines, cars and lorries, fertilisers and sewage farms. | 250 times more powerful | +16% |
| Halocarbons | 1% | Used in industry, solvents and cooling equipment. | 3000 times more powerful | Not natural |

The extra greenhouse gases which pollute the atmosphere are produced by humans. In the UK we use lots of fossil fuels. Burning these produces carbon dioxide, which ends up in the atmosphere as pollution. The main source of this pollution is power stations that produce our electricity (see below).

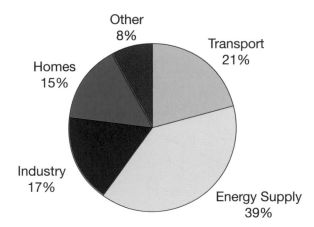

▲ Percentage of carbon dioxide emissions from different sources in the UK (2006).

On a global scale, there are big differences in carbon dioxide production. Most is produced by the developed world.

- The EU, USA and Japan emit over 40% of all carbon dioxide.
- Adding Russia, China and India brings the total to 70% of all carbon dioxide emissions.

Most people in the developing world produce less than 1 tonne of carbon dioxide per person per year, compared to 10–25 tonnes per person in the developed world. The map above right shows the differences.

Many scientists are becoming concerned about greenhouse gas emissions and their effect on our climate. Some issues we need to think about are:

- how to reduce emissions in the developed world, where we use a lot of fossil fuels
- how to persuade big developing countries like China and India to slow down the growth in their carbon dioxide emissions
- how to protect vulnerable people from the future impacts of climate change.

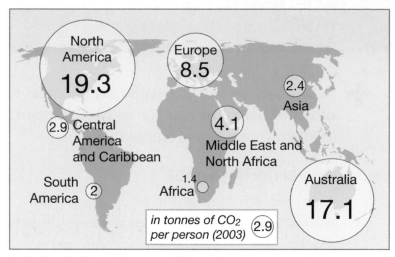

▲ Carbon dioxide emissions around the world.

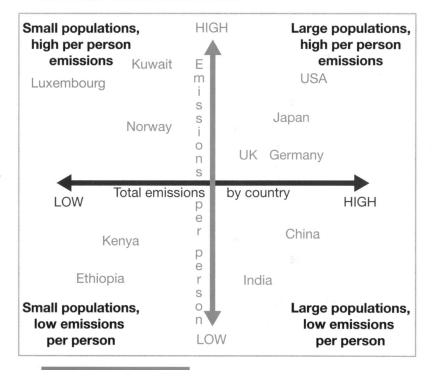

## your questions

1 Why is carbon dioxide in the atmosphere so important?
2 What are the main greenhouse gases?
3 a Draw your own version of the diagram showing the greenhouse effect.
  b Label it to explain how the greenhouse effect works.
4 a Make a list of human activities which add extra greenhouse gases into the atmosphere.
  b Explain why people in the developing world produce only small amounts of greenhouse gases. Think about their lifestyles and activities compared to the developed world.

In this section you will learn how pollution of the atmosphere with greenhouse gases has led to the enhanced greenhouse effect, also known as global warming.

## A warming planet

Today our climate seems to be changing. This is known as global warming. Global warming means a warming of the Earth's temperatures, and is caused by the **enhanced greenhouse effect**. 'Enhanced' simply means 'working more strongly'. Because humans have polluted the atmosphere with carbon dioxide and other greenhouse gases, the natural greenhouse effect has been given a boost. The graph shows the increase in carbon dioxide in the atmosphere in Hawaii. More heat is trapped in the atmosphere by the greenhouse gases, and temperatures are rising.

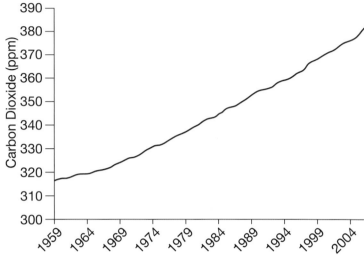

▲ Carbon dioxide concentrations, Hawaii 1959-2005.

Global warming has been measured:
- Global temperatures rose by 0.75°C between 1905 and 2005.
- Sea levels rose by 195 mm from 1870 to 2005. They are rising because the sea expands as it warms up. This is called **thermal expansion**. In future, if the glaciers and ice sheets continue to melt, sea levels could rise significantly.

Since 1980, global warming seems to have been happening more quickly:
- 19 of the 20 warmest years ever recorded have been since 1980.
- Floating sea ice in the Arctic shrank from 7.6 million square kilometres in 1980 to only 4.4 million in 2007.
- Over 90% of the world's valley glaciers are shrinking. On the right are photos of the Upsala glacier in Argentina in 1928 and 2004.

*On your planet*
+ Worldwide, 2005 was the warmest year on record, with 1998 and 2007 tied in second place.

▲ The Upsala Glacier, Argentina, in 1928 (top photo) and 2004.

## What do scientists think?

Many scientists and politicians are worried about global warming. In 2007, over 2500 scientists and climatologists from 130 countries wrote a report called 'Climate Change 2007'. They work for the Intergovernmental Panel on Climate Change, which is part of the United Nations. Their report said:

> 'Most of the increase in global average temperatures since the mid-20th century is very likely due to the increase in human greenhouse gases.'

In other words, they blamed humans for the increasing temperatures. But there are some scientists who believe that humans are not the main cause of global warming, or who think most of the warming is natural.

## Future climate?

Scientists do not know exactly how global warming might affect our planet. All they can do is try to estimate future changes. Their estimates are that:

• temperatures will rise between 1.1 °C and 6.4 °C by 2100
• sea levels will rise by between 30 cm and 1 metre by 2100.

A 'best guess' might be a warming of 3.5 °C and a sea level rise of 40 cm by 2100. Floods, droughts and heatwaves could become more common, and storms and hurricanes stronger (see the photo). Predicting future global warming is very difficult because we don't know:

• what the world's future population will be.
• if we will continue to use fossil fuels, or change to cleaner energy like wind or solar.
• if people will change their lifestyles and recycle more, or use public transport.

The graph on the right shows how temperatures might increase in three different situations. Also, we don't really know how the climate might react if we continue to pollute the atmosphere. We could be in for some big surprises.

▼ The aftermath of Hurricane Katrina.

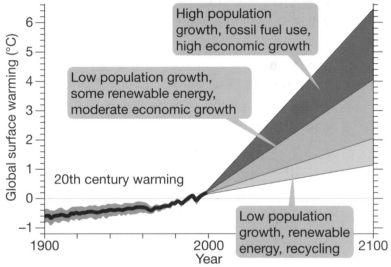

High population growth, fossil fuel use, high economic growth

Low population growth, some renewable energy, moderate economic growth

20th century warming

Low population growth, renewable energy, recycling

*Global surface warming (°C)* — *Year*

### your questions

1 Copy and complete the paragraph using the words below:
ENHANCED    ATMOSPHERE    WARM
CARBON DIOXIDE    GLOBAL    GAS
POLLUTE

_____ _____ is a greenhouse ____.
These gases in the _____ keep the planet comfortably _____. Human activities are producing more greenhouse gases which _____ the atmosphere. This is causing the _____ greenhouse effect and seems to be leading to _____ warming.

2 Use the graph on the opposite page to describe how carbon dioxide levels have risen since the 1950s.

3 Use data to explain what global warming is.

4 Why is it hard to predict how temperatures might rise with global warming? Write down four reasons.

✚ In this section you will learn how the UK might be affected by global warming.

## Moving south?

There is not much doubt that global warming will make the UK hotter, but precipitation will also change. The maps on the right show estimated changes by 2050. In summer, if London was warmer by:

- 2 °C - it would be as warm as Paris
- 4 °C - it would be like the south of France
- 6 °C - it would be like Madrid

A warmer UK would have costs as well as benefits (as the table below shows).

Many people think that the costs of global warming might be greater than the benefits. There could be some big changes to physical geography if sea levels rise, because:

- low-lying coasts like East Anglia, Essex and London could flood
- some coastlines could erode more rapidly, such as Holderness.

Trying to hold back rising sea levels would be very expensive. Sea defences and flood barriers (such as the Thames Barrier in the photo) cost millions of pounds.

Annual temperature change

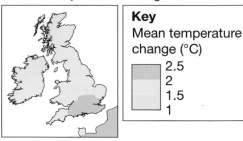

**Key**
Mean temperature change (°C)
2.5
2
1.5
1

Winter precipitation

**Key**
Precipitation change (%)
25
20
15
10
0

0
−10
−20
−30
−40

Summer precipitation

▼ The Thames Barrier was raised four times between 1984–90, 35 times between 1991–2000, and 67 times between 2001–07.

| Costs of a warmer UK | Benefits of a warmer UK |
|---|---|
| In the summer, drought and water shortages could become more common, especially in the south. | In winter, heating costs and road gritting costs could fall. |
| There could be more illnesses such as heat-stroke and skin cancer. | More people could take their holidays in the UK, which would be good for the tourism industry. |
| Very hot temperatures can melt road surfaces and buckle railway lines. | There could be fewer deaths of old people in the winter from cold. |
| Farmers might have to change crops to those that need less water and more sunshine. | New crops might mean new sales opportunities for farmers. |
| Some plants and animals might die out in the UK if it gets too hot. | More land could be farmed at higher altitudes. |

## Wild weather

Some scientists think that extreme weather might become more common in the UK as a result of global warming. This could mean more:

- heatwaves, like summer 2003, when temperatures reached 38 °C
- flooding, like summer 2007, when parts of the Midlands had a month's rainfall in 1-2 days
- storms, like the 'great gales' in 1987 and 1990, which caused millions of pounds of damage, as shown in the photo.

Extreme weather is hard to predict. It is also costly. Insurance companies have to pay out, and the Government has to help. To stop floods destroying homes and businesses, we would have to spend billions of pounds on flood defences.

## A Stern warning

In 2005, Sir Nicholas Stern wrote a report on global warming and its impacts. This was called the 'Stern Review'. It warned that we should act now to reduce global warming. Stern said:

- we should spend 2% of our GDP reducing greenhouse gas pollution now
- if we don't do this, the effects of global warming could reduce our GDP by 20%.

He is basically saying 'Spend now, or pay later'.

## What can we do?

Could we reduce greenhouse gas pollution? The answer is 'yes'. However, it would only really work if the UK and other countries all acted together.

- We could reduce our use of fossil fuels and switch to 'green' energy, like wind, solar and tidal power.
- We could recycle more.
- We could use cars less and public transport more.

The 1997 Kyoto Protocol is an international agreement to reduce carbon dioxide emissions. Some countries, like the UK and Germany, have reduced their emissions. Others, like China and the USA, have not. In future we need a much tougher agreement that all countries must sign up to.

▲ The aftermath of the great gale in 1987.

**On your planet**
+ By 2030, heatwaves like the one in 2003 could happen once every three years

### your questions

1 Describe how global warming in the UK could affect temperatures and precipitation.
2 How might the UK's weather become more 'wild' due to global warming?
3 What was the Stern Review and what did it say?
4 Write a letter to the Prime Minister saying what you think the UK should do about global warming.
5 **Exam-style question** Using examples, describe how global warming in the UK could have both costs and benefits. (5 marks)

## 2.8 Egypt on the edge

**+** In this section you will learn about some of the possible impacts of global warming in the developing world, by exploring the situation in Egypt.

> ### On your planet
> **+** People who are forced to move due to changes to the environment (including changes caused by global warming) are called environmental refugees.

### Egypt's contribution

As a developing country, Egypt's greenhouse gas emissions are low. They are 2.6 tonnes of greenhouse gases per person per year. The world average is 6.8 tonnes per person, compared with 11 tonnes in the UK. Egypt's 75 million people produce less than 1% of all greenhouse gases.

### Challenges

Egypt is unusual. 99% of its people live on only 5% of the country's land. Much of the country is desert (see the satellite photo).

The River Nile is a very important water supply. It is Egypt's only reliable source of water. Egypt's geography makes it very vulnerable to global warming. It could be on the edge of a climate disaster.

If sea levels rise by only 50 cm, over one third of the city of Alexandria will be under water. The same sea level rise would flood 10% of the Nile delta. This would mean over 7 million people having to find somewhere else to live. The loss of land would also hit farming. Less food would be produced, possibly leading to famine. Rising sea levels, and more frequent storms, are already eroding the delta coastline by over 5 metres per year in some areas.

> **+ Desertification** is the gradual change of land into desert.

Global warming could have wider impacts on Egypt, such as:

- temperature rises of 8 °C by 2080, double the global average
- less and more unreliable rainfall
- the spread of the Sahara Desert (**desertification**) onto areas of farmland
- falling crop yields as temperatures rise and water shortages increase
- heatwaves bringing more illness and death
- the spread of diseases like malaria.

Rainfall in Egypt averages under 10 mm per year.

Egypt depends on the River Nile for its water supply.

Egypt has 6 million acres of farmland, almost all of it is irrigated using River Nile water.

Egypt grows large amounts of maize, wheat and cotton in the fertile Nile Delta.

## Water wars?

Water is already in short supply in Egypt. The amount of water available per person is far below the world average, as the graph shows. Also, the rainfall that feeds the River Nile does not fall on Egypt but in mountainous areas to the south. 86% of the Nile's water starts its journey in Ethiopia.

If climate in Africa changes, water could be in short supply. Countries south of Egypt (see the map) are starting to take more water from the Nile. Many are building large dams and reservoirs.

- Uganda is building a $400 million dam near Lake Victoria.
- Sudan is building the $1 billion Merowe dam near Khartoum.
- Ethiopia is building the $225 million Tekeze dam on a tributary of the Blue Nile.

All of these dams will supply hydro-electric power (HEP) and water for irrigation and drinking. However, they could also have a serious impact on the amount of water reaching Egypt – leading to conflict.

## The cost of global warming

As global warming continues, Egypt could be facing the loss of farmland to rising sea levels, and also water shortages. In the developed world it might be possible to build sea defences and use water more efficiently. But this costs money. Egypt is a developing country, with debts of $30 billion in 2007 and 44% of its people living on less than $2 per day. Like many developing countries, it may not be able to afford to cope with global warming.

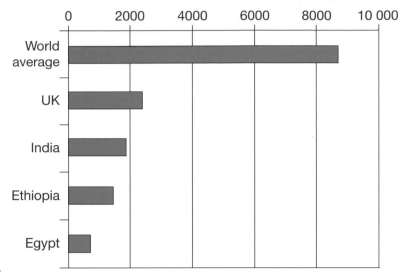

▲ Water resources (m³ per person per year).

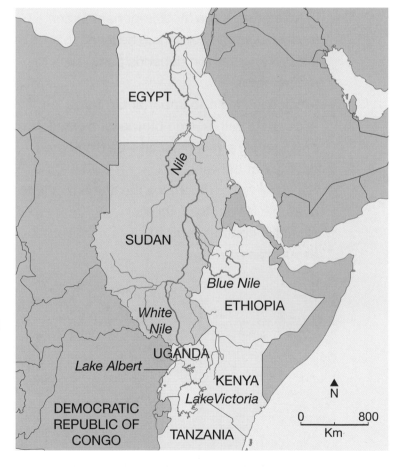

### your questions

1 Why might many Egyptians think global warming is 'unfair'?
2 Give three changes that might happen in Egypt due to global warming.
3 Explain how some Egyptians could become 'environmental refugees' due to global warming.
4 How could water cause conflict between Egypt and other countries?

➕ **In this section you will learn about the distribution of global biomes.**

The **biosphere** is the part of the Earth's surface inhabited by living things. A **biome** is a world-scale ecosystem. It covers a huge area. The world can be divided into nine major biomes. Each one has its own type of vegetation and wildlife. The location and characteristics of each biome are mainly determined by climate. This is because climatic factors affects the growth of plants.

- Temperature is the most important factor. It varies with the seasons. The length of the growing season depends partly on temperature.
- Precipitation is also important. A forest ecosystem with a large **biomass**, needs lots of rainfall. The rain must also be distributed throughout the year.
- Sunshine hours determine the amount of light available for photosynthesis.
- Humidity controls rates of **evapotranspiration**.

## Mapping biomes

The map below shows the distribution of global biomes. It shows natural vegetation. For instance, you can see that the whole of Great Britain is classified as temperate deciduous forest. There are still forests in Great Britain, but most of the land is now farmed or built on, because most native British forests were chopped down many years ago.

➕ **Evapotranspiration** is when water evaporates from the pores of leaves into the atmosphere. This results in water being drawn up plant stems.

**Key**

- Tundra
- Coniferous forest
- Temperate deciduous forest
- Temperate grassland
- Mediterranean
- Hot desert
- Tropical rainforest
- Tropical grassland (savanna)
- Other biomes (e.g. ice, mountains)

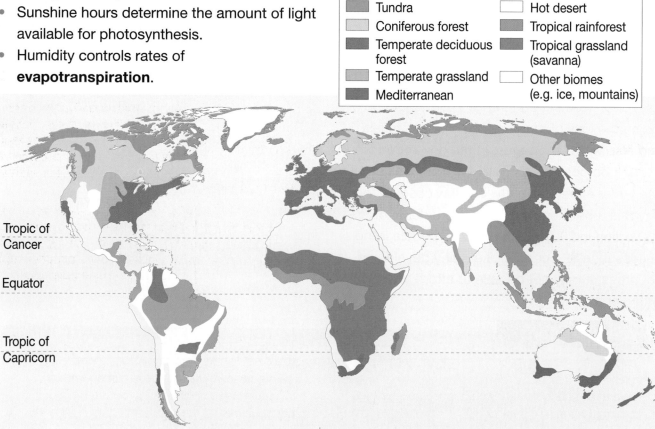

Tropic of Cancer

Equator

Tropic of Capricorn

The diagram below shows a more complex picture. You can see how the biomes gradually change as you move away from the tropics towards the North and South Poles. Take, for example, the tropical region. As you move away from the Equator, tropical rainforest develops into tropical grassland and finally into hot desert.

Altitude and distance from the coast also affect vegetation patterns. At a high altitude few plants will grow. Look again at the tropical region on the diagram. You can see that tropical rainforest develops into coniferous forest and tundra as you gain height and move inland.

*On your planet*

✛ Did you know that, even today, forests cover 30% of the world's land surface? While tropical forests are decreasing, temperate forests are actually increasing.

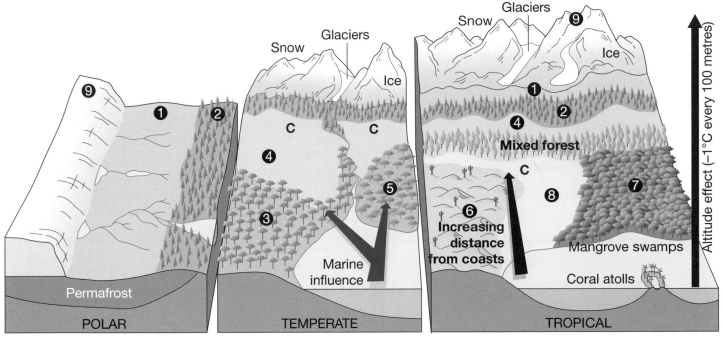

Latitude influences biomes by altering temperature and precipitation

Arctic · · · · · · · · · · · · · · · · · · · · · · · · · · · · · · · · · · · Equator

**Key**

**Biomes seasonally lacking in heat and/or water**

❷ Coniferous forest
❸ Temperate deciduous forest
❹ Temperate grassland
❺ Mediterranean
❽ Tropical grassland (savanna)

**Biomes permanently lacking in heat and/or water**

❶ Tundra
❻ Hot desert
❾ Mountain or Alpine

**Biomes promoting growth all year round**

❼ Tropical rainforest

Marine influence The sea cools nearby land in the hot season and warms it during the cold season. This reduces annual temperature range and increases precipitation.

Continentality **C** Away from the sea, the land heats up in the hot season and cools quickly in the cold season. This increases the annual temperature range and reduces precipitation.

## your questions

1  What is the difference between the biosphere and a biome?
2  Explain the following words – tropical, temperate, polar, tundra, deciduous.
3  **a**  What is the difference between altitude and latitude?
   **b**  How do these affect how plants grow?
4  Draw and complete a table to compare marine and continental climates. Use the diagram above to help you.

In this section you will learn more about the effect of climate and local factors on vegetation.

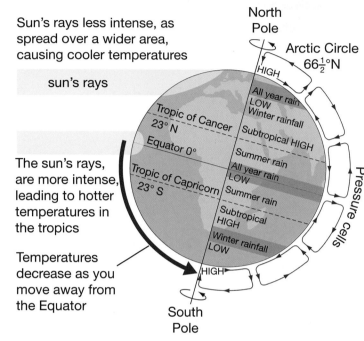

**Key**

Rainfall zones

## Temperature

Average temperature is the main factor affecting plant growth. Temperature gradually decreases as you move away from the Equator. As latitude increases, so temperature decreases.

In the tropics, the sun's rays are at a high angle in the sky for the whole year. These rays are concentrated over a smaller area than at the poles. Concentrated rays provide a lot of heat and sunlight. Plants grow well, so there is dense vegetation in the tropics.

In polar areas, the sun's rays are less concentrated. The lack of heat and light limits vegetation growth. Plants are stunted and low growing.

Sun's rays less intense, as spread over a wider area, causing cooler temperatures

sun's rays

The sun's rays, are more intense, leading to hotter temperatures in the tropics

Temperatures decrease as you move away from the Equator

North Pole

Arctic Circle 66½°N

HIGH

All year rain
LOW
Winter rainfall

Subtropical HIGH

Tropic of Cancer 23° N

Equator 0°

Summer rain

All year rain
LOW

Tropic of Capricorn 23° S

Summer rain

Subtropical HIGH

Winter rainfall
LOW

HIGH

Pressure cells

South Pole

▲ How latitude affects temperature and rainfall patterns.

## Precipitation

Around the world, precipitation is more likely in some places than others. Precipitation happens in **low-pressure belts**, where air masses **converge** (meet) and air rises. The two main areas of year-round rainfall occur at the Equator and at mid latitudes, such as in the UK. Forests grow in both of these areas.

In polar and desert areas, high-pressure zones occur, causing very dry conditions.

An added complication is that the whole pattern of pressure belts changes with the seasons. Mediterranean and tropical areas sometimes become low-pressure zones and experience rainy seasons for nearly half the year.

## Local factors

As well as global factors such as temperature and precipitation, local factors affect plant growth:

- Altitude: temperatures decrease by 1°C for every 100 metres in height. So the top of Mount Kilimanjaro is covered in snow, although it is near the Equator.
- **Continentality** is the effect of distance from the sea and is also important. See the diagram on page 41.
- Nutrient-rich environments encourage the growth of ecosystems. Nutrients are supplied by the soil or from upwelling ocean currents.
- Geology, **relief** and drainage are also important.

+ **Pressure belts** are regions of the atmosphere which run around the Earth. They are parallel to the Equator. Some are high-pressure areas. Others are low-pressure areas.

## your questions

**1** Look at the photos and climate graphs for three biomes.

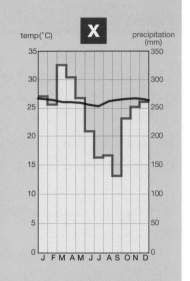

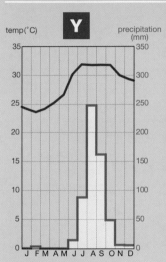

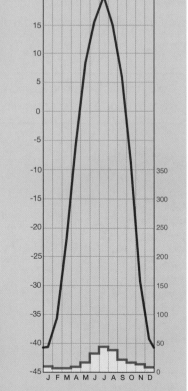

**1** Match the photos above (A,B and C) with the climate graphs (X, Y and Z).

**2** Which photo is a tropical grassland b northern coniferous forest c tropical rainforest?

**3** Design a table to compare the climates, using the 3 graphs. Include maximum and minimum temperatures, total annual rainfall, and the number of months where rainfall is over 50mm.

# A life-support system

**On your planet**

+ In 100 km² of rainforest, you could find 1500 types of flowering plants, 30 000 insect species, 150 butterfly species, 100 kinds of reptile, and 750 species of tree.

+ **In this section you will learn about the value of the biosphere as a provider of goods and services.**

The biosphere is a life-support system. It provides us with a wide range of **goods**, both for survival and for commercial use.

The biosphere also provides many vital **services**, such as:

- regulating the composition of the atmosphere
- maintaining the health of the soil
- regulating water within the hydrological cycle.

The problem is that different people want to use the same biome in different ways. If we overexploit forests or overharvest marine life, we aren't using the biosphere in a **sustainable** way. If the biosphere is damaged, it may fail to provide us with services. This can be disastrous.

+ **Sustainable** means a process that does no lasting harm to people or the environment.

| Green lungs | • Forests remove carbon dioxide from the atmosphere (carbon sinks). This reduces global warming. |
| | • Forests give out oxygen – purifying the atmosphere. |
| Water control | • Forests protect watersheds from soil erosion and intercept precipitation – preventing flash flooding. |
| | • By trapping silt, forests keep water pure. |
| | • Reefs and mangroves provide protection from coastal storms. |
| Nutrient cycling | Forests provide leaf litter which forms humus. This makes the soil more fertile for growing crops. |
| Providing habitats for wildlife/ biodiversity | Rainforests and reefs are very biodiverse. They provide 'homes' for a huge range of organisms, including some very rare animals. |
| Recreation | Reefs and rainforests provide attractive scenery for tourism. |

▲ How the biosphere serves you.

**On your planet**

+ Sometimes we use up biosphere resources so quickly that they are lost to future generations. For example, the **gene pool** found in rainforests might one day yield cures for cancer. But we won't find this out if we destroy rainforests.

+ The **gene pool** is the genetic information contained in living organisms.

▼ The biosphere provides us with goods and services.

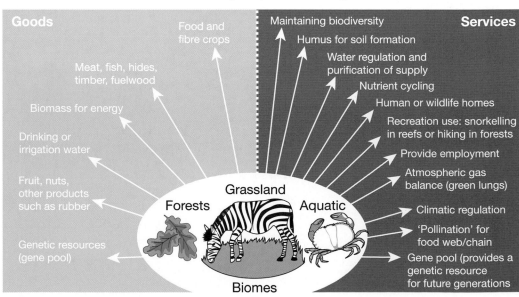

# Delivering the goods

For **indigenous peoples** in the tropical rainforest biome, the biosphere provides almost everything: fuelwood for cooking, timber for building, herbs for medicine, and foods such as nuts, fruit, meat and fish. They can grow subsistence crops, such as yams and millet, using the 'slash and burn' method. Many see this type of farming as being sustainable. The patch of forest is only used for 5-6 years. And once the soil is exhausted, the plot recovers with new forest gradually developing.

> **+ Indigenous peoples** are peoples who have originated in and lived in a country for many generations.

The biosphere also provides many goods for commercial use. But there is a problem. Rival commercial users can destroy the rainforest biome for short-term gain. **Transnational companies** exploit the forest by logging for timber or paper manufacture. They deforest the land to grow commercial plantations of rubber, cocoa, and palm oil. Commercial farmers also cut down trees to graze cattle or grow soya beans for biofuels. Drug companies search the forest for plants to provide the ingredients for new medicines. Mining companies search for minerals or oil. And governments may want to develop hydroelectric power.

> **+ Transnational companies** are giant companies operating in many countries.

| | Forested area | Deforested area |
|---|---|---|
| **Use** | Subsistence farming by indigenous local communities. | Ranching, mainly for poor-quality hamburger meat. |
| **Soil** | Soil is protected from heavy rain and is nutrient rich. Water moves slowly through the soil, preventing flooding. | Nutrients lost from soil due to heavy rainfall, and surface soil is washed away, blocking rivers and reservoirs. Rapid surface runoff leads to flooding. |
| **Trees** | Provide important habitat for wildlife. Tree roots bind the soil, preventing landslides. Wood provides fuel for local communities. | Loss of wildlife because of habitat destruction. Without roots to hold the soil together, landslides can occur. Source of wood for fuel is lost. |
| **Water** | Clean river water is fit for drinking. | Water is muddy and unsuitable for drinking. |
| **Economic and environmental gain** | Little economic gain, though the forest can support local indigenous communities without suffering permanent damage. | Reasonable economic gain, but only short-term. Deforestation can cause desertification. The natural environment would struggle to recover. |

▲ The effect of deforestation on the services offered by the forest.

## your questions

1 Write sentences to show how the biosphere keeps
   a air clean
   b water protected
   c biodiversity rich
   d tourists happy.

2 a Write 200 words explaining why indigenous peoples believe that they maintain the forest, and TNCs cause nothing but damage.
   b Write about 200 words to show how a TNC might reply.

3 Exam-style question Explain how one biome is being threatened by human interference. (4 marks)

✚ **In this section you will learn more about the different demands made on the biosphere.**

As you learnt in section 3.3, the biosphere provides a wide range of goods and services. But different people and organisations (known as **players**) want to use the biosphere in different ways.

Transnational companies and governments often have completely different ideas about the value of the biosphere. Their plans can conflict with the needs of local people. The table shows some of the conflicts which occur between the various players.

| | |
|---|---|
| Soufrière reef, St Lucia | Environmentalists wished to conserve the reef for tourism. But local fishermen needed to catch fish. Creating a marine reserve meant that fishermen were not allowed to operate in certain areas of the reef. Also see pages 115-117. |
| Temagarni region, Northern Ontario, Canada: an ancient red and white pine forest | There were many groups with conflicting interests:<br>• Forestry operations clashed with ecotourism developments.<br>• Commercial forestry conflicted with native land claims from the local people.<br>• Mining company developments clashed with forest conservation.<br>Therefore areas had to be developed for protection/conservation and production. |
| The Guyana mountains rainforest | The Guyanan government was short of money, because it had soaring foreign debts. It wanted to develop the forest for timber and mining. Environmentalists and local people were opposed to this. |

▲ Conflicting interests – commercial logging is carried out by transnational companies (left); local people collect fuelwood from the forest (right).

# How is the rainforest used?

The diagram here shows an **ecological tree**.
It shows some of the uses of a rainforest.
These include commercial and industrial uses,
ecological uses or services, providing for the
subsistence needs of local people, and possible
genetic uses.

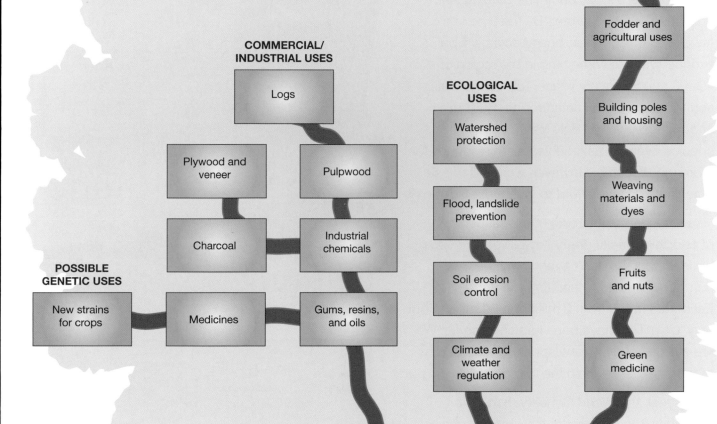

**COMMERCIAL/ INDUSTRIAL USES**

Logs

Plywood and veneer

Pulpwood

Charcoal

Industrial chemicals

**POSSIBLE GENETIC USES**

New strains for crops

Medicines

Gums, resins, and oils

**ECOLOGICAL USES**

Watershed protection

Flood, landslide prevention

Soil erosion control

Climate and weather regulation

**SUBSISTENCE NEEDS**

Fuelwood and charcoal

Fodder and agricultural uses

Building poles and housing

Weaving materials and dyes

Fruits and nuts

Green medicine

## your questions

**1 a** On a double page draw two copies of the ecological tree above. Include the boxes, but leave them empty.

**b** Complete the boxes of the left hand tree with all the things that would happen if TNCs and governments got rid of most of the rainforest.

**c** Complete the boxes of the right hand tree with all the things that would happen if the rainforests were well protected.

**2 Exam-style question** Explain the value of one biome you have studied. (4 marks)

### On your planet

➕ Chewing gum (as we know it) was first made from the resin of the chicozapote tree, found deep in the tropical rainforest.

On your planet
+ But it's not all doom and gloom: in 2007 conservationists saved 16 species of beautiful birds from extinction.

+ In this section you will learn how the biosphere is being degraded by human actions, both directly and indirectly.

## Species under threat

Norman Myers in his book *The Singing Ark*, published in 1979, wrote: 'By the time you have read this chapter, one species will be extinct. We lose something in the region of 40 000 species every year – 109 a day'. This sounds like a global catastrophe. But put another way, we will only lose 0.7% of all species over the next 50 years. This sounds far more manageable. However, we need to ask ourselves, can we afford to lose these species?

Headlines talk about extinctions, especially of cuddly animals, pretty flowers or gorgeous birds. In fact we don't even know the true number of species inhabiting the Earth. Estimates range from 2 million to 80 million, but only 1.4 million have been identified so far. We are still discovering new species in the depths of the oceans or within the rainforest.

Every year, the World Conservation Union publishes *The Red List* of threatened species. Currently over 16 000 known animal and plant species are in danger of extinction. Since records began, 30 years ago, 784 known species have been declared extinct. Increasingly habitats are being damaged and destroyed, and that increases the threat to species survival.

▲ Four species on the brink of extinction:
**a** Mexican black howler monkey
**b** Yellow-breasted capuchin (found in Brazil)
**c** Grey whale (Pacific Ocean)
**d** Riverine rabbit (South Africa).

▶ A success story: numbers of the white rhino, found in Africa, are on the rise again.

# Causes of biosphere threats

The causes of threats to the biosphere can be classified in two ways:

- Immediate causes, such as logging, overfishing, pollution.
- Root causes, such as rapidly expanding populations of people who use fuelwood, etc.

A further root cause is economic development, for example in China and India, which have a huge need for industrial raw materials. Living standards are also improving in these countries. This means that people now consume more food and fuel. All this puts a strain on the biosphere.

Certain species and parts of the world are particularly under threat. This includes 25 **hotspots** which have the greatest concentration of **biodiversity**. These areas cover less than 2% of the Earth's land area, yet account for 44% of all plants and 38% of all vertebrates. Most hotspots are in developing countries.

> **+ Biodiversity** is the range of animal and plant life found in an area.

| Threats to the biosphere | Impacts | Examples |
|---|---|---|
| Deforestation | Commercial logging destroys forest unless sustainable forestry principles are used. It affects rates of flooding, soil erosion and humus formation. | Logging in the rainforests of the Amazon and Indonesia. |
| Conversion to farmland or urban use | Commercial intensive farming destroys or alters the ecosystem. Urban sprawl destroys ecosystems and encourages wildfires. | Wheat farming in the American Prairies has removed natural grassland where bison once grazed. Soya beans are grown for biofuels. The urban sprawl of Los Angeles. |
| Overharvesting/ overfishing | Overharvesting causes wild animals to be hunted to extinction. Overfishing of some species such as krill, needed as food for fish farms, destroys food chains. | Big game, such as tigers (India) and rhino (Africa). The krill population (Antarctica and North Sea) is now managed by quotas. |
| Mining and energy | Mining cuts away whole hillsides. Opencast mining destroys the surface and restoration is only partly successful. Drilling for oil causes surface damage. Oil tankers cause oil pollution. | Removal of forests for mining iron ore in the Brazilian Amazon. Tin mining in Malaysia. Oil drilling in Alaska – a very fragile environment. |
| Pollution | Water pollution from sewage, fertilisers and industry. Toxic fumes emitted into the atmosphere, destroy species and damage ecosystems. | Too much silt and nutrients kill fish or coral. Acid rain kills forests. |
| Introduction of alien species | Sometimes we introduce new species deliberately. More usually they arrive by accident, e.g. via ships or aircraft. Alien species often breed well and take over. | Pheasants for shooting. Rhododendrons are alien species which poison the soil so that other plants won't grow there. Mink escaping from fur farms. |
| Tourism and recreation | Eco-tourism has little impact. But high-density mass tourism in fragile environments disturbs wildlife. | The Galapagos Islands are under threat due to tourism. |

## your questions

1 Draw a table with 3 columns, using the headings Environmental, Economic and Social. Using examples from this chapter, list the impacts of losing so many plant and animal species (a loss of biodiversity). Place them under the correct heading in your table.

2 'The economic gain is worth the environmental loss.' How far do you agree with this statement about the rainforest? Explain your view in 250 words.

**✚ In this section you will learn about species extinction and climate change.**

Over geological time (millions of years) ecosystems change. Species have become extinct as a result of natural events, such as long-term climate change. But scientists are now worried that destruction and degradation of the biosphere are happening at an accelerating rate. As a result, some species could be wiped out a hundred times faster than first feared. Another **mass extinction** may be on its way. And we may be playing a key role in this.

> **✚ Mass extinction** refers to the extinction of a large number of species within a short period of geological time.

## Our role in species extinction

Our impact on the planet is growing and we are threatening the ability of the biosphere to provide goods and services. Human impacts include:

- population growth. In the developing world, average population growth is 3% per year.
- rising consumption of resources, such as food, oil, water and minerals.
- human-induced climate change (**global warming**). This is happening now, and is probably the biggest single threat to the biosphere, and the species that live in it.

**On your planet**

✚ Some people say that animals should be helped to migrate to avoid the impact of global warming. Should we interfere in this way?

▶ Some impacts of climate change:

**a** Krill: rising ocean temperatures mean that numbers have fallen by 25% in 25 years. This affects Antarctic food chains.

**b** Emperor penguin: numbers fell by 50% in Adelie Land, Antarctica, when winter temperatures rose by 3 °C.

**c** Polar bear: some turn to cannibalism because the reduction in the amount of sea ice makes hunting difficult.

**d** Glaciers retreat as temperatures climb.

# Climate change

A recent headline suggested that half of all known species could become extinct due to future climate change. Global warming is occurring too rapidly for species to adapt. Already:

- there are fewer fish in African lakes
- plants flower earlier
- there are new patterns of bird migration.

Reports from seven continents suggest that rising temperatures are also altering vegetation belts. These are beginning to shift towards the poles by 6 km a decade.

Climate change will have many impacts on the biosphere, including these:

- Habitats will become increasingly fragmented (broken up).
- Habitats will change due to rising temperatures, changing rainfall patterns, and rising sea levels.
- Extreme weather events such as storms, floods and droughts may become more common.
- Species face extinction as they can't migrate to new habitats quickly enough.
- Pests and diseases will thrive in rising temperatures.

The table above right shows the possible effects of a 1-3°C rise in temperature.

| Temperature rise | Impact on species | Impact on environment |
|---|---|---|
| 1 °C | 10% of land species facing extinction | Disappearance of glaciers and mountain ecosystems leads to loss of alpine plants. |
| 2 °C | 15%-40% of land species facing extinction | Oceans become more acidic as more freshwater is added. Bleaching of coral kills reefs. |
| 3 °C | 20%-50% of land species facing extinction including the many hotspots, e.g. Kenyan coast | Sea level rises causing flooding of mangroves. Drought in the Amazon kills the rainforest. |

▲ The impact of rising temperature on the biosphere.

▲ The white dead nettle grows in Great Britain. It now flowers 55 days earlier than it did in the 1950s. This has led to a rise in leaf-eating grubs which benefit blue tits.

## your questions

1 Explain why the world's polar regions could lose a lot of their species because of climate change.
2 Explain the possible impact on farmers, gardeners, and you, of:
   a plants flowering earlier than usual
   b a greater number of storms, floods and drought
   c a continued rise in temperatures.
You might want to present your answer in a table.

*On your planet*

+ All the Arctic ice could disappear by 2030 due to global warming. There would only be ocean left.

In this section you will explore strategies for conservation.

## What should we conserve?

Sections 3.5 and 3.6 showed that human actions are threatening the biosphere. But, as the diagram shows, maintaining the biosphere and reversing its devastation will require vast sums of money and massive international effort. However, it's not really a question of can we afford the money, but one of can we afford *not* to spend it?

How do we decide which habitats and species to conserve? Here are some questions that we need to ask:

- Do we save the hotspots – the best bits which are most under threat? Or do we save representative samples of all the world's biomes?
- Is it best to get value for money by conserving areas in developing countries? Money goes further in these countries because of lower costs.
- Is it worth restoring completely devastated areas, because here the costs would be highest?
- What about species? Should we save high-profile animals like pandas and tigers? Or should we conserve the gene pool and **keystone species** like bees, which are so important for the whole food web? It's much easier to raise money to save whales or elephants than obscure insects.

Global, national and local initiatives are all very important in the battle for the biosphere. But there are often tensions over how to conserve it.

> **+** A **keystone species** is one which has a particularly large effect on other living organisms.

> **On your planet**
> **+** Should we only protect wildlife in its natural habitat? Or can zoos, seed banks and botanical gardens play an important role?

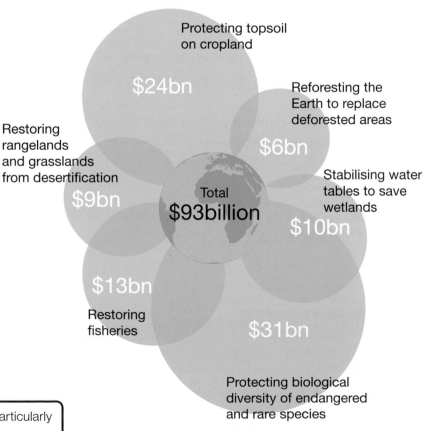

Protecting topsoil on cropland $24bn

Reforesting the Earth to replace deforested areas $6bn

Restoring rangelands and grasslands from desertification $9bn

Total **$93billion**

Stabilising water tables to save wetlands $10bn

$13bn

Restoring fisheries

$31bn

Protecting biological diversity of endangered and rare species

▲ The price of conserving the biosphere.

## Act global

Countries can get together to develop wildlife conservation treaties. Two good examples are:

- the RAMSAR Convention on Conserving Wetlands, signed in 1971 and adopted by 147 countries.
- CITES – the Convention on International Trade in Endangered Species, signed in 1973 and adopted by 166 countries. The CITES treaty lists the endangered species. The aim is to stop the trade in products such as elephant ivory or handbags made from crocodile skins.

International treaties, such as CITES, are very difficult to manage, there are so many conflicting interests. However, they do provide a useful legal framework for conservation.

## Act local

National policies can be delivered in a local area, involving local people. For example, Biodiversity Action Plans (BAPs) protect natural vegetation in Great Britain. These plans act at a local level. They arose from the Convention of Biodiversity at the 1992 World Summit in Rio de Janeiro.

## Act national

At a national scale, governments can set up protected areas, which help to conserve, manage and restore biodiversity.

- The map below shows the distribution of **National Parks** in England and Wales. Here the demands of recreation are managed in some of the most attractive, yet fragile, coastal and upland landscapes.
- **Community forests** have been established to provide new areas of trees near major cities.
- There are many different types of conservation area, all with different conservation aims.

Governments can also pay farmers to maintain and replant hedgerows and become more environmentally friendly.

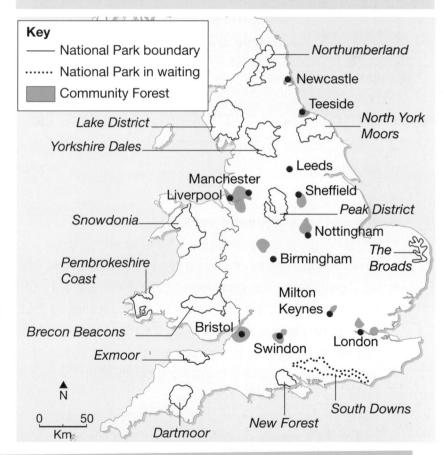

**Key**
— National Park boundary
······ National Park in waiting
▨ Community Forest

### your questions

1 Look at the diagram opposite. In pairs, imagine you have a budget of US$ 30 billion. What would your priorities be for conserving the biosphere? What would you save, and how might you do it?

2 What is the difference between National Parks and Community Forests? Use some named examples in your answer.

3 Go to www.ukbap.org.uk and look up your region. Use the website link to find out about plans for your area. Design a poster to explain the aims, and some protected species and areas.

4 **Exam-style question** Using examples, explain some ways of conserving threatened species.

✚ **In this section you will learn about sustainable management of ecosystems.**

In the 1980s, people realised that 'closing off' great areas of rainforests and reefs to try to protect them was not a success. People still carried out illegal poaching and harvesting. Another form of managing these areas was needed, and this was **sustainable management**.

Sustainable management of ecosystems can be thought of as a middle way between total protection of ecosystems (where no-one has access) and total exploitation where there is no protection.

*On your planet*

✚ In parts of southern Africa local people within reserves can make money from game hunting. Is shooting wildlife really sustainable management?

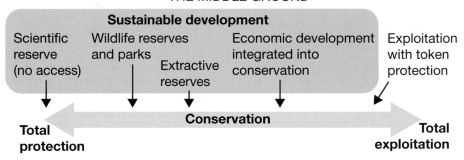

THE MIDDLE GROUND

**Sustainable management involves:**

- conserving the ecosystem for future generations, by ensuring that it isn't used faster than it can be renewed. You can do this by zoning.
- local people, so the ecosystem still provides them with resources.
- schemes which train and educate local people, so that they can be involved in decision making – as the most important **stakeholders**.
- helping local people living in poverty. Sustainable schemes allow local people to make a living from ecotourism, or by carrying out activities in the buffer zone.
- being environmentally friendly. It avoids practices like clear cutting, where the forest is completely destroyed. Trees are left to protect watersheds. Harvesting is selective, with only large trees logged. Only adult fish are caught.

**Key**

Core conservation area

Buffer zone – light use on rotational basis. This surrounds the core

▼ A sustainable forest reserve – Kilum in the Cameroon Republic. The land is divided into zones. These are used for different purposes and have different levels of protection.

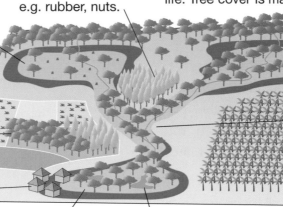

Area of selective logging. No clear cut felling so a tree cover is maintained.

Long-term leases/selective small-scale clearance with replanting.

Tree nurseries to replace cut down forest, i.e. afforestation.

Ecotourism

Multiple zoning, e.g. hunting, tourism, conservation.

Extractive reserve, e.g. rubber, nuts.

Size of reserve is large enough to support wild life. Tree cover is maintained on watershed.

Forest reserve protected area with minimum human interference.

Reserves linked by natural corridors for migration.

Agroforestry – maintains biodiversity of agricultural land. Crops can be grown beneath the shade of banana trees.

Tree cover in watersheds reduces flood risk and improves water quality and quantity for villagers.

▲ Local people plant new trees to replace the cut down forest.

## your questions

1 Look at the diagram above of the sustainable forestry reserve. Say how sustainable management of forests will work. Use these headings: Looking to the future, Community involvement, Eco-friendly and Pro-poor.
2 From what you have read here, how far do you think that the battle to conserve the biosphere has been won?
3 What do you think needs to be done next? Explain your answer.

✚ In this section you'll learn about the hydrosphere and hydrological (water) cycle.

## Water and life on Earth

Water is what makes Earth – the **Blue Planet** – unique and different from other planets in the solar system. Without it, life could not exist. It is more important than anything else on Earth. You can live for three weeks without food, but without water you'll be dead in three days!

The **hydrosphere** consists of all the water on the planet – in seas, oceans, rivers and lakes, in rocks and soil, in living things and in the atmosphere. Water exists on the Earth's surface and in the atmosphere in three states: as a liquid (water); as a solid (ice); and as a gas (water vapour).

Water can change states by:
* evaporation – liquid changes to vapour
* condensation – vapour turns into liquid
* melting or freezing.

## Water – a continuous cycle

Water flows in a never-ending cycle between the atmosphere, land, and oceans. The **hydrological cycle**, or water cycle, is a **closed system** – a bit like a central heating system in a house. The water goes round and round, but none is added or lost from the system as a whole, so the Earth gets neither wetter nor drier.

Think of the global water cycle as having a number of water **stores**, such as in rocks and soil, or lakes and oceans.

The table below shows how important the oceans are as a store, but remember that oceans store salt water which cannot be used by people unless it is converted into fresh water. This can be done but it is hugely expensive. Most of the fresh water is stored in ice sheets and glaciers (especially in Antarctica), and these stores are gradually being reduced as a result of global warming. Relatively small amounts of water are stored in rocks as groundwater, and in lakes and rivers, and these are in huge demand as sources of water.

Water stays in the stores for varying amounts of time, from around a day to thousands of years! Just think what happens after a heavy storm – water will be dripping off trees, and the grass will be wet for about 12 hours afterwards. The puddles will dry up after a day or so but water which finds its way underground can stay there for years. If rainfall is heavy and goes on for a long time, some of the smaller stores fill and we end up with floods.

| Store | Size (km³ x 10 000 000) | % of all water |
|---|---|---|
| Oceans | 1370.0 | 97.0 |
| Polar ice and glaciers | 29.0 | 2.0 |
| Groundwater | 9.5 | 0.7 |
| Lakes | 0.125 | 0.01 |
| Soils | 0.065 | 0.005 |
| Atmosphere | 0.013 | 0.001 |
| Rivers | 0.0017 | 0.0001 |
| Living things | 0.0006 | 0.00004 |

The photos show some water stores:
**A** Lakes
**B** Oceans and atmosphere
**C** Polar ice and glaciers
**D** Trees and vegetation

## your questions

**1 a** Make a copy of the diagram below.
   **b** Label A and B to show water changing state.
   **c** Label the water stores C, D, E and F.

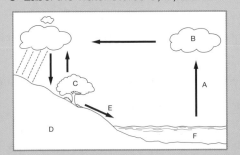

**2 a** Explain what is meant by a 'closed system.'
   **b** Why is the hydrological system on Earth a closed system?
**3** Draw an annotated pie chart to show why so little of the water on Earth is drinkable.

✛ In this section you'll find out about the role of the biosphere and lithosphere in the hydrological (water) cycle.

## The global hydrological cycle

The diagram on the right shows how the global hydrological (water) cycle works. The stores of water are linked by processes which transfer water into and out of them. These processes regulate the water cycle.

- Evaporation from oceans and rivers, and evapotranspiration from trees, condenses to cause precipitation (rainfall).
- This precipitation follows a number of routes:
  - Some runs off over the surface.
  - Some seeps into the soil or rock.
  - Some collects as snow or ice.

### The biosphere and lithosphere

The **biosphere** and **lithosphere** play a vital role in the water cycle, and act as sub-cycles. The diagram opposite shows these sub-cycles in a **river basin system** (part of the water cycle which operates on land). In the biosphere, trees intercept precipitation, and over half of it is then evaporated and transpired without ever reaching the ground. (This water is known as **green water**.) If the storm or rainfall is very heavy, or goes on for a long time, precipitation drips from the leaves and stems and slowly makes its way into the river system. Precipitation infiltrates into the soil, where it flows down slope as throughflow, or if the underlying rock is permeable, into the ground, to be stored as groundwater – a vital supply of water. Only after many hours is the water released into the river basin. Both the biosphere and the lithosphere help to regulate the water cycle.

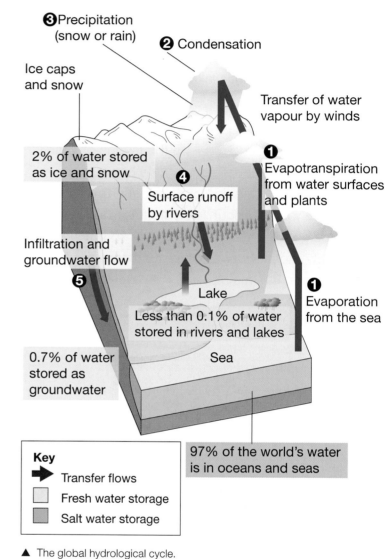

**❸** Precipitation (snow or rain)

**❷** Condensation

Ice caps and snow

Transfer of water vapour by winds

2% of water stored as ice and snow

**❶** Evapotranspiration from water surfaces and plants

**❹** Surface runoff by rivers

Infiltration and groundwater flow

**❺**

Lake

Less than 0.1% of water stored in rivers and lakes

**❶** Evaporation from the sea

0.7% of water stored as groundwater

Sea

**Key**
→ Transfer flows
☐ Fresh water storage
☐ Salt water storage

97% of the world's water is in oceans and seas

▲ The global hydrological cycle.

✛ The **biosphere** is the part of the Earth and atmosphere in which living organisms exist.

✛ The **lithosphere** is the outer layers of the Earth's surface (the crust and upper mantle).

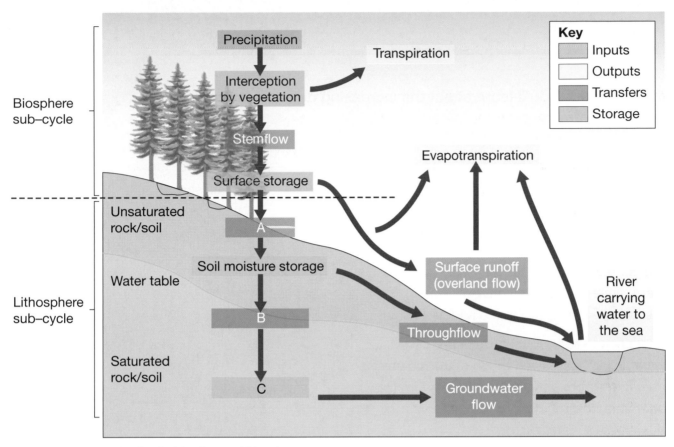

▲ A river basin system showing the biosphere and lithosphere.

**Infiltration** – movement of water into the soil from the surface

**Percolation** – movement of water into underlying rocks

**Groundwater storage** – water stored in rocks following percolation

**Saturation** – when soil is full of moisture

**Water table** – the level at which saturation occurs in the ground or soil

**Inputs** – things which enter the system

**Outputs** – things which leave the system

**Transfers or flows** – movements within the system

**Stores** – held within the system

On your planet
+ Did you know that some of the groundwater deep down in the Earth has been stored there for over 10 000 years?

## your questions

1 Make a copy of the river basin system above, and complete the terms in boxes A, B and C.

2 Define these words: evaporation, evapotranspiration

3 Using the diagram and definitions, explain the passage of water from the time it falls as precipitation, to the time it reaches the river.

4 **Exam-style question** Explain why the biosphere and lithosphere are important to the hydrological cycle.
   (4 marks)

+ In this section you'll learn about the increasing demand for water and water stress.

## Water crisis

The world is currently facing a freshwater crisis. Demand is soaring as population increases, and supplies are becoming increasingly unpredictable. Many economists and experts predict that in the future we will have **water wars**, where countries fight over water resources – especially in the Middle East.

As you can see from the graph, our use of water is increasing and much of the increasing demand is from agriculture. Modern farming often requires irrigation, which uses vast quantities of water.

The increasing use of water for agriculture is having a major impact on the amount of water remaining for other uses. A rapidly rising global population and constant economic development, especially in countries such as China and India, means that industrial and domestic demand has also grown. More water is needed for manufacturing industries, and rising living standards mean people use more water at home for showers, washing machines and so on.

Key:
- Reservoir losses
- Industrial
- Municipal/Domestic
- Agriculture

▲ Increasing global water use.

**Population growth**
World population grew from 2.5 to 6.2 billion between 1950 and 2000. Water supplies per person decreased by one third from 1970-1990. Seven billion people are likely to have insufficient water by 2050.

**Agricultural demand**
Rising population increases the demand for food and the water needed for farming. The area of irrigated land doubled in the twentieth century. Future water shortages could threaten food supplies in many developing countries.

**Energy**
Energy consumption has increased with population growth and industrial development. Developing countries have vast, untapped HEP resources. But storing water in reservoirs increases the amount of water lost through evaporation, and the risks of water-related diseases.

**Decline in water availability and water quality**

**Urbanisation**
By 2025 nearly 60% of the world's population will live in urban areas. The water supply and sanitation infrastructure won't be able to cope.

**Tourism**
There has been a massive increase in tourism in developing countries. But tourism developments (including hotels and golf courses) take a huge share of a region's water resources.

**Climate change**
Global warming and climate change will affect rainfall, evaporation and water availability. Places already suffering from water shortages are likely to experience lower rainfall. The dry areas of developing countries will be hardest hit.

**Industrial development**
The growth of manufacturing industry depends on water supplies. Industries such as steel and paper are major water users. They also use rivers and seas to get rid of waste.

▶ Pressure on water supplies.

# Water stress and scarcity

So far, freshwater isn't actually scarce.

- Globally, only half of the annual freshwater runoff (known as **blue water**) available for human use is currently used in agriculture, industry or for domestic use.
- But much freshwater is either inaccessible (i.e. it falls in the wrong place), or is available only at certain times of the year.

However, many parts of the world are now experiencing **water stress**. Many lakes, rivers and groundwater supplies are drying up from overuse. Water stress takes the form of shortages of water supplies, especially for irrigation. Turkmenistan and Uzbekistan in Central Asia are currently the most water-stressed countries in the world, as they use huge quantities of water for irrigating crops such as cotton.

The map shows that currently most of North and South America and Northern Eurasia have plenty of water, but other areas suffer from **water scarcity**. Areas, such as the south-west USA and Central Asia, are experiencing **physical water scarcity**, where demand exceeds local availability.

Other areas, such as sub-Saharan Africa, experience **economic water scarcity**. Here, there are sufficient supplies available, but people cannot afford to exploit them. They lack the money to build water storage facilities to provide water for dry seasons, or to distribute water to rapidly growing cities.

**What do you think?**
+ Could we really have water wars?

**On your planet**
+ The minimum amount of water each person needs for drinking, hygiene and growing food is 1000 cubic metres per year — equivalent to 40% of an Olympic-size swimming pool.

+ **Water stress** occurs when the demand for water exceeds the amount available during a certain period, or when it is not good enough quality to use.

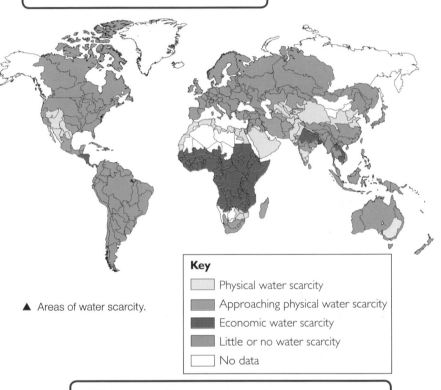

▲ Areas of water scarcity.

**Key**

- [ ] Physical water scarcity
- [ ] Approaching physical water scarcity
- [ ] Economic water scarcity
- [ ] Little or no water scarcity
- [ ] No data

+ **Water scarcity** comes in two types:

- **Physical scarcity**
  Shortages occur when demand exceeds supply.

- **Economic scarcity**
  When people simply cannot afford water, even if it is readily available.

## your questions

1 Look at the spider diagram opposite. Which two of the pressures on water supply are greatest? Explain your choice.

2 In groups of three produce a wall display about water scarcity, using both text and images. You should cover one area of physical water scarcity, one approaching physical water scarcity, and one of economic water scarcity.

In this section you'll find out about chronic water shortage in the Sahel, and how climate change may affect water supplies in other parts of the world.

## The Sahel

The graph shows some of the countries suffering from water stress and scarcity. Worldwide, over 2 billion people live in areas of water stress and 500 million live in a state of chronic water shortage. The Sahel region is one of those areas.

The Sahel is a narrow belt of semi-arid land immediately south of the Sahara desert, as shown on the map. Rain falls in only 1 or 2 months of the year, and both the total amount of rainfall (usually between 250 and 450 mm) and the length of the rainy season are very variable. The graph below the map shows that, since 1970, rainfall has more often than not been below average – and in some cases up to 25% below average. Sometimes the rain comes in torrential downpours and is then lost as surface runoff, causing flooding. In other years the rains fail completely – recent years have seen several lengthy droughts.

- Drought causes seasonal rivers and water holes to dry up and the water table to fall.
- Drought spells disaster for the **nomads** who graze animals, and for **subsistence farmers** who rely on rain to grow millet and maize.
- Grasses die, and soil erosion and desertification follow, due to overgrazing by animals.

Many of the countries in the Sahel are developing countries, such as Chad, Niger, Sudan, and Ethiopia, and are among the poorest in the world. They have rapidly growing populations, which puts pressure in drought years on failing food supplies. These semi-arid lands are a very fragile environment and water stress soon causes humanitarian crises. Famine happens regularly.

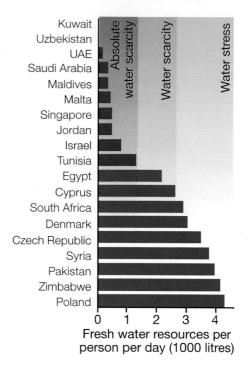

▲ Countries that are water-stressed, water-scarce or facing absolute water scarcity.

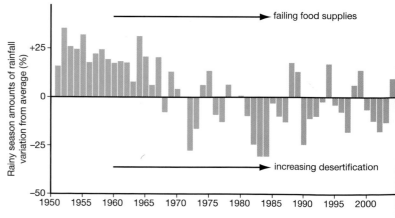

▲ The Sahel, and variations in rainfall.

# All change

The World Bank estimates that, by 2025, 50% of the world's population will face water shortages.

The top map on the right shows the impact of climate change on water scarcity by 2025.

Global warming will lead to:

- changes in climate patterns, so rain-bearing winds will reach some areas less regularly
- an increase in the rate of glacier melting
- more extreme weather events, with floods and storms as well as droughts.

The second map shows that the global effects of population growth and climate change combined could be devastating (in terms of water scarcity).

Richer countries can buy their way out of water stress – for example, Kuwait and Saudi Arabia can afford to build desalination plants (to convert sea water to drinking water) with the profits from their oil industries. However, rapid changes in weather patterns could spell disaster for poor subsistence farmers in developing countries, who rely on rain to water their crops. These farmers are increasingly vulnerable to unstable weather patterns and, for them, water insecurity will almost certainly lead to food insecurity and famine.

▼ Climate change will influence water scarcity ...

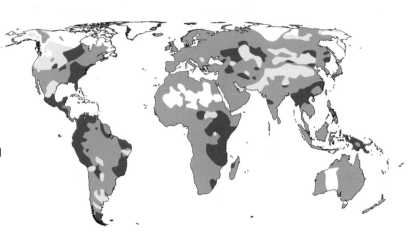

▼ ... but population growth with climate change could be devastating.

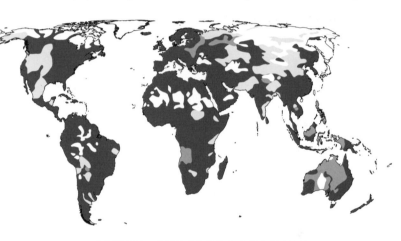

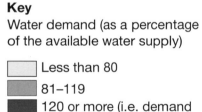

**Key**

Water demand (as a percentage of the available water supply)

- Less than 80
- 81–119
- 120 or more (i.e. demand is at least 20% more than the available supply)

## your questions

1 Explain what the graph opposite shows about rainfall in the Sahel since 1950.
2 Place these in order to show how drought leads to soil erosion: drought, roots can't hold the soil together, wind blows, grass dies, soil blows away, soil dries out.
3 Explain how and why farmers and cattle can make this worse.
4 Draw a table with two columns to compare the effects of climate change in wealthy and poor countries.

+ In this section you'll explore the human threats to water quality.

## Quality not quantity

Not only do we need enough water for everyday use, but the quality of our water is just as important as the quantity. People can suffer from water stress if their water isn't safe, or is contaminated.

The diagram below shows a wide range of sources of pollution. These sources can be classified as:

- domestic
- industrial
- agricultural
- transport-based
- other – those which don't fit into any of the above categories.

### Pollution in emerging and developing economies

Levels of water pollution can be related to economic development, as the graph opposite shows. The highest levels of water pollution are usually linked to the most rapid rates of economic growth – found today in countries such as India and China. These countries are industrialising and developing their energy sources rapidly. They tend to put economic growth before environmental protection. Countries like this are also experiencing very rapid urban growth, and cities are growing faster than the infrastructure for piped water and waste disposal systems can be installed. As a result, streams flowing through the slums and shanty towns of megacities are badly affected by pollution (see the photo).

Many countries with rapidly developing economies are also developing commercial agriculture, which relies on pesticides and fertiliser to increase yields. Runoff of those chemicals increases water pollution.

Developing countries generally lack the concentrations of industry found in more-developed countries. Up to 70% of people live and work in rural areas as subsistence farmers. While pollution exists widely, it tends to be less concentrated, e.g. many streams are polluted by raw sewage.

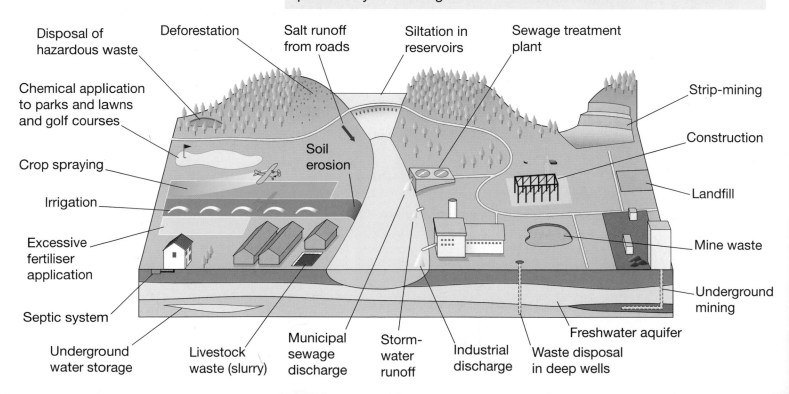

Disposal of hazardous waste

Deforestation

Salt runoff from roads

Siltation in reservoirs

Sewage treatment plant

Chemical application to parks and lawns and golf courses

Strip-mining

Construction

Crop spraying

Soil erosion

Landfill

Irrigation

Excessive fertiliser application

Mine waste

Septic system

Underground mining

Freshwater aquifer

Underground water storage

Livestock waste (slurry)

Municipal sewage discharge

Storm-water runoff

Industrial discharge

Waste disposal in deep wells

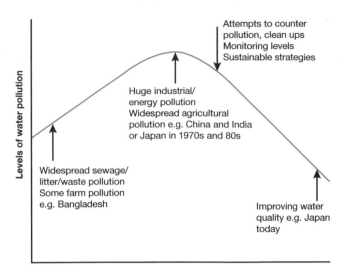

Attempts to counter pollution, clean ups
Monitoring levels
Sustainable strategies

Huge industrial/ energy pollution
Widespread agricultural pollution e.g. China and India or Japan in 1970s and 80s

Widespread sewage/ litter/waste pollution
Some farm pollution e.g. Bangladesh

Improving water quality e.g. Japan today

Levels of water pollution

Developing    Emerging economies    Developed
**Economic development**

## Pollution in developed economies

Developed countries, such as Japan or the UK, have taken big steps to control pollution. Their economies are heavily based on **tertiary** and **quaternary** activities, which cause less water pollution than **primary** and **secondary** industries.

In the late 1960s, Japan's lakes, rivers and seas were all polluted. The pollution caused major health problems and damaged ecosystems, for example mercury poisoning at Minamata. During the 1970s, the Japanese government produced standards to improve water quality, and tackle pollution. Lake Biwa and the Inland Sea were polluted which now have high water quality, with swimming areas and fish farms. The rivers entering Tokyo and Osaka Bay – two industrial areas – are now almost free of pollutants.

▼ The main sources of water pollution and their impacts on people and the environment.

| Type of water pollution | Impact |
|---|---|
| Organisms such as bacteria, viruses, worms, rats | Cause a range of diseases such as cholera, dysentery, etc. |
| Organic wastes such as domestic and farm sewage | Consumes oxygen in the water and kills many organisms |
| Inorganic solid compounds, such as toxic materials and cyanide from mines | Can poison stretches of river completely, killing all life |
| Plant nutrients from fertilisers leaching from the fields, e.g. nitrates and phosphates | Leads to **eutrophication** (water is deprived of oxygen by the rich nutrients) |
| Chemicals such as plastics, oil, pesticides, PCBs | Highly toxic – even at low concentrations – causes the death of most life in the river |
| Suspended solids | Affects the colour of the water and kills fish/shellfish |
| Radioactive substances | Can lead to cancer |
| Thermal pollution from hot waste water from cooling towers in power stations | Increases rate of decomposition of biodegradable waste, reducing the water's ability to hold oxygen |

## your questions

**1 a** Draw a table to classify all of the pollutants from the diagram on the opposite page. Use the headings domestic, industrial, agricultural, transport and other.

**b** Make another copy of the table, this time classifying all of the types of pollution, shown in the table above.

**c** Which category has the worst impacts? Explain your answer.

**2** Why do you think countries such as India and China do not have standards for water quality, like Japan now does?

**3** Exam-style question Using examples explain why pollution threatens water quality. (4 marks)

# Interfering in the hydrological cycle

+ In this section you'll look at some of the impacts of human interference in the hydrological (water) cycle.

## Human interference

Water isn't just for drinking and washing. It is used for industry, farming (irrigation), power generation (HEP), and also for waste disposal – plus we use it for recreation. Water creates wetland habitats which are important environmentally for their **biodiversity**. Many wetlands are protected as **Ramsar Sites**.

The wide range of uses reflects the processes and links within the water cycle. The diagram on the right shows how people intervene in the water cycle. While some of these have positive impacts, others have negative impacts on water supplies. And the downside of the links within the cycle means that over-use for one purpose, or disruption of the cycle, can have serious knock-on effects elsewhere in the system.

## Overabstraction

Overabstraction of water in the Thames Valley in Southern England in recent years has led to a dramatic drop in river flow, with some tributary streams drying up completely – damaging the river ecosystem, home to many plants and animals. Droughts in Southern England, and rising demand from increasing numbers of homes, has led to increased use of groundwater supplies. This has lowered the water table across the Thames Basin so the aquifer (water store) is not being used sustainably.

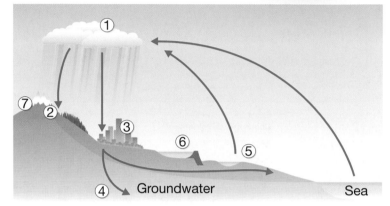

Groundwater        Sea

**Interventions**

① Cloud seeding to make rain

② Deforestation and changes in land use, leads to loss of interception capability and possible flooding (Himalayas)

③ Widespread urbanisation (cuts evapotranspiration)

④ Overabstraction of groundwater leads to falling water table

⑤ Overabstraction from rivers and lakes leads to conflict between users and loss of evaporation

⑥ Building of dams and larger reservoirs

⑦ Impacts of global warming melts glaciers

+ **Overabstraction** means too much water is being taken from the river, lake or other water source.

Most water companies now have strict policies called CAMS (Catchment Abstraction Management Strategies) for managing local water resources. Water levels are managed to keep the competing demands of the area in balance – sufficiently high for all the users, but not so high that there is an increased flood risk. The pie chart shows water abstraction in the Test-Itchen CAMS area in Hampshire.

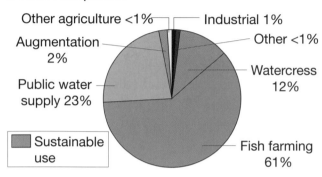

Other agriculture <1%
Augmentation 2%
Public water supply 23%
Industrial 1%
Other <1%
Watercress 12%
Fish farming 61%
■ Sustainable use

## Reservoir building

In some parts of the world, natural lakes are drying up, such as Lake Chad in Africa, but in other areas artificial reservoirs can add new stores to the water cycle.

There are several types of reservoir, including those that are used for HEP generation (which usually have high dams in order to get sufficient water to generate power) and those that are used for water storage (which usually have a much less spectacular earth dam). Reservoirs can be very useful, but they can also bring problems:

- Loss of land. In the UK, nearly 275 km$^2$ are covered by reservoirs, some of which have drowned whole villages and large areas of valuable farmland.
- In some countries they can be a source of disease, because they are home to insects such as mosquitoes.
- Vegetation drowned by the lakes decays and releases methane and carbon dioxide (greenhouse gases).

Reservoirs are not all bad news. Sometimes they are designed to be multi-purpose – like Grafham Water in the diagram on the right.

## Deforestation

Deforestation affects the water cycle:

- Removing the trees reduces evapotranspiration, so less green water is recycled. This can lead to a reduction in rainfall and the possibility of desertification.
- It exposes the soil surface to intense heat, which hardens the ground – making it impermeable and increasing runoff.
- It leads to a loss of soil nutrients (because of a loss of biomass). Nutrients are quickly flushed out of the system.
- Raindrop splash washes out the finer soil particles, leaving behind a coarser, heavier sand surface.
- It cuts out the process of interception. If trees are removed from watershed areas, this can increase the siltation in rivers and lead to increased flood risk.

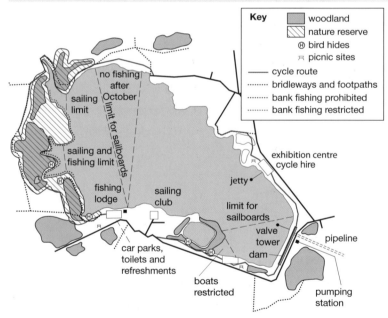

**Key**
■ woodland
▨ nature reserve
⊕ bird hides
⋔ picnic sites
— cycle route
···· bridleways and footpaths
···· bank fishing prohibited
···· bank fishing restricted

no fishing after October
sailing limit
limit for sailboards
sailing and fishing limit
fishing lodge
sailing club
car parks, toilets and refreshments
boats restricted
exhibition centre cycle hire
jetty
limit for sailboards
valve tower dam
pipeline
pumping station

Grafham Water is a water supply reservoir covering 600 hectares near Huntingdon in the UK. It has been zoned to allow wildlife conservation and different leisure activities, such as windsurfing, sailing, fly-fishing, cycling and birdwatching.

## your questions

1 Draw a flow diagram for each of the following, to show the knock-on effects: deforestation, urbanisation, dam building, over-abstraction.
2 Which of these has the worst effects? Explain your reasons.
3 **Exam-style question** Using examples, show how different water uses can have unintended effects. (4 marks)

In this section you'll evaluate large-scale solutions to managing water supplies.

## Sustainability

Water – demand is rising and supplies can be unpredictable. Can we manage our water supplies sustainably? If we use water faster than it can be replenished, its use is not sustainable. Not only that, but we need to store and distribute it efficiently – and this can be expensive.

There is still a huge amount to be done to provide access to reliable and safe water supplies for billions of people around the world. Strategies designed to use water more sustainably range from large-scale schemes which this Section looks at, through to small-scale solutions which Section 4.8 deals with.

## Large-scale water management projects

Large-scale projects to increase water supply tend to involve big dams. They are usually complex, multi-purpose projects. The Big Dam age began with the building of the Hoover Dam on the River Colorado in the USA in the 1930s. Since then, dams have increased in scale (because of improvements in concrete-making and earth-moving technology). Up to the 1980s the rate of building increased, until most of the best sites in North America and Europe had been developed, in the so-called '**Blue Revolution**'. Protests about dam building have slowed down the rate of development.

Currently dams supply:

- 40% of the world's irrigated water
- 20% of the world's electricity
- and 15% of all **blue water**.

A few countries, like New Zealand, get nearly all of their electricity from hydroelectric power, and so are very concerned about global warming's impact on their rainfall!

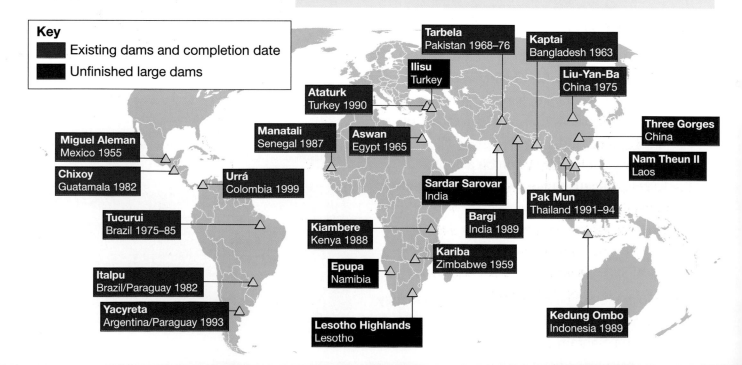

**Key**

Existing dams and completion date

Unfinished large dams

Tarbela Pakistan 1968–76
Kaptai Bangladesh 1963
Ilisu Turkey
Liu-Yan-Ba China 1975
Ataturk Turkey 1990
Manatali Senegal 1987
Aswan Egypt 1965
Three Gorges China
Miguel Aleman Mexico 1955
Nam Theun II Laos
Chixoy Guatamala 1982
Urrá Colombia 1999
Sardar Sarovar India
Pak Mun Thailand 1991–94
Tucurui Brazil 1975–85
Bargi India 1989
Kiambere Kenya 1988
Italpu Brazil/Paraguay 1982
Epupa Namibia
Kariba Zimbabwe 1959
Yacyreta Argentina/Paraguay 1993
Lesotho Highlands Lesotho
Kedung Ombo Indonesia 1989

Dams may bring benefits in terms of increased water supply but they can have negative impacts too (see below). Their construction plays havoc with fragile aquatic ecosystems, destroying fisheries and wildlife. Both natural and cultural resources are submerged. Environmental groups and human rights activists campaign to halt and restrict dam construction.

Around 45 000 large dams worldwide affect 6 out of 10 major rivers and have caused about 80 million people to be forcibly relocated (including around 2 million for China's Three Gorges Project).

Loss of farmland and villages

Habitat for water birds

Scenic asset – recreational use

Increase in humidity

Fish stocking

Water for domestic use and irrigation – can reduce quality

Sedimentation in lake

Dam interferes with logging, navigation and fish migration

Hydroelectric power attracts industry

Dam acts as a knickpoint – energy is reduced and deposition results

Less sediment means more energy, leading to 'clear water' erosion

Regulated flow – floods are controlled

## China's big schemes

China's centralised government enables it to develop huge schemes. Water is plentiful in the south of China, but scarce on the parched Northern Plains, and likely to get scarcer. China's answer to its water problems has been to develop major schemes such as the South-to-North Water Diversion Project and the Three Gorges Project (see above right).

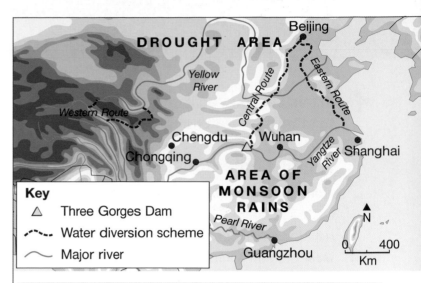

### South-to-North Water Diversion Project

This huge water diversion project will transfer water to the drier north of the country. Planned for completion in 2050, the scheme will eventually divert 44.8 billion m³ of water annually. The work will link China's main rivers and requires the construction of three diversion routes, stretching south-to-north across the eastern, central and western parts of the country. The complete project is expected to cost more than twice as much as the Three Gorges Dam. The diversion scheme has caused many environmental concerns, regarding the loss of ancient sites, the displacement of people and the destruction of pasture land. Plans for further industrialisation along the routes of the project pose a serious risk of pollution to the diverted water.

### Three Gorges Project

The Three Gorges Project is due for completion in 2009. The project will have 26 generators capable of generating 84.7 billion kWh of electricity a year. When the reservoir is filled in 2012, water will rise to a height of 175 metres, extend 600 kilometres and inundate 632 km² of land. the impact on local biodiversity has been devastating.

As well as generating electricity and helping navigation, the Three Gorges Dam will help to prevent seasonal flooding that annually threatens the lower reaches of the Yangtze River. The project has required the relocation of an estimated 1.4 million people from 1200 villages. It is believed that the water quality of Yangtze River tributaries has deteriorated after the reservoir began to store water, and some scientists fear the dam could significally change the salt content of the Sea of Japan, affecting the climate of the region.

## your questions

1 In pairs, draw a large table with social, environmental and economic down the left hand column, and benefits and problems across the top. Complete the table to show the advantages and disadvantages of a big dam.

2 How far do big dams produce economic benefits, but social and environmental problems? Write 300 words to explain your opinion.

3 Discuss as a class – are big dams sustainable?

4 **Exam-style question** Using examples show how big dams can bring either benefits or problems. (4 marks)

✚ In this section you'll look at small-scale sustainable solutions to managing water supplies.

## Small-scale solutions

**Non-governmental organisations** (NGOs), such as WaterAid or Practical Action, often develop small-scale sustainable solutions to local problems in developing countries. Local communities are involved in projects to develop safe and reliable water supplies. NGOs set up low-cost projects using **appropriate** or **intermediate technology**. This means that it is appropriate to the geographical conditions of the local area, and within the technical ability of the local community so that they can operate and maintain it themselves. Local people are trained to take responsibility for the development and management of the schemes. The aim is that when the NGO ends its involvement, local people can continue with the projects, and even replicate them elsewhere.

The schemes include rainwater harvesting (see the diagram on the right), protecting springs from contamination, developing gravity-fed piped schemes and building hand-dug or **tube wells** for villages. Water can be obtained from hand-dug wells using buckets or hand/treddle pumps. Recently, scientists at Bristol University have developed a simple hand-held device for communities to check the safety of their water supplies. Other NGO schemes aim to purify water supplies.

> ✚ **Tube wells** are built where the water table is too deep to be reached by a hand-dug well.

> ✚ **Appropriate** or **intermediate technology** – development schemes that meet the needs of local people and the environment in which they live.

Gutters collect rainwater

Tank is made from clay covering a simple bamboo frame

Local people make the taps and dig the collection pit

Each tank takes 3 days to make and can be maintained by the family

▲ A rainwater harvester for use in rural villages.

▶ A handpump at a hand-dug well.

Many of the NGO projects are in rural areas but others have been developed to supply fresh water for the shanty towns in urban areas. They have also developed projects to help with sanitation problems, such as low-cost pit or composting toilets, which will prevent water supply contamination.

### Dhaka, Bangladesh

Old Zhimkhana, a slum community built on the site of a disused railway station in Dhaka, Bangladesh, had no safe water or toilets. But with the help of an organisation called Prodiplan (one of WaterAid's partners) things are changing. Prodiplan is working with the community to help deliver water, sanitation, and hygiene education.

Six deep tube wells have been constructed, saving people time and energy in collecting water, and two new sanitation blocks provide toilets and water for washing.

People in Old Zimkhana are no longer continually ill. People run the facilities themselves and the project has helped them begin to move out of poverty.

### Large scale or small scale?

Are there any snags to small-scale intermediate technology schemes? Some don't succeed because they can be inefficient. In many African countries, huge numbers of people suffer from HIV/AIDS and people may be too ill to operate them. But, in general, intermediate technology works well and is often more sustainable than the large-scale schemes. Access to clean water can make a huge difference to people's lives. In the past, people (usually women) often had to walk long distances to get water. Now they can have safe drinking water, and water for cooking, washing, and personal hygiene.

*What do you think?*

✚ Large-scale schemes or small-scale projects – which is the best way to get sustainable water supplies?

## your questions

1 Explain how a hand-dug well works. Draw a diagram to help your answer.
2 **a** In one column of a table, write out the main benefits and problems with hand-dug wells. In a second column, compare the main advantages and disadvantages of big dams. (Look back to section 4.7.)
  **b** Which is more sustainable? Explain your answer.
3 Look at the pie chart below showing use of water in England and Wales.
  **a** Add up percentages to show total water used for **i** personal hygiene **ii** sanitation **iii** food and cooking **iv** leisure.

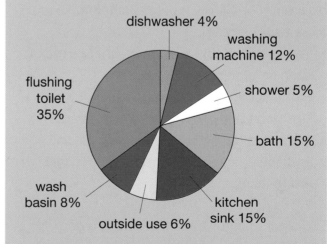

- dishwasher 4%
- washing machine 12%
- shower 5%
- bath 15%
- kitchen sink 15%
- outside use 6%
- wash basin 8%
- flushing toilet 35%

  **b** For each of these, suggest ways of reducing the amount of water used.
4 **Exam-style question** Using examples, explain how water use could be made more sustainable.
(4 marks)

➕ In this section you will learn about the coastal zone, and investigate how rock type influences coastal landforms.

## The coastal zone

The coastal zone is the transition zone between the land and the sea. Coasts are very dynamic places, and are always changing. The way a coast looks, and how it changes, is a result of the land and the sea working together. Coastlines are very popular places for people. Being on a coast has many advantages:

- Access to the sea for fish and other resources, as well as access to farmland.
- Good access for trade, and for connecting to other places.
- Recreation and tourism opportunities.

## Geology and rock type

The most important feature of a coast is often the type of rock in the area. Some rocks are resistant to **erosion**, whereas other rocks are more easily eroded.

- Very resistant rocks are hard igneous rocks, such as granite and basalt.
- Some sedimentary rocks, like sandstone, limestone and chalk, are fairly resistant.
- Weaker sedimentary rocks, such as clay and shale, are the least resistant and will erode fastest.

The diagram shows the three main types of coastal erosion.

> ➕ **Erosion** is the process of wearing away and breaking down rocks. There are three main types of coastal erosion: abrasion, attrition and hydraulic action.

▲ Resistant rocks tend to produce cliff coastlines, while less resistant rock types produce much gentler features

1. Water is forced into cracks in the rock. This compresses the air. When the wave retreats the compressed air blasts out. This can force the rock apart. This is called **hydraulic action**.

2. Loose rocks, called sediment, are thrown against the cliff by waves. This wears the cliff away and chips bits of rock off the cliff. This is called **abrasion**.

Cliff

Waves crashing against cliff

3. Loose sediment knocked off the cliff by hydraulic action and abrasion is swirled around by waves. It constantly collides with other sediment, and gradually gets worn down into smaller, and rounder sediment. This is called **attrition**.

# Coastal landforms

On coasts with hard, resistant rocks, erosion is slow. It may only be a few millimetres or centimetres a year. Most erosion happens during big storms, when waves are very powerful. Gradually, erosion produces certain characteristic landforms that give the coast its shape.

- Wave power is concentrated at the base of a cliff, where abrasion forms a **wave-cut notch**.
- Above this notch there is a cliff **overhang**.
- As the notch grows, the overhanging cliff become unstable and eventually collapses.
- The resulting pile of rock debris at the base of the cliff protects the cliff from further erosion.
- Over time, the loose rock is eroded by attrition, exposing the cliff to erosion again.

Over thousands of years, a succession of wave-cut notches form, and the cliff collapses again and again. Gradually the cliff retreats inland. You can tell where the cliff line once was, because a level area of smooth rock is left (called a **wave-cut platform**), which stretches out into the sea.

## Hydraulic action

Hydraulic action produces coastal landforms by eroding weaknesses in the rock – cracks, joints and fissures. This produces the classic sequence of landforms shown in the diagram below. It starts with a large crack, which grows into a cave, then forms an arch, then a stack and finally a stump.

Hard rock coasts erode slowly. But, from time to time, a slab of cliff, or an arch, collapses spectacularly.

*On your planet*

+ In 1990, the London Bridge Arch in Victoria, Australia suddenly collapsed. Two people were left stranded on the newly formed stack and had to be rescued by helicopter!

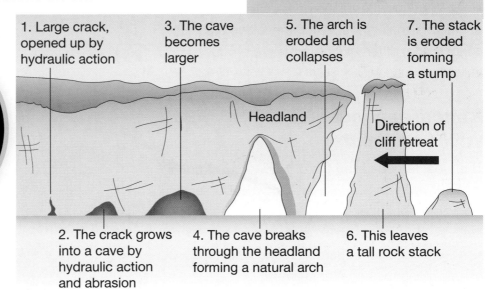

1. Large crack, opened up by hydraulic action

2. The crack grows into a cave by hydraulic action and abrasion

3. The cave becomes larger

4. The cave breaks through the headland forming a natural arch

Headland

5. The arch is eroded and collapses

6. This leaves a tall rock stack

7. The stack is eroded forming a stump

Direction of cliff retreat

## your questions

1 Make a list of the advantages of living on a coast.
2 What does erosion mean?
3 Describe the three ways cliffs erode.
4 Look at the diagram showing how arches, stacks and caves form. Imagine you could visit the coast in 50 years' time. Explain:
   **a** how the large crack might have changed
   **b** what could have happened to the arch
   **c** what might have happened to the stump.

✚ In this section you will explore how rock structure influences the shape of the coast.

## Rock structure

Rock structure simply means the way different rock types are arranged. Rocks are generally found in layers, called strata. This means there may be several types of rock in one cliff (see right). The cliff will only be as resistant as its weakest layers. Rock strata can be arranged in two ways along coastlines:

- If the layers are parallel to the coastline, the coast is **concordant**.
- If the layers are perpendicular (at 90°) to the coast, the coast is **discordant**.

Concordant coasts have the same type of rock all along the coastline. Discordant coasts have lots of different rock types. When these two types of coast erode, different landforms are produced, as the diagram shows.

## Concordant coasts: coves and cliffs

The most famous example of a concordant coast is at Lulworth on the Jurassic Coast in Dorset (see the photo opposite). This is a World Heritage Site, because the geology here is so important.

- A resistant layer of hard Portland limestone runs along the coast at Lulworth.
- Hydraulic action and abrasion have eroded this and 'punched' through, exposing the less resistant rock behind.
- Where the waves have been able to reach the softer rock, a **cove** has quickly widened.
- The erosion at Lulworth slows when the waves reach the more resistant chalk at the back of the cove.
- A steep chalk cliff has formed at Lulworth.

This resistant sandstone layer forms a very large overhang

Coal, a very weak layer. A Wave cut notch has formed.

Ironstone, a resistant layer

Shale, a weak layer

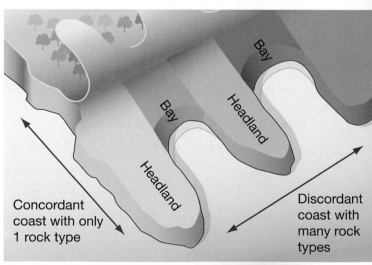

Concordant coast with only 1 rock type

Bay

Bay

Headland

Headland

Discordant coast with many rock types

✚ A **cove** is an oval-shaped bay with a narrow opening to the sea.

*On your planet*
✚ Because coves have narrow entrances from the sea, but sheltered beaches hidden by steep cliffs, they were often used by smugglers in the past.

Close to Lulworth Cove is Stair Hole. This is a cove that is beginning to form. The sea has punched an arch in the resistant limestone, and is widening the cove behind.

## This cliff is faulty!

Another factor which influences erosion is weakness in the rock forming the cliffs. There are two types of weakness:

- Joints are small, natural cracks, found in many rocks.
- Faults are larger cracks caused in the past by tectonic movements.

The more joints and faults there are in a cliff, the weaker the cliff will be. Hydraulic action attacks faults and joints, causing erosion.

## Discordant coasts: headlands and bays

In south west Ireland there is an unusual coastline, with very long headlands and bays. This is a discordant coast and is shown below. Layers of resistant sandstones and less resistant limestones are found along the coast. Waves have eroded the limestone to form bays, leaving the harder sandstone as headlands.

Resistant chalk

Steep cliff

Less resistant sands and clays

Resistant Portland limestone

Fairly resistant limestone and shale

Lulworth Cove

Stair Hole

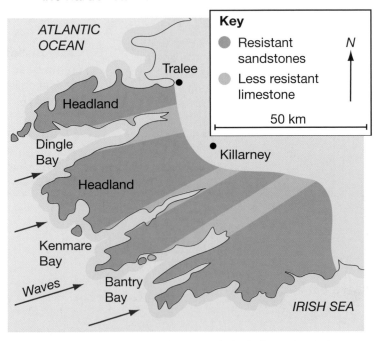

ATLANTIC OCEAN

Tralee

Headland

Dingle Bay

Headland

Kenmare Bay

Waves

Bantry Bay

Killarney

IRISH SEA

**Key**

Resistant sandstones

Less resistant limestone

N

50 km

## your questions

1 Draw a sketch of the cliff photo on page 74 and label it to show the most and least resistant rock.
2 Explain in a sequence of diagrams how it got to look like this.
3 Study the photo of Lulworth Cove and put these changes into a correct sequence:
 - erosion by sea
 - less resistant sand and clays are eroded
 - cuts through resistant limestone
 - cove forms
 - cliffs of resistant limestone
 - sea can't erode resistant chalk so widens cove
 - forms a break in the cliff
 - sea reaches resistant chalk
4 Describe how Stair Hole will change in the future as it continues to erode.

+ In this section you will learn how waves form, and how different types of waves affect beaches.

## What causes waves?

If you blow across the surface of a glass of water, ripples will form. The same process forms waves in the sea. When wind blows across the sea, friction between the wind and water surface causes waves (see below). The size of the waves depends on:

- the strength of the wind
- how long the wind blows for
- the length of water the wind blows over – called the **fetch**.

Some waves have a fetch of thousands of miles. Waves can start near Florida and travel right across the Atlantic Ocean before hitting Cornwall, a fetch of about 6000 kilometres.

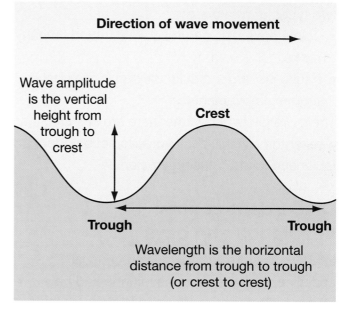

Direction of wave movement

Wave amplitude is the vertical height from trough to crest

Crest

Trough          Trough

Wavelength is the horizontal distance from trough to trough (or crest to crest)

▲ Wavelength and amplitude

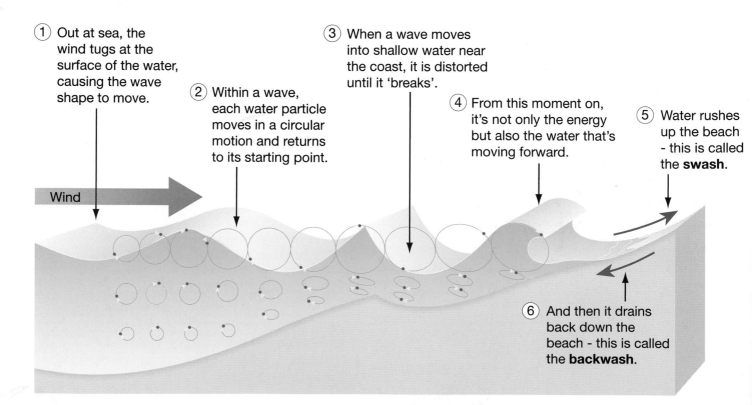

① Out at sea, the wind tugs at the surface of the water, causing the wave shape to move.

② Within a wave, each water particle moves in a circular motion and returns to its starting point.

③ When a wave moves into shallow water near the coast, it is distorted until it 'breaks'.

④ From this moment on, it's not only the energy but also the water that's moving forward.

⑤ Water rushes up the beach - this is called the **swash**.

⑥ And then it drains back down the beach - this is called the **backwash**.

Wind

# Summer waves and winter waves

Not all waves are the same. The shape of a beach, or **beach profile**, is a result of how waves break on the beach.

In the summer, waves are small. They are called spilling waves. They have long wavelengths and low amplitudes. These are the type of waves everyone runs away from on summer holiday; when they break they spill up the beach!

- They have a strong swash.
- This transports sand up the beach.
- The sand is deposited as a bank or beach berm.

Because these waves build up the beach by depositing sand on it, they are called constructive waves (see top right).

In the winter, when storms and strong winds are more common, waves are different. They are taller (larger amplitude) and closer together (shorter wavelength). These are called plunging waves.

- They have a strong backwash.
- This erodes sand from the beach.
- This creates a steep beach profile.
- The sand is carried offshore by an underwater rip current.
- The sand is deposited out at sea, forming an offshore bar.

Plunging waves are dangerous, because they arrive one after another, very quickly. After a wave breaks, the next incoming wave arrives so quickly that the backwash has to flow under the incoming wave. This flow is called a rip current (see bottom right). These currents can be very strong and can drag weak swimmers out to sea, under the water.

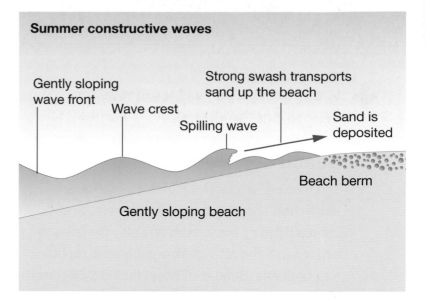

**Summer constructive waves**

Gently sloping wave front

Wave crest

Spilling wave

Strong swash transports sand up the beach

Sand is deposited

Beach berm

Gently sloping beach

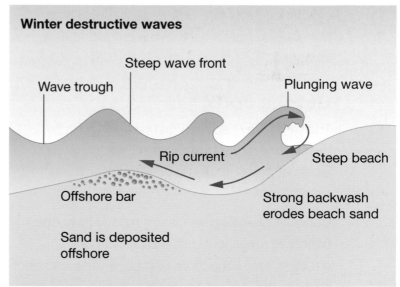

**Winter destructive waves**

Steep wave front

Wave trough

Plunging wave

Rip current

Steep beach

Offshore bar

Strong backwash erodes beach sand

Sand is deposited offshore

## your questions

1 a Using an atlas, measure the fetch in kilometres:
- from Calais to Dover
- from Denmark to the Yorkshire coast
- from Greenland to the west coast of Scotland.

b Which of the three places in the UK will get the biggest waves?

c Explain why this is so.

2 What is meant by swash and backwash?

3 Explain why:

a beaches in the summer are flat

b beaches in the winter are steep.

➕ In this section you will learn how coastal processes can create depositional landforms.

## Beach sediment

The material eroded from cliffs by hydraulic action and abrasion is called sediment. Sediment comes in many sizes, from tiny clay particles to larger sand and silt, right up to pebbles, cobbles and boulders. Sand is material that is 0.06-2 mm in size. Over time, attrition will make sediment smaller and rounder. Beach sediment is also transported from where it was eroded to new locations.

## Get the drift?

The main way sediment is transported is by **longshore drift**. This happens when waves break at an angle to the coast, rather than parallel to it.

Because prevailing winds are mostly from one direction, longshore drift is usually in one direction too. Longshore drift transports sediment along coastlines, as the top right diagram shows – sometimes for hundreds of kilometres before it is eventually deposited.

## Depositional landforms

When rocks are eroded, creating sediment, it will first be deposited very close to where it eroded. In a partly sheltered area such as a cove or bay, a beach will form, because the sediment is trapped in the bay.

Sediment transported by longshore drift will create new landforms where it is deposited – as the bottom right diagram shows.

Many **beaches** are simply rivers of sand and shingle (pebbles) slowly moving along the coast,

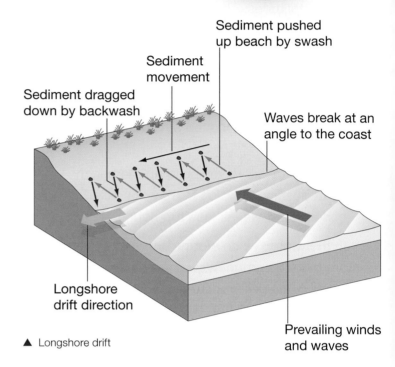

Sediment pushed up beach by swash

Sediment movement

Sediment dragged down by backwash

Waves break at an angle to the coast

Longshore drift direction

Prevailing winds and waves

▲ Longshore drift

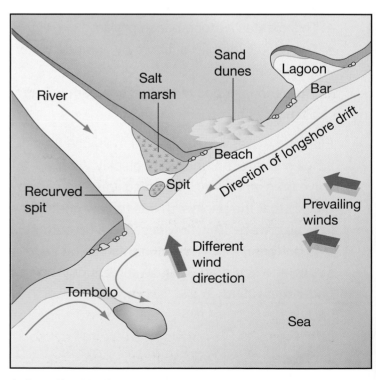

River

Salt marsh

Sand dunes

Lagoon

Bar

Direction of longshore drift

Beach

Recurved spit

Spit

Prevailing winds

Different wind direction

Tombolo

Sea

▲ Depositional landforms

as sediment is transported by longshore drift.

- Strong onshore winds can blow sand inland, forming **sand dunes** parallel to the shoreline.

- Small bays on the coast can sometimes be blocked by a **bar** of sand which grows across the bay. This is pushed across the mouth of the bay by longshore drift. Behind the bar, a shallow **lagoon** forms. These are often important habitats for birds.

- At a river mouth, longshore drift pushes sediment out into the river. It is deposited forming a long neck of sand and shingle called a **spit**. The spit will stop growing when the deposition of sand by longshore drift is balanced by erosion from the river. In the calm water behind a spit, **salt marshes** form.

- Many spits have a hooked or **recurved** end. This is caused by waves from a different direction pushing the end of the spit upriver, curving its end.

- An unusual landform called a **tombolo** can form when a beach grows out to meet an island just offshore. This is often produced when two longshore drift currents from different directions meet.

Depositional landforms are made of loose sediment, so they are not very stable. Sand dunes form when plants grow on the sediment helping to stabilise it. Plants that grow on beach sand need to be tough.

- They have long roots to hold them in place in the strong winds.

- They have tough, waxy leaves to stop them getting sandblasted.

- They can survive being sprayed by salt water.

**Marram grass** is one of the tough plants that colonises sand dunes. But even when depositional landforms are strengthened by the presence of plants, storms can often cause damage. Any damage is repaired by the processes of longshore drift and deposition.

✚ In this section, you will examine how climate change might increase the risk of erosion and flooding on coastlines.

## Rising sea levels

Many scientists fear that global warming will cause sea levels to rise. How much they will rise by is not known, but there are estimates of between 30 cm and 1 metre by the year 2100.

- Sea level is rising today, as the sea is warming up and expanding.
- Melting ice sheets are likely to speed up the rise.

For people who live on very low-lying land next to the sea, this could spell trouble. There are many areas around the world at risk:

- In Bangladesh, if sea levels rose by 1 metre, up to 15% of the country might be flooded.
- In the UK, London and Essex are at risk, because they are low-lying.
- Many small coral islands in the Pacific and Indian Oceans, like the Maldives and Tuvalu (pictured), could disappear underwater.

## Flood risk

Sea levels are constantly changing. Twice a day, due to the gravity of the moon, high tides cause raised sea levels. A few times every year there are exceptionally high tides, called spring tides. During spring tides the flood risk rises.

If spring tides occur when there are large waves, the sea will be even higher. Worse, if spring tides and waves combine with low air pressure, then a **storm surge** can form. Storm surges are caused by hurricanes and depressions, which are both low-pressure weather systems.

It is possible that global warming could make hurricanes and depressions more powerful. They might also become more frequent. This would mean storm surges would happen more often. If melting ice sheets raise sea levels as well, the combined results could be very serious indeed.

"There is a threat we could become the world's first global warming refugees"

Spokesperson for the Maldives Government, 2006

"Thousands flee to higher ground as stormy seas gather momentum"

The Times, Nov 2007

✚ During a **storm surge**, sea level rises. This is because the air pressure falls. Sea level rises by 10 mm for every 1 millibar drop in air pressure.

### On your planet

✚ In 1953, when there was a 5 metre storm surge in the North Sea, over 300 people were killed in the UK and 1800 in the Netherlands in devastating flooding.

◀ An exceptionally high tide washes waves into a home on the island of Tuvalu in the Pacific Ocean.

The diagram on the right shows how vulnerable some people could be to the combination of spring tides, storm surges and rising sea levels.

As the newspaper headlines show, coastal flooding is a real risk. In the UK in November 2007, a 3 metre storm surge was predicted, and thousands of people were evacuated from homes in North Sea coastal areas. In the event, very little flooding occurred, but storm surge floods may become increasingly common.

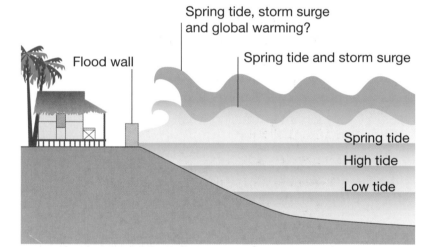

## Erosion

With higher sea levels in the future, and possibly increased storms, some coastlines will experience faster erosion rates.

This is a real problem for the east and south coasts of the UK, where the rocks are less resistant. Clays, shales and mudstones are soft and easily eroded. Sea level rises of 50 cm would increase erosion rates and make existing sea defences useless. The only choice would be to try and build new, expensive defences, or abandon some areas to the sea.

### your questions

1 What are the estimates of sea level rise by 2100?
2 Use an atlas to find 10 major cities in the world that are on the coast, and that could be at risk from rising sea levels.
3 Explain how a storm surge forms.
4 **Exam-style question** Using examples, explain how sea level rise could threaten people and their property. (6 marks)
5 a Use the Met Office website http://www.metoffice. gov.uk to research the causes and impacts of the 1953 North Sea floods.
  b Make brief case study notes on the 1953 flood under the headings Causes and Impacts.

In this section you will investigate why some coasts are eroding very rapidly, and what impact this has on people.

## A complex problem

Some coastlines are eroding quickly by over 2 metres a year. In the UK, the Holderness coast in Yorkshire, the North Norfolk coast and some coastal areas of Hampshire and Dorset have very high erosion rates. All of these locations have weak rocks that easily collapse. They all suffer from erosion (hydraulic action and abrasion), but **weathering** and **mass movement** are making the problem much worse. Weathering and mass movement are called **sub-aerial** processes. Water movement is often important in causing weak cliffs to collapse.

> **+ Weathering** is the breakdown of rocks *in situ*. This means it happens where the rock is. Rocks are weakened by being chemically attacked and mechanically broken down.

> **+ Mass movement** is the movement of materials downslope, such as rock falls, landslides or cliff collapse.

## Christchurch Bay

Christchurch Bay on the UK's south coast has a severe problem. Without any management, the cliffs could erode by over 2 metres a year.

This would threaten many of the towns and residential areas along the coast, such as Barton-on-Sea (see the map opposite).

## Why cliffs collapse

- **Marine processes**
1 The base of the cliff is eroded by hydraulic action and abrasion (cliff foot erosion), making the cliff face steeper.

- **Sub-aerial processes**
2 Weathering weakens the rock. This can be mechanical weathering, like freeze-thaw action, or chemical weathering, like solution.
3 Heavy rain saturates the permeable rock at the top of the cliff. Rainwater may also erode the cliff as it runs down it or emerges from the cliff at a spring (cliff face erosion).
4 The water flows through the permeable rock, adding weight to the cliff.

- **Human actions**
5 Building on top of the cliff adds a heavy load, which pushes down on the weak cliff.

During a big storm, heavy rain saturates the permeable rock, and erosion by the sea undermines it. Eventually a large chunk of cliff gives way and slides down the cliff – a rotational slide (see the photo opposite).

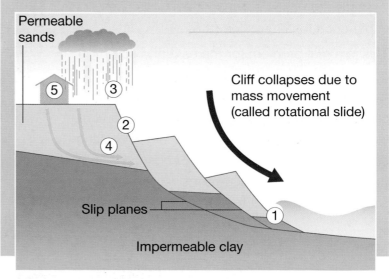

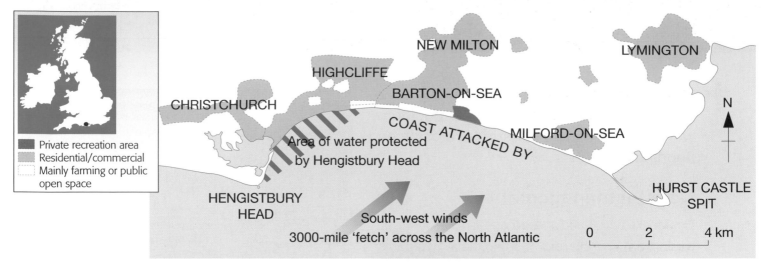

Private recreation area
Residential/commercial
Mainly farming or public open space

NEW MILTON

LYMINGTON

HIGHCLIFFE

BARTON-ON-SEA

CHRISTCHURCH

COAST ATTACKED BY

MILFORD-ON-SEA

N

Area of water protected by Hengistbury Head

HENGISTBURY HEAD

South-west winds
3000-mile 'fetch' across the North Atlantic

HURST CASTLE SPIT

0    2    4 km

At Barton-on-Sea, mass movement is the major cause of cliff erosion. But bad weather, cliff foot erosion, weathering, and water movement in the cliff all play a part. Even human activity, like building at the top of the cliff, increases the risk. This makes managing this type of erosion very difficult.

## Impacts

Erosion in Christchurch Bay affects many people:

- Homeowners could lose their homes to the sea. House values will fall, and insurance may be impossible to get.
- Rapid cliff collapses are dangerous for people on the cliff top, and on the beach.
- Sometimes roads and other infrastructure are destroyed.
- Many people would say the erosion makes the area unattractive.

People who live in Christchurch Bay have argued that they need sea defences to protect their coast. These are very expensive, and there is no agreement about which type of defence works best.

*On your planet*
+ 1 metre of sea defences costs about £5000 to build!

### your questions

1 What does the term mass movement mean?
2 Explain how weathering is different to erosion.
3 Look at the map of Christchurch Bay. Make a list of the type of people who might live and work here.
4 Read the section on why cliffs collapse, and draw a spider diagram to show all the different factors.
5 People who live in Christchurch Bay live in fear of a major storm. Why is this?

# Managing the coast

+ In this section you will learn what coastal management is, and examine the traditional ways of protecting a coast from erosion and flooding.

## Coastal management

Engineers who build sea defences to protect the coast from erosion and flooding, have two basic choices:

- **'Hard' engineering** – using concrete and steel structures, such as sea walls, to stop waves in their tracks.
- **'Soft' engineering** – using smaller structures, sometimes built from natural materials, to reduce the energy in waves.

The 'hard' method is the traditional method of coastal management. However, it has two major problems:

- It is very costly.
- It usually makes the coast look unnatural, and often ugly.

On the positive side, large, very strong sea defences may be the best way to stop erosion. Engineers have many different types of sea defences they can use, as the table shows. Often they use several types of defences together.

| Type of defence | Cost | About the defence |
|---|---|---|
| Sea wall | £2000 per metre | • Reflects waves back out to sea.<br>• Can prevent easy access to the beach.<br>• Suffers from wave scour, where plunging waves erode the beach and attack the wall's foundations. |
| Sea wall with steps and bullnose | £5000 per metre | • Steps **dissipate** wave energy, the bullnose throws waves up and back out to sea. |
| Revetments | £1000 per metre | • Break up incoming waves.<br>• Restrict beach access and look ugly.<br>• Can be destroyed by big storms. |
| Gabions | £100 per metre | • A cheap type of sea wall.<br>• Absorb wave energy as they are permeable.<br>• Not very strong. |
| Rock armour (rip-rap) | £300 per metre | • Easy to build.<br>• More expensive if built in the sea.<br>• Dissipate wave energy and look 'natural'. |
| Groynes | £2000 per metre | • Prevent longshore drift, trapping sand and shingle.<br>• Larger beach dissipates wave energy, reducing erosion.<br>• May increase erosion downdrift. |

+ **Dissipate** means to reduce wave energy, as some of it is absorbed as waves pass through, or over, sea defences.

Flood gates are closed during very high tides and storms

Concrete bullnose sea wall deflects waves back out to sea

Wooden groyne field prevents longshore drift and builds up the beach

Rip-rap dissipates wave energy before it can hit the sea wall

▲ Coastal management at Hornsea, East Riding of Yorkshire.

◄ Erosion near Naish Holiday Village in Christchurch Bay showing how groynes have protected one part of the coast but caused erosion elsewhere. This is called 'terminal groyne syndrome'.

*Image labels: Longshore drift; Beach grows; Stone groynes trap sand; No new sediment reaches here; Rapid erosion; Beach cannot form*

## Groynes

Beaches are nature's sea defences. When a wave breaks on a wide beach, the energy of the wave is dissipated by the beach before it hits the cliff. On coasts with rapid erosion, there are often very narrow beaches or no beach at all.

The traditional way of solving this problem is to 'grow' a beach. This is done by building wooden or stone groynes across the beach at right-angles to the coast. Each groyne costs £200 000–250 000. Groynes stop longshore drift by trapping sediment, building up the beach.

But building groynes in one place stops sediment reaching other places further downdrift, so beaches elsewhere can disappear. Solving the beach problem in one place creates problems somewhere else. This can lead to conflict.

## Sea walls

It is likely that sea walls built today will have to be made higher in future to cope with rising sea levels. On low-lying coasts, settlements could end up in a race to build higher defences in the future. This will be very expensive.

## Conflict at the coast

Many people on eroding coasts are in favour of hard engineering. It looks like something 'serious' is being done to protect their coast. But it's not quite that simple, because other people have conflicting views (see the table).

It can be even more complex. Some local businesses that rely on tourism might think hard engineering is ugly and would reduce visitor numbers. It's hard to please everyone.

| In favour of hard engineering | Against hard engineering |
|---|---|
| Local residents whose homes are in danger | Local taxpayers who do not live on the coast |
| Local businesses, like caravan parks and hotels | Environmentalists, who fear that habitats and natural beauty will be affected |
| Local politicians, who want the support of residents and businesses | People who live downdrift, and might lose their beach |

### your questions

1 List all of the types of defences mentioned, in order of cost (highest to lowest).
2 How can building sea defences in one place increase erosion in another place nearby?
3 Why do people who live on eroding coasts often like hard engineering defences?

+ In this section you will learn how coasts today are often managed in a more holistic, sustainable way.

## Managing the whole coast

The modern way of managing a coast is called **holistic management**. This means managing a whole stretch of coast, like all of Christchurch Bay, not just one place such as Barton-on-Sea. Holistic management takes into account the:

- needs of different groups of people
- economic costs and benefits of different strategies today, and in the future
- environment, both on land and in the sea.

**Integrated Coastal Zone Management (ICZM)** is the name for this approach to managing coasts. For long stretches of coast a plan is drawn up, called a **Shoreline Management Plan (SMP)**, which sets out how the coast will be managed. This should prevent one place building groynes, if this will then cause more erosion downdrift.

## Soft engineering

On many coasts, soft engineering is replacing hard engineering. This works with natural processes, and tries to stop erosion by stabilising beaches and cliffs and reducing wave energy. Soft engineering solutions can be cheaper than hard, and are often less intrusive.

## The choices

In the UK, local councils pay for sea defences. They may get some money from the Government, or the Environment Agency if there is a flood risk. There are four choices that councils can make about how to manage the coast:

1. **Hold the line** – use sea defences to stop erosion, and keep the coast where it is today. This is expensive.
2. **Advance the line** – use sea defences to move the coast further into the sea. This is very expensive.
3. **Strategic retreat** – gradually let the coast erode, and move people and businesses away from at-risk areas. This may involve financial compensation for people when their homes are lost.
4. **Do nothing** – take no action at all, and let nature take its course.

These choices can cause conflict. Choices 3 and 4 may mean some people lose land, businesses or homes. The diagram below shows the choices that have been made along part of the North Norfolk coast.

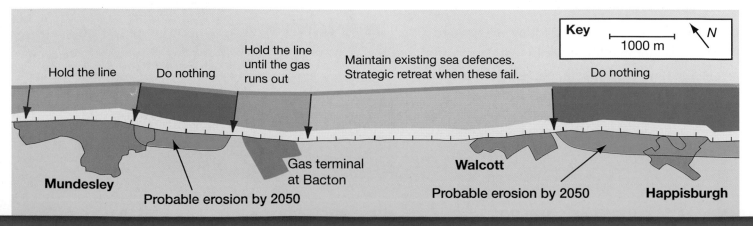

Key    1000 m    N

Hold the line    Do nothing    Hold the line until the gas runs out    Maintain existing sea defences. Strategic retreat when these fail.    Do nothing

Mundesley
Probable erosion by 2050
Gas terminal at Bacton
Walcott
Probable erosion by 2050
Happisburgh

The diagram on the right shows some soft engineering solutions:

- Planting vegetation: £20-50 per square metre.
- Beach nourishment: £500-1000 per square metre.
- Offshore breakwaters: £2000 per metre.

Soft engineering may not work in all places. Where rocks are very weak, hard engineering or 'do nothing' may be the only choices.

# Here comes the sea

Holistic coastal management means that some places are not protected from flooding or erosion. This is because:

- protection would be too expensive, as the value of land and buildings does not justify the cost
- building defences might cause more erosion elsewhere
- it might become impossible very soon, because of global warming and rising sea levels
- it might be better for the environment to create new areas of marsh, for instance.

This type of management is seen as sustainable. In some places in the UK, sea defences have been abandoned and nature is taking its course.

In the next few decades, the UK will face many difficult decisions about how best to protect the coast.

- At the moment the Government thinks it is too expensive to protect farmland and isolated houses.
- Residents, councils and businesses often disagree.
- It is very hard to persuade people who have lived by the coast all their lives that protecting their property is not sustainable.
- We don't know exactly what the impact of rising sea levels will be, so planning new defences is difficult.

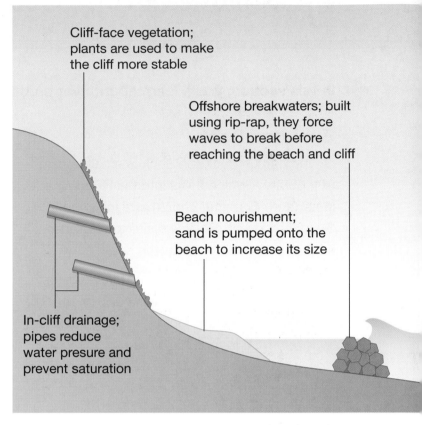

Cliff-face vegetation; plants are used to make the cliff more stable

Offshore breakwaters; built using rip-rap, they force waves to break before reaching the beach and cliff

Beach nourishment; sand is pumped onto the beach to increase its size

In-cliff drainage; pipes reduce water presure and prevent saturation

▲ Soft engineering solutions.

## your questions

1 What does 'holistic' coastal management mean?
2 Who pays for most sea defences in the UK?
3 a Look at the map of North Norfolk. Make a list of land uses and features that will be lost to the sea by 2050.
  b Make a list of people this could affect.
  c Why is the gas terminal at Bacton being protected?
  d Explain how people in this area might have very different views about the shoreline management plan.
4 **Exam-style question** Using named examples, explain how coastal management choices can cause conflict at the coast. (6 marks)

➕ In this section, you'll learn about river processes in upland areas.

## Buckden Beck

On the right is Buckden Beck, a small stream – or tributary – which flows into the River Wharfe in northern England. The Wharfe joins the River Ouse, and eventually flows into the Humber Estuary – where it reaches the sea.

Look at the small rapids and waterfalls in the photo. This is typical of a river's **upper course**. The river loses height rapidly, making the stream look very fast. In fact, the stream is flowing slowly, because so much of the river's energy is lost through **friction** with the stream bed.

However, rainstorms increase the river's energy a lot, and allow it to carry large boulders and stones, which wear away – or **erode** – the channel. Therefore, most erosion is done during periods of wet weather.

> ➕ **Erosion** means wearing away the landscape.

▶ Buckden Beck – the **gradient** (slope) of the river course is steep.

### How a river carries its load

➕ The material carried by a river is called its **load**.

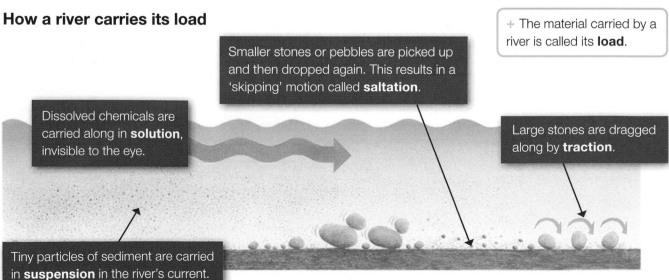

Smaller stones or pebbles are picked up and then dropped again. This results in a 'skipping' motion called **saltation**.

Dissolved chemicals are carried along in **solution**, invisible to the eye.

Large stones are dragged along by **traction**.

Tiny particles of sediment are carried in **suspension** in the river's current.

## How a river erodes its channel

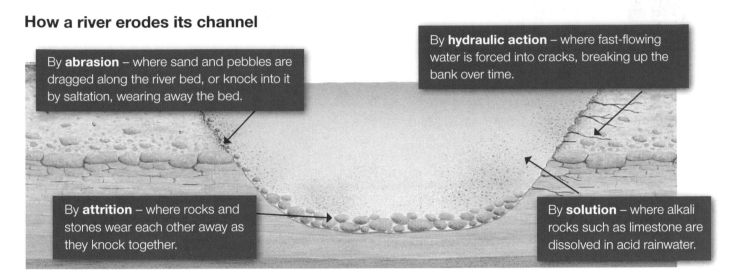

By **abrasion** – where sand and pebbles are dragged along the river bed, or knock into it by saltation, wearing away the bed.

By **hydraulic action** – where fast-flowing water is forced into cracks, breaking up the bank over time.

By **attrition** – where rocks and stones wear each other away as they knock together.

By **solution** – where alkali rocks such as limestone are dissolved in acid rainwater.

In the upper course, most erosion is vertical. Whenever the river meets a hard – or **resistant** – rock, a step is formed. This can eventually become a waterfall. The river gains a great deal of energy where it falls over the lip of the waterfall, allowing it to erode rapidly.

## How waterfalls are formed

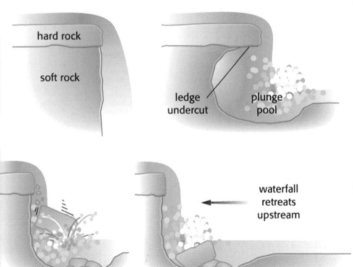

1 Waterfalls occur when a river crosses a bed of more resistant rock.

hard rock

soft rock

ledge undercut

plunge pool

2 Erosion of the less resistant rock underneath continues – undercutting the hard rock above it. The river's energy creates a hollow at the foot of the waterfall, known as a **plunge pool**.

3 The less resistant rock beneath is eroded more rapidly by abrasion and hydraulic action. This creates a ledge, which overhangs and collapses.

waterfall retreats upstream

4 The waterfall takes up a new position, leaving a steep valley or **gorge**.

### your questions

1 For each set of words **a** say which is the odd one out, **b** explain your choice.
   * stream, tributary, river, waterfall
   * solution, attrition, plunge pool, abrasion
   * suspension, gorge, traction, load
   * hydraulic action, saltation, suspension, traction

2 Explain how the following might change during wet weather:

   * The amount of water in the stream, and its energy.
   * The amount of erosion that a stream can do.
   * A waterfall.

3 Explain the following features of a river's upper course:
   * It usually flows slowly, but sometimes flows rapidly.
   * There are several large stones and boulders on the stream bed.

In this section, you'll learn how rivers and their valleys develop in upland areas and what causes this.

▼ Buckden Beck is typical of a valley in its upper course – it forms a V-shape, with interlocking spurs.

## The valley of Buckden Beck

Look at the steep valley sides in the photo. This is typical of a river's upper course in upland areas. It has two main features:

- It has steep sides and the valley bottom is narrow (you can see why valleys like this are called V-shaped).
- The river winds gently around **interlocking spurs** – or ridges of land – which jut into the river valley, looking as though they're interlocked.

Although the stream is vital in eroding the valley, what happens on the valley sides is also important. Two things happen: **weathering** and **mass movement**.

## Weathering on the valley sides

The valley sides are pale grey limestone, and have been attacked and broken down by weathering. The cliffs of rock are known as **rock outcrops**, where rock comes to the surface. Below them, the valley is covered in limestone rubble, called **scree**. The scree has broken away from the cliffs above because of weathering. There are different types of weathering – physical, chemical and biological, as shown in the photo.

+ **Weathering** is the breakdown of rocks *in situ*. This means it happens where the rock is. Rocks are weakened by being chemically attacked and mechanically broken down.

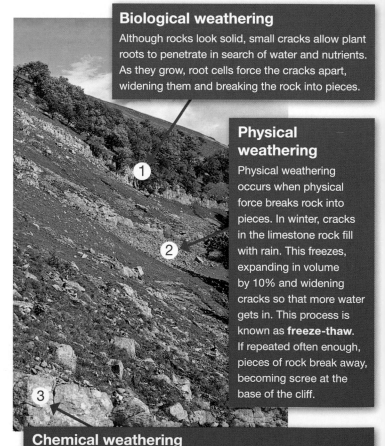

### Biological weathering

Although rocks look solid, small cracks allow plant roots to penetrate in search of water and nutrients. As they grow, root cells force the cracks apart, widening them and breaking the rock into pieces.

### Physical weathering

Physical weathering occurs when physical force breaks rock into pieces. In winter, cracks in the limestone rock fill with rain. This freezes, expanding in volume by 10% and widening cracks so that more water gets in. This process is known as **freeze-thaw**. If repeated often enough, pieces of rock break away, becoming scree at the base of the cliff.

### Chemical weathering

Chemical weathering is any chemical change or decay of solid rock. Rainwater mixes with atmospheric gases, e.g. $CO_2$, to form weak acids which dissolve alkaline rocks such as limestone.

# Mass movement

Once rock is broken up, the fragments move down the slope towards the stream. Some move quickly, others slowly. This is known as mass movement, but there are different processes:

- Rapid – such as **landslides** and mudflows. These are less common in the UK, although they do occur on railway cuttings and along cliff coastlines.
- Slow – the most common of which is **soil creep**. Soil creep occurs really slowly – perhaps 2 cm a year – but over decades it has many effects, like those in the top diagram.

# The shape of the valley

Valley shape is affected by three things, as shown in the bottom diagram:

- The speed of weathering. If scree piles up, weathering is taking place rapidly.
- The speed of mass movement.
- How quickly the river can remove the material brought by mass movement.

If the river has plenty of energy, it takes the material away and uses it to help erode the valley, making it steeper. However, if it is slow and cannot cope with all the material, weathered rock collects at the bottom of the slope, making the valley gentler and flatter. You'll see in the middle course how this alters the valley shape.

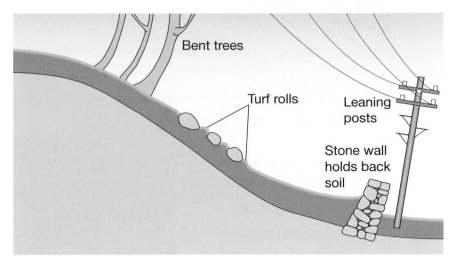

▲ Different ways in which soil creep can be recognised in the landscape. Although it is slow, soil creep can cause walls or telegraph poles to lean.

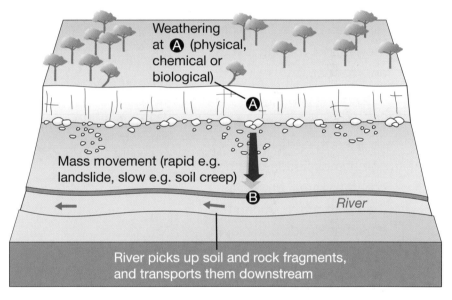

▲ Weathering takes place at A on the valley sides to break down solid rock into fragments. These move down the slope to the river at B, where they are removed. The river then uses these to wear away – or erode – the river bed.

## your questions

1 Draw labelled diagrams to explain these features of a valley in the upper course:
- How scree is formed.
- Why valley slopes get covered in scree.
- Why hedges and stone walls can fall over.
- How trees can grow out of solid rock cliffs.
- Why stream beds contain large rock fragments.

2 Explain how the following might change the valley side:
- A cold frosty spell.
- Prolonged rain.

✚ In this section, you'll learn about how both the river and its valley change in the middle course.

## The River Wharfe in its middle course

The photo shows the River Wharfe between Kettlewell and Starbotton, downstream from Buckden Beck. By now, the river is in its **middle course**. Several streams like Buckden Beck have joined the Wharfe, making it wider and deeper.

During wet weather, the volume of water can be so great that the river sometimes floods over a broad flat area, known as the **flood plain**. During this time, fine sands and clays brought down by the river from upstream, settle over the flood plain and form layers of **alluvium**. Flood plains are at risk of flooding, but alluvium is fertile and attracts farmers.

The valley shape has also changed. From a V-shape, the flood plain has broadened the valley floor, so it has become a U-shape. The edge of the flood plain in the far distance is steep where the flood plain ends and the valley side rises.

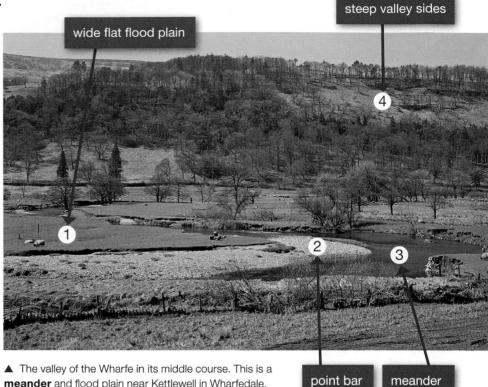

wide flat flood plain

steep valley sides

point bar    meander

▲ The valley of the Wharfe in its middle course. This is a **meander** and flood plain near Kettlewell in Wharfedale.

## The river's energy

By the time the Wharfe reaches its middle course, the gradient is gentle, but the river actually flows faster. The increase in the volume of water leaves more energy, once friction is overcome. Now, instead of eroding down, the river's energy is directed towards the sides, or **laterally**.

### On your planet
✚ Generally, the greater the volume of water, the greater the river's energy.

### What do you think?
✚ Should the fertile alluvium on flood plains always be farmed?

# How meanders change the valley

**1**

The river bends in its middle course; each sharp bend is called a meander. Meanders are natural; rivers almost never flow in a straight line. Water flows naturally in a corkscrew pattern. This is called **helical flow**.

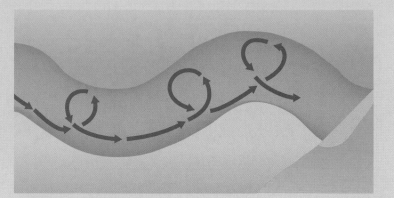

**2**

Helical flow sends the river's energy laterally – i.e. to the sides. The fastest current (called the **thalweg**) is forced to the outer bend (A), where it undercuts the bank.

The helical flow of the river shifts sediment across the channel to the inner bank (B). Here the sediment is **deposited** by the slower moving water, to form a **point bar**.

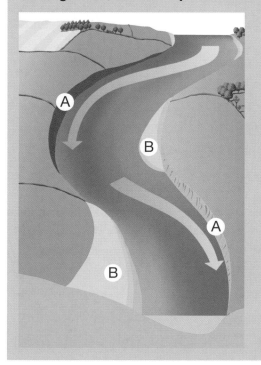

**3**

Continued erosion can create a narrow neck between two meanders (X). Eventually, the neck will be breached at (Y), cutting off the meander to create an **ox-bow lake** (Z).

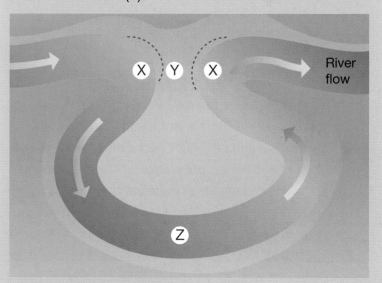

## your questions

1 Classify each of these features by whether they are part of **a** the river in its middle course, **b** its valley in its middle course:

| | | |
|---|---|---|
| flood plain | alluvium | U-shape |
| lateral erosion | meander | helical flow |
| thalweg | river cliff | point bar |
| meander neck | ox-bow lake | |

2 Draw a sketch of the photo on page 92, adding the following labels: flood plain, alluvium, lateral erosion, meander, thalweg, undercutting, point bar.

3 Using the information above to help, draw 3–4 labelled diagrams to show how an ox-bow lake is formed.

In this section, you'll learn about how both the river and its valley change in the lower course.

## The lower course of the Wharfe

By the time the Wharfe has reached its lower course, the differences with its upper course are really obvious! The river – once narrow, with waterfalls – is now wide and deep, flowing over a gentle gradient. Its meanders are large, the volume of water is also larger, and there is a wide, flat flood plain. This is low-lying and floods easily.

Beside the river, there are embankments. These can be either natural or artificial, and are known as **levées** (see right).

## Towards the sea

The Ouse River (joined by the Wharfe) eventually joins the Humber and forms an estuary – where the river meets the sea. Two directions of flow take place here – **outwards** by the river, taking water out to sea, and **inwards** by incoming high tides. Twice a day, incoming tides meet the outgoing river and the flow stops, forcing the river to deposit its sediment. This forms a broad wide area where the river deposits mud – hence the name **mudflats** (see below).

▲ The lower part of the Wharfe at Tadcaster. By this stage, the river is flowing over an almost flat gradient.

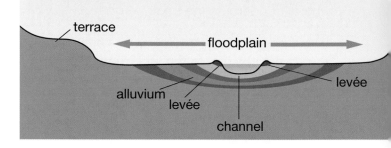

Natural levées form beside the river's bank where it first floods. As a river reaches **bankful** – i.e. before it spills on to the flood plain – it deposits sand and clay particles where the flow is slower. These build up beside the river as a bank.

Artificial levées are built by engineers to protect farms or towns from flooding. These are common in the UK along rivers like the Severn and the Ouse, where flooding is common, and along some of the world's giant rivers, e.g. the Mississippi in the USA.

◀ Mudflats and salt marshes on the River Humber, near its estuary. Note the artificial levée in the top right of the photo.

Because the estuary is tidal, submerged by the sea twice a day, **salt marsh** plants have to be able to stand salt water as well as fresh. Salt marshes are valuable for wildlife; migrating birds use them to shelter during stormy weather, and the mud is rich in shellfish and worms. But salt marshes are under threat from ports and industry – the photo shows oil refineries along the Humber Estuary.

## Conclusion – spot the changes!

Two kinds of changes occur during the upper, middle and lower courses of a river.

- Changes in the **long profile** of the river. As the first diagram shows, the long profile is the way that the gradient of the river changes from its upper to lower course. Put simply, it's steep in upland areas, and gentle in the lowlands.
- Changes in the **cross profile**, or valley shape. Put simply, the valley is V-shaped in the upper course and almost flat by the lower course.

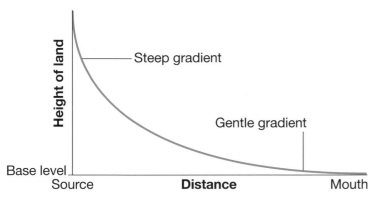

▲ The long profile of a river – showing how river gradient changes as the river moves from its upper to its lower course.

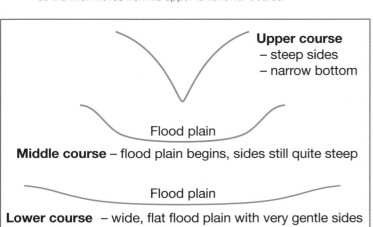

**Upper course**
– steep sides
– narrow bottom

Flood plain

**Middle course** – flood plain begins, sides still quite steep

Flood plain

**Lower course** – wide, flat flood plain with very gentle sides

▲ An aerial view of the Humber Estuary. Estuaries have enormous value economically, as shelter for shipping and flat land for industrial development. However, this can conflict with their ecological value.

### your questions

1 Put these phrases into the correct order to explain why mud is deposited in estuaries to form salt marshes:
*river brings sediment downstream, incoming tide, plants colonise mud, outgoing river flow, mud deposited on the estuary floor, mudflats formed, plants immersed at high tide, flow stops, plants help to trap more mud, mudflats become permanent*

2 Now write out a similar sequence to show how levées are formed.

3 Make a large copy of this table and compare the river in its upper, middle and lower courses.

| | Upper course | Middle course | Lower course |
|---|---|---|---|
| **River features** | | | |
| **River processes** | | | |
| **Main landforms** | | | |
| **Processes on valley sides** | | | |

4 Why is the lower course most at threat from climate change?

5 **Exam-style question** Explain how river valleys alter their shape as a river moves downstream. (4 marks)

◄ The cross profile of a river – showing how valley shape changes as the river moves from its upper to its lower course.

✚ In this section, you'll learn how sudden floods affected Sheffield in the summer of 2007.

## A June to remember

In June and July 2007, several periods of extreme rainfall gave rise to widespread flooding in parts of England and Wales. It was the wettest May to July period since 1766, when reliable data were first collected. Nationally, 49 000 households and 7000 businesses were flooded. Major transport links, schools, power and water supplies were disrupted. South Yorkshire suffered record-breaking floods.

## What caused the Sheffield floods?

### 1 Prolonged rain

Sheffield's flood story began when extreme wet weather hit the north of England. Heavy and prolonged rain fell across South Yorkshire as a result of a depression over northern England.

- On 15 June, 90 mm of rain fell over Sheffield; more than one month's normal rainfall!
- This was exceeded by even heavier rain on 25 June, when almost 100 mm fell in just 24 hours! This was the most rain Sheffield had ever had in one day.

With almost 190 mm in those two days alone, June was the wettest month recorded in Yorkshire since 1882.

### 2 Soil saturation

After the rains of 15 and 16 June, the soil was **saturated**, causing localised flooding. The ground had not dried out by 25 June, when more extreme rain overwhelmed the rivers and drains in Sheffield. All the extra rain just ran straight off into rivers as **surface runoff**.

# Two swept to death in Sheffield

Hundreds of families in South Yorkshire have been moved to safety amid severe flooding, and a man and a teenage boy have been drowned in Sheffield.

About 900 people in Sheffield are using emergency shelters, and about 700 have left villages near Rotherham, fearing the nearby Ulley dam could collapse.

Police closed the M1 completely between junctions 32 to 34, because of the risks posed by the dam, and rail companies announced cancellations to services.

The village of Catcliffe near Sheffield was under water.

London Fire Brigade sent two 'high volume' pumps to help with the flooding problems.

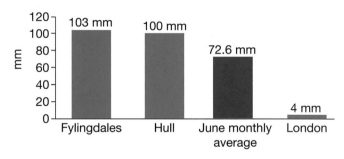

▲ Rainfall in Yorkshire on 25 June, compared to a) June's average monthly rainfall total and b) the London area.

▼ Club Mill Bridge in Sheffield was washed away.

## 3 The confluence of several rivers

The map shows where the worst Sheffield flooding occurred, in the Hillsborough area near the football ground. This was because of a combination of circumstances.

- The River Rivelin joins the River Loxley at confluence X. The volume of water increased hugely at this point, and flooding became a risk straight away.
- Only a short distance away, confluence Y occurs – this time with the River Don. At this stage, the Don was already almost up to its banks.
- A 'backlog' of water therefore occurred along both rivers, causing water to back up and overflow the banks.

## 4 The physical landscape

Sheffield lies at the foot of the Pennines, at the point where three rivers meet. Several storage reservoirs at the heads of these rivers provide water for Sheffield, and normally help to store water in rainy periods. This time, however, they filled rapidly and overflowed. The slopes near Sheffield are steep (see below), so water runs off rapidly during wet spells, quickly filling the rivers. The heavy rains raised river levels, and soon blocked drains in parts of the city.

▼ The slopes of the Don valley on the outskirts of Sheffield.

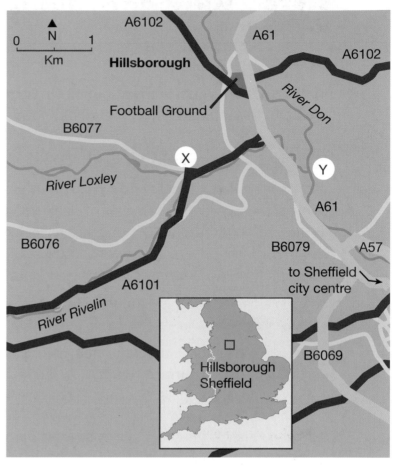

▲ The area around Hillsborough, north-west Sheffield, where some of the worst flooding occurred. Also see section 6.7.

### your questions

1 Copy the diagram below. On it, label how each of the factors shown helped to produce the floods in Sheffield.

```
        1. Prolonged                    4. The physical
           rain                            landscape
                      Sheffield
                      floods June
                      2007
        2. Soil                         3. Confluence
           saturation                      of rivers
```

2 Now add other points to show how human factors helped to cause the flood event.
3 Draw a flow diagram to show all the stages that cause roads to be flooded; start with 'Rain falls' and end with 'Roads flooded'.

# Why does flooding occur?

+ In this section, you'll learn about storm hydrographs, and how they can help to explain the Sheffield floods.

## Stage 1 – from rain to soil

When it rains, very little falls directly into rivers; most falls elsewhere. As the diagram below shows, leaves and branches of plants trap a lot of the rain that falls. This is known as the **interception zone**. The amount intercepted depends on the vegetation, and also the season. For example, deciduous plants (those which lose their leaves in winter) intercept more rain in summer, when they're in leaf.

Some intercepted water is **evaporated** into the atmosphere. The rest drips from leaves to the soil and soaks in – this is called **infiltration**.

Eventually, the soil becomes **saturated**, and cannot take any more. Any extra rain flows overground – called **surface runoff**. How quickly this happens depends on three factors:

1. How much rain has fallen recently – known as **antecedent rainfall**. If the weather has been wet, the soil may already be saturated.
2. How permeable the soil is. Sandy soils are **permeable** – they absorb water easily, and surface runoff rarely occurs. Clay soils are more closely packed, so are **impermeable** – water can't infiltrate easily.
3. How heavily the rain falls. Heavy storms cause rapid surface runoff; the greater the runoff, the greater the flood risk.

## Stage 2 – from soil to river

Once water enters the soil:

- some is taken up by plants and **transpired** through leaves into the atmosphere
- some seeps into the river through soil air spaces – known as **throughflow**
- some continues into solid rock, and saturates it. The upper limit of saturated rock is known as the **water table**. From here, water seeps slowly towards the river as **groundwater flow**, which keeps a river flowing even when there is no rain.

▼ A cross-section of a river valley to show how water moves down valley sides to a river.

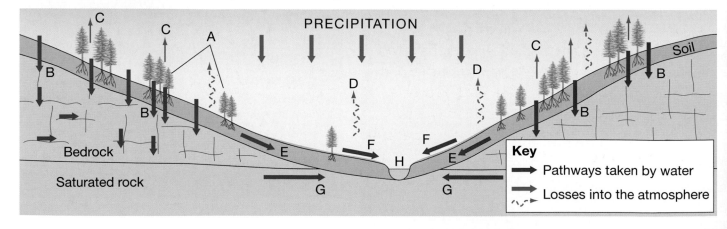

# Understanding storm hydrographs

A **storm hydrograph** shows how a river changes as a result of rainfall. Because most rain falls on the valley sides, it takes time to reach the river. The amount of water in the river (the line graph), known as the discharge, rises gradually (unless the rain falls very heavily, as it did in Sheffield). Notice the way the graph rises and falls – there is a delay between the rain and the discharge.

+ A **storm hydrograph** is a graph which shows how a river changes as a result of rainfall. It shows two things: rainfall and changing river volume or discharge.

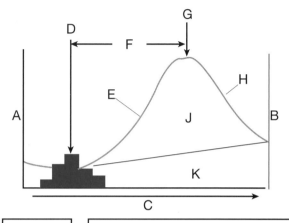

Discharge – in cubic metres per second – or cumecs

The amount of water in the river that has come from the storm

The amount of water in the river that comes from baseflow

Rainfall in millimetres

Time in hours

Peak rainfall – when rain is falling most heavily

Time lag – the length of time between peak rainfall and peak discharge of the river

peak discharge – where river flow reaches a peak

Falling limb – where river discharge decreases

Rising limb – as water reaches the river, the discharge increases

# How do cities change rivers?

Urban areas change everything about the hydrological cycle. Natural landscapes (e.g. forests) allow rainwater to soak slowly into the ground. By contrast, urban surfaces, like roads, roofs and buildings, are impermeable and don't let runoff seep into the ground. Water remains on the surface, then runs off into drainage systems that are designed to take the runoff.

Runoff gathers speed once it enters the drains, joining rainwater from other drains, and eventually spilling into the river. Sometimes drains become blocked when rivers are high and block the exit. Once the drains are blocked, water flows down the street – which, in a hilly city like Sheffield, can mean very fast-flowing water. The two people who died in Sheffield were caught unawares – one drowned walking across a park, while another was trying to cross a flooded street.

## your questions

1 Read the text, then match the letters on the diagram of the valley opposite with these terms: groundwater flow, evaporation, infiltration, interception zone, river channel, surface runoff, throughflow, transpiration.

2 Match the letters on the storm hydrograph above to the text boxes around it.

3 In your own words, explain:
  a why water normally takes so long to reach a river. Use the words in question 1.
  b why streams in rural areas are less likely to flood than in urban areas.
  c why many urban floods can take local people by surprise.

4 Would the flooding have been worse, or less serious, if the rain had happened in January instead of June? Explain.

5 **Exam-style question** Explain the time lag between a period of rain, and the peak discharge in a river. (4 marks)

6 **Exam-style question** Describe what happens to rain (from street to river) when it fall on an urban area. (4 marks)

✚ In this section, you'll learn about the impacts of Sheffield's floods in 2007.

## The causes of flooding

Most of Sheffield's flooding was caused by drains, river channels and flood defences being overwhelmed by extreme flows of water. Sudden downpours of rain made everything happen so quickly, and made it difficult to predict where flooding might occur.

- Fallen trees along the rivers Sheaf, Loxley and Don quickly blocked river channels.
- As surface runoff raced down Sheffield's hills, drains blocked and overflowed.

## The impacts of flooding on Sheffield and South Yorkshire

### A  Hillsborough and north-west Sheffield

- Hillsborough Football Stadium (on the right) was flooded up to 8 metres deep. It cost several million pounds to repair the damage.
- Of 300 homes on one estate, 128 (43%) were flooded. All council tenants returned home within 9 months; some owner-occupiers had to wait much longer as builders were so inundated with work.
- Displaced people suffered stress. Some families were moved into caravans for the winter.
- There were severe health risks from raw sewage escaping into the floodwaters.

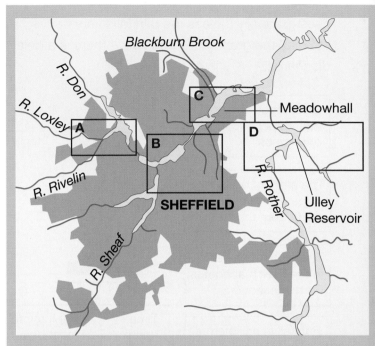

The floods had a devastating impact on people living and working in Sheffield.

- Two people drowned in the floodwaters.
- Over 1200 homes were flooded across the city, and more than 1000 businesses were affected.
- Roads were damaged and a bridge collapsed, blocking roads and travel for several days.
- 13 000 people were without power for two days.

## B The city centre and the Don flood plain

In Sheffield city centre, on 25 June, motorists abandoned their vehicles as roads flooded and traffic gridlocked.

- As the floodwaters rose, many people were caught unawares and had to be evacuated from flooded buildings.
- So many trains and buses were cancelled that many people were unable to get home. Some were trapped overnight; over 900 people spent the night in their offices.
- 20 people were airlifted to safety from one building, a further 3 from the roof of another. 200 people were stranded on the first floor of a Royal Mail distribution centre.

## C The Lower Don Valley

- The deepest flooding occurred in areas near the River Don, where some industries were badly affected.
- The tool-making company Clarkson Osborn suffered £15 million in flood damage to its works, with equal damage to Sheffield Forgemasters International and Cadbury Trebor Bassett.
- Meadowhall Shopping Centre was flooded (see below), causing millions of pounds' worth of stock losses and damage, and closing the centre for a week. Some shops were closed for three months. Meadowhall's flood defences, built to withstand a 1-in-100-years flood event, were overwhelmed.

## D The Ulley Reservoir area

On 26 June, there were fears that the Ulley Dam might collapse, following the rains and damage to its structure (see above).

- 700 residents were evacuated from the villages of Whiston, Canklow, and Catcliffe and Treeton. Some were allowed back 2 days later; others were away for 2 weeks.
- 100 people took shelter at Dinnington Comprehensive School, where the Salvation Army provided food, clothing and bedding.
- About 100 people had to be rehoused for up to a year where flood damage was severe.
- The M1 motorway was closed for two days between junctions 32 and 36 because of the risk of the dam bursting.

## your questions

1 Classify the effects of the flooding in Sheffield as shown in the table. Use different colour pens to write in the effects for the four areas of Sheffield shown.

|  | Short-term | Medium-term | Long-term |
|---|---|---|---|
| Social |  |  |  |
| Economic |  |  |  |
| Environmental |  |  |  |

2 Which were the greatest effects – social, economic or environmental? Explain your answer.

3 Which area was worst affected – A, B, C, or D? Explain your answer.

In this section, you'll learn about whether more should be spent on flood protection in Sheffield.

Since 2007, Sheffield has debated what it can do to avoid flooding again. Engineers have to choose between 'hard' and 'soft' solutions.
- 'Hard' solutions are structures built to defend from floodwater.
- 'Soft' solutions adapt to flood risks, and allow natural processes to deal with rainwater.

## Hard engineering in Sheffield

### Drains and culverts

Sheffield's drains and culverts (which carry away rainwater) were designed to deal with rainfall amounts that might only occur once in 30 years. But the 2007 floods were a 1 in 400 year event, and people wonder whether anything could have protected them. The drains had two problems.
- They could not cope with all the rain, so streets flooded.
- Where drains met the rivers, water could not escape as the rivers were already so high.

### The River Sheaf

The River Sheaf has an enlarged concrete-lined channel where it joins the River Don in central Sheffield. It aims to speed up the flow of water away from the city. In June 2007, it worked, and there was no flooding in this part of Sheffield.

### Meadowhall shopping centre

Meadowhall has its own flood defences, constructed when the centre was built, consisting of an embankment around it to hold off water. The photo on the right shows that the bank simply wasn't high enough to contain all the floodwater!

Sheffield City Council denies that the city's drainage system was to blame for the extensive flooding. The council says the rain was so severe that no amount of planning could have prevented it. They said: 'The rain was the worst for 35 years. The rivers, drains and streams were already full, due to heavy rainfall for many days, and couldn't cope. Most floods were caused by rivers overflowing, and pipes and drains simply weren't big enough to cope with the deluge.'

▲ Adapted from the *Sheffield Star*, 19 June 2007.

### What do you think?

+ The UK's Environment Agency believes that:
- hard defences are not the solution
- buildings at risk must be flood-proofed (e.g. on raised sites or protected by walls), or companies relocated to safe locations
- councils should increase maintenance – streams and drains block easily with debris or tree branches.
Are they right?

▼ Flood defences at Meadowhall shopping centre just weren't high enough.

## Flood storage reservoirs

In Rother Valley Country Park, east of Sheffield, lakes have been shaped from old quarries along the River Rother to take floodwater. They hold it until it is released into the River Don a few miles away. In 2007, they prevented flooding in this area. The lakes are now also used for boating within the park.

# What further hard defences could Sheffield build?

The table below shows some hard engineering methods that could be considered for Sheffield. But they all have both costs as well as benefits.

▲ Creating a river diversion in Rotherham.

## 1 Hard engineering

| Method | How it works | Comment |
|---|---|---|
| **a. Build flood banks** | Raise the banks of a river to increase its capacity. | These are fairly cheap, one-off costs. However, they disperse water quickly and increase flood risk downstream. |
| **b. Increase the size of the river channel** | Dredge the river to increase its capacity, or line it with concrete to speed up the flow of water. | Lining with concrete is expensive, and dredging needs to be done every year. Speeding up the water can increase the flood risk downstream. |
| **c. Divert the river away from the city centre** | Create a diversion for excess water to avoid flooding the city centre. | This was done in 2008 in Rotherham, 10 km away (see above right). It costs £14 million for a 1km stretch. |
| **d. Increase the size of the drains** | Dig up every major road into Sheffield and enlarge the major drains. | Gets runoff away from the city, but it is only as good as the capacity of the river to take all the water from the drains. |
| **e. Increase the maintenance budget** | Clear rivers, drains and sewers to remove debris or vegetation. | This needs to be done every year, so it can be costly over a 10- or 20-year period. |

**What do you think?**

+ Should cities just build bigger drains?

## your questions

1 Explain what is meant by 'hard engineering'.
2 In pairs, complete the table on the right show the advantages and disadvantages of hard flood protection in Sheffield.
3 Now award points up to 5 for each advantage and subtract points to -5 for each disadvantage. Add up a total for each method. Which is best?
4 **Exam-style question** Select **two** methods which you think should be used to protect Sheffield from future floods. Justify your choices.

| Method | Advantages | Disadvantages |
|---|---|---|
| 1. Build flood banks | | |
| | | |
| | | |
| | | |
| | | |
| | | |
| | | |
| | | |
| | | |

# What about soft engineering?

In this unit you will learn about flood management through soft engineering.

## Is hard engineering the only way?

After 2007, the Environment Agency considered the situation in Sheffield.

- Increased building in Sheffield since the 1980s had increased surface run-off, but nothing more was spent on flood protection.
- The reservoirs in the upper Don valley outside the city were originally designed to store water for Sheffield – it was an added advantage that, by storing water, they helped to prevent flooding. However, in 2007, they were so full that they were of no use when heavy rains fell.

The UK's Environment Agency now believes that hard defences are not the solution. They cost a great deal and can rarely be made big enough to cope with the largest floods. Instead they suggest:

- upstream, upland areas should be planted with trees to reduce surface run-off
- buildings at risk would be better protected using flood-proofing (e.g. protecting with walls, or building on raised land)
- planning permission should not be given for building near rivers
- companies should relocate, if flood proofing is not possible.

So far, Sheffield has almost no 'soft' methods. But the Environment Agency believes that these would make flood management more sustainable.

▶ The Environment Agency monitors rivers, to help them predict possible floods.

## 2 Soft engineering

| Method | How it works | Comment |
|---|---|---|
| a. Flood abatement | Change land use upstream, e.g. by planting trees. | This delays the passage of water into rivers by increasing interception and transpiration. |
| b. Flood proofing | Design new buildings or alter existing ones to reduce flood risk. | Only affects new buildings; can be expensive to alter existing buildings. |
| c. Flood plain zoning | Refuse planning permission where flood risk is high | It phases out development in risk areas, or restricts permission to certain uses, e.g. leisure centres. |
| d. Flood prediction and warning | The Environment Agency monitors rivers (see below), and uses forecasts from the Met Office. | Accurate predictions help to reduce flood damage and evacuate people. |

▲ Some soft engineering methods.

## Sustainable management along the River Skerne

The Environment Agency is responsible for the environment around rivers as well as the rivers themselves. In Darlington, the River Skerne has recently been improved to create an amenity for local people AND to prevent flooding.

In the 19th century, its meanders were straightened to allow industries to build on the flood plain. This decreased its length by 13% and increased the flood risk, so the flood plain was raised with industrial waste. This made the river environment poor, especially when industries closed and left derelict land.

So could it be restored, without increasing the flood risk in Darlington? Meanders could not be restored to their former length because pipes containing gas, electricity and sewage followed the new course.

In the end, the Environment Agency restored 2km of river, creating a riverside park as an outdoor space.
- Some meanders were rebuilt, lengthening the river and slowing water down.
- Banks were lowered to make the river flood the park instead of Darlington.
- The flood plain was lowered to increase its ability to store floodwater.

## Has it worked?

The River Skerne did flood in June 2007, but rainfall that year was exceptional. Despite this the damage caused was still less than in the floods of 2000, before the work was done. 50 homes were flooded, all due to the backlog of rainwater entering the city's drains.

Another result has been large increases in species with a 30% increase in birds and insects, within one year. Locals also liked the changes – in a survey, 82% of people liked it either 'mostly' or 'strongly'.

### your questions

1 Explain what is meant by 'soft engineering'.
2 In pairs, copy the table to show the advantages and disadvantages of 'soft' flood protection. Base it on what you have learnt about the River Skerne.
3 Now award up to 5 points for each advantage and up to 5 minus points for each disadvantage. Add up a total for each method. Which is best?
4 Based on this, decide whether hard or soft methods are best to protect Sheffield from flooding. Justify your choices.

| Method of 'soft' protection | Advantages | Disadvantages |
|---|---|---|
| 1. Flood abatement | | |
| | | |
| | | |

**On your planet**
+ The most threatened ecosystems include coral reefs, mangrove forests, coastal wetlands, sea grass beds and continental shelves.

+ In this section you will learn how and why the oceans are threatened with destruction.

## Global patterns

In 2008, scientists completed the environmental damage map below. It shows the combined impact of 17 types of human activities on 20 types of ocean ecosystems.

The map shows that many of the world's oceans were experiencing serious environmental damage in 2008, with only a small percentage in good condition. Enclosed seas in densely populated, industrialised areas (North Sea, Mediterranean, South and East China Seas) are most at risk. Polar areas are at present the least damaged — but for how long? As global warming melts polar ice, especially in the Arctic, future exploitation could threaten even these ocean environments.

The increasing number of **dead zones** is very worrying. In dead zones, ecosystems have collapsed completely. The biggest one, in the Baltic Sea, is half the size of the UK. Section 7.3 will tell you more about ocean dead zones.

### Think about this:
- Our supplies of seafood (fish, crabs, lobsters, shellfish) could die out by 2050 if we don't farm them.
- Every year, hundreds of dolphins are killed by deep sea trawling.
- Vast areas of coral reefs and mangrove forests are at risk.
- The depletion of individual species could upset **food chains** and destroy entire ocean **ecosystems**. This is especially likely near crowded coastal areas, where nearly 60% of people live.

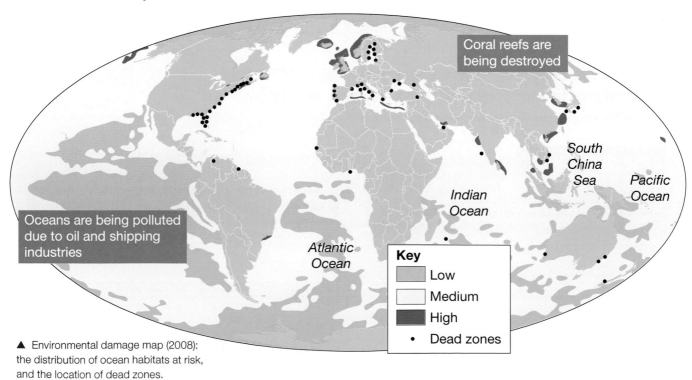

Coral reefs are being destroyed

Oceans are being polluted due to oil and shipping industries

South China Sea

Pacific Ocean

Indian Ocean

Atlantic Ocean

**Key**
- Low
- Medium
- High
- • Dead zones

▲ Environmental damage map (2008): the distribution of ocean habitats at risk, and the location of dead zones.

# The value of coral reefs

Coral reefs are the 'rainforests of the oceans', because of their high levels of **biodiversity** (biological richness of species and habitat).

**On your planet**

+ We have probably only discovered about 10% of all the species of fish and corals. Many species could become extinct without even being discovered!

**Key**

- Coral reefs
- Reefs at greatest risk

Red Sea

Tropic of Cancer

Caribbean

Equator

South Pacific Islands

Tropic of Capricorn

River mouths avoided because of sediment.

Shallow water of 25 metres or less to ensure abundant light.

Tropical water temperature of 24–26°C. It must be saline water.

▲ The distribution of the world's coral reefs.

Coral reefs are of great value, as this table shows.

| Exploitation for fish | 4000 species of reef-living fish (25% of all known marine fish) provide food for local communities. 25% of the world's total commercial fish catch comes from coral reefs. |
|---|---|
| Shoreline protection | Reefs provide shoreline protection from storms, tsunami and wave erosion. Reefs can grow with rising sea levels and protect against the impacts of climate change. |
| Aquarium trade | Reefs supply tropical fish, sea horses, and 'plants' for the aquarium trade. |
| Tourism | Reefs are a magnet for the world's tourists. Many countries in the Caribbean get over half of their income from reef tourism. |
| Education and research | Reefs can be visited easily to learn about marine life. |
| Other uses | As a medicine source (some drugs originate from reef organisms). To make decorative objects, such as jewellery. A source of lime for cement and building. |

▲ A coral reef supports a fantastic range of life.

## your questions

1  Why are enclosed seas in densely populated areas the most at risk?

2  Why is it difficult to prevent over-fishing?

3  In pairs, find 3-4 pictures that show the value of the coral reef. Produce an A3 colour poster that illustrates the reef's value.

4  Does it matter if we run out of seafood? Explain your answer.

In this section you will learn about what threatens the world's coral reefs.

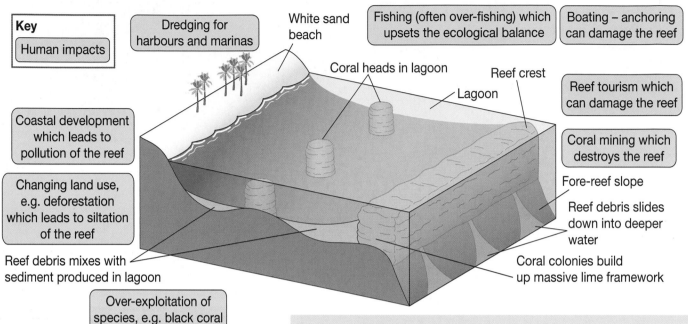

**Key**

Human impacts

Dredging for harbours and marinas

White sand beach

Fishing (often over-fishing) which upsets the ecological balance

Boating – anchoring can damage the reef

Coral heads in lagoon

Reef crest

Lagoon

Reef tourism which can damage the reef

Coastal development which leads to pollution of the reef

Coral mining which destroys the reef

Fore-reef slope

Changing land use, e.g. deforestation which leads to siltation of the reef

Reef debris slides down into deeper water

Reef debris mixes with sediment produced in lagoon

Coral colonies build up massive lime framework

Over-exploitation of species, e.g. black coral

▲ 'The web of risk' – how we impact on coral reefs.

The world's reefs are under increasing risk. This is because of an increase in destructive factors – both natural and human (see right).

Population growth in coastal areas is a root cause. Most coral reefs are located in developing countries, with populations growing at 3% a year. Also, people migrate to coastal areas, where there are jobs in tourism or fishing.

Land development causes problems. Building work disturbs the land, which causes soil erosion. This gets washed to sea, which clouds the coral. Pollution from cars and industry causes further damage..

## Destructive factors

- Global warming will devastate reefs by **bleaching** them. Rising temperatures lead to coral stress. The algae that live within corals are expelled. This makes the coral white.
- Local warming of oceans by **El Niño** has the same effect.
- Blast fishing with dynamite, poisoning with cyanide, and trawling reefs, all cause damage.
- Parrot fish graze the algae on coral reefs. Overfishing leads to smothering by algae which destroys reefs.
- Diseases such as black band coral disease destroy reefs. Siltation and pollution make disease more likely.
- Siltation damages the way in which corals breathe and feed.
- Hurricanes produce huge waves and heavy rainfall, which increases siltation. Waves also cause damage themselves.
- Coral mining for sand and lime for urban development destroys reefs. Often coral is the only local building material.
- Tourism causes both direct and indirect damage (see opposite).
- Pollution from sewage (nutrient enrichment) and from oil and toxic chemicals can kill reefs.
- Predators, such as crown of thorns starfish (COTS), breed in nutrient-enriched environments and kill the coral. Overfishing removes their natural predators, so COTS are increasing in number.

## Tourism

Tourism benefits the economy and gives people jobs. But the diagram on the right shows that tourism can have a wide range of direct and indirect impacts on the very reefs the tourists come to see.

## Fishing

Fishing has damaged reef ecosystems for a number of reasons:

- Overfishing of reef species, especially to meet huge demand in the Far East, is widespread.
- People living on the coast near reefs are often very poor. They need fish as a source of protein. They fish at a subsistence level using spears and traps. Some use blast fishing with dynamite, which damages the coral and kills many species – not just the fish being hunted.
- Many breeding grounds for reef fish are being destroyed.
- Many developing countries have sold fishing rights to nations with large trawler fleets, who have exhausted their own fisheries. Trawling methods are very efficient. Their small-mesh nets catch young fish and therefore harm the reef ecosystem by preventing its replenishment.
- Research at Australia's Great Barrier Reef suggests that just five trawls across a reef can lead to 90% coral damage.
- Fishing is often species selective. If you remove particular types of fish, this upsets the balance of the reef's food web.

Construction – impact of building tourist facilities, hotels, jetties. Loss and degradation of coastal habitats, e.g. mangroves.

Heavy demand for reef seafood, e.g. snappers and lobsters.

Coastal infilling – increased sedimentation due to increased runoff from a concrete coast.

Demand for reef curios, e.g. shells and black coral jewellery.

Beach enhancement – surplus sand can ruin local reefs.

**Tourism issues or concerns**

Damage from boat anchors, especially in busy areas such as yacht harbours.

Extraction of fresh water for use at hotels.

Recreational fishing by tourists.

Sewage and other pollutants come from hotels, e.g. sun screen!

Use of pontoons (floating hotels for day trippers).

Problems caused by feeding fish.

Trampling by snorkellers and divers breaks corals.

### On your planet

+ The UN's Food and Agriculture Organisation claims that fish stocks are being depleted faster than any other resource. By 2050, the fish protein needs of only half the world will be met. What can we do about this?

### your questions

1 Draw a table with 3 columns, with the headings 'Natural', 'Human' and 'Both'. List the destructive factors opposite under the appropriate headings.
2 'The real threats to the oceans are land based.' Do you agree with this statement?
3 Research a reef tourist destination, such as the Maldives or the Great Barrier Reef. Weigh up the costs and benefits of developing reef tourism.
4 **Exam-style question** Explain why the world's oceans are under threat. (4 marks)

In this section you will learn about the impacts of unsustainable use on marine ecosystems.

**Unsustainable use** includes:
- overfishing – where oceans are fished at a greater rate than the stock of fish can be replenished by breeding
- habitat destruction
- pollution.

All of these accelerate the decline of ocean environments, which has an impact on food webs.

**What do you think?**
+ Fish farms are seen as the great hope for providing sustainable supplies of fish! Do you see fish farms as a sustainable way ahead?

## Food web disruption

Krill are under threat. Scientists used to blame global warming. But now they believe the threat has come from intensive 'suction' harvesting. This is the method used by fishermen to meet our growing demand for krill. It gathers up huge quantities of the tiny creatures. Onboard processing and fast-freeze technology enable ships to take ever bigger catches. This will soon begin to damage the food web (see below).

Interestingly, until intensive harvesting by humans began, supplies of krill had been increasing. Many whale species (baleen whales) eat vast amounts of krill. However, when whale numbers decreased due to overhunting by whalers, krill numbers rose. It's a fine balance.

Krill are tiny shrimp-like animals. They are a keystone species, feeding on algae and plankton at the bottom of the food chain. In turn, they are eaten by whales, penguins and fish. However, the krill population has declined by 80% since the 1980s. Krill is now in growing demand for use in omega 3 health supplements, and above all as food in fish farms.

▶ The role of krill in the food web for the Southern Ocean in Antarctica.

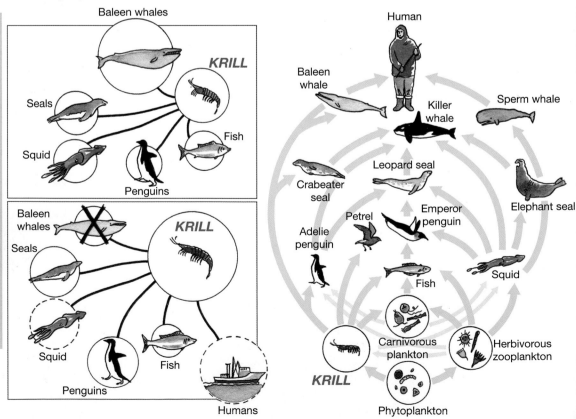

# Nutrient overloading

Nutrient cycles can be disrupted by pollution. Polluted dead zones could be the greatest emerging threat to life in the world's oceans. Nutrient overloading is called **eutrophication**. And it's happening on a huge scale and at an ever increasing pace. **Dead zones** are ocean areas which are critically low in oxygen. These zones lack the ability to support life. Once rare, they are now commonplace offshore from developed and overpopulated countries.

Climate change could be adding to the dead zones. Extreme storms could lead to more flooding and more flushing out of nitrates and phosphates into the oceans.

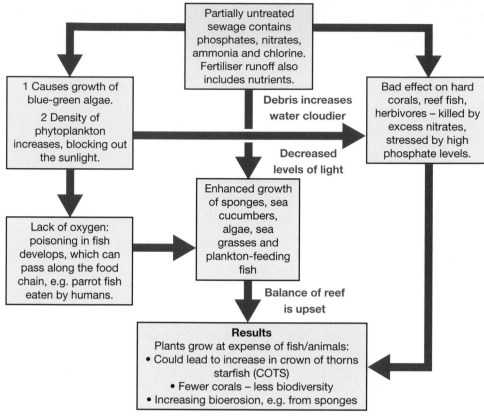

Partially untreated sewage contains phosphates, nitrates, ammonia and chlorine. Fertiliser runoff also includes nutrients.

1 Causes growth of blue-green algae.
2 Density of phytoplankton increases, blocking out the sunlight.

Debris increases water cloudier

Decreased levels of light

Bad effect on hard corals, reef fish, herbivores – killed by excess nitrates, stressed by high phosphate levels.

Lack of oxygen: poisoning in fish develops, which can pass along the food chain, e.g. parrot fish eaten by humans.

Enhanced growth of sponges, sea cucumbers, algae, sea grasses and plankton-feeding fish

Balance of reef is upset

**Results**
Plants grow at expense of fish/animals:
• Could lead to increase in crown of thorns starfish (COTS)
• Fewer corals – less biodiversity
• Increasing bioerosion, e.g. from sponges

▲ The impacts of eutrophication.

▲ An algal bloom in the Baltic Sea.

Fertilisers and sewage wash into oceans. Toxic 'blue algal blooms' grow on all the extra nutrients. Marine bacteria feed on algae in the blooms. When they die they sink to the bottom and decay. This uses up oxygen dissolved in the oceans. Highly poisonous conditions result, which asphyxiate bottom-dwelling organisms like worms and shellfish. In turn this affects fish, and even us who may eat the poisoned fish.

*On your planet*
✛ Did you know about the dead zones? There is no other threat of such ecological importance that has changed so dramatically in such a short time.

## your questions

1 Draw a simplified food web from the one shown opposite. Show the importance of krill in the Southern Ocean food web. Also show the factors which affect its supply.
2 What is the greatest threat to the oceans? Give reasons to support each of the following.
 • habitat destruction
 • overfishing
 • pollution and siltation.
 Give each of your reasons a score from 0-5. Add up the scores to determine the greatest threat.

✚ **In this section you will learn how climate change impacts directly and indirectly on oceans.**

Short-term climate change includes the impacts of both global warming and El Niño events (see page 133). If average temperatures rise by 3 °C by the end of this century, oceans could suffer irreversible destruction, which has been called the 'tipping point'.

## Direct impacts

Direct impacts of climate change relate to increases in temperature. These will affect oceans as much as the land. Many ocean ecosystems, such as coral, are vulnerable to changes in ocean temperatures.

Climate change can also lead to extreme weather, such as storms and floods. These damage ocean ecosystems by increasing pollution and siltation.

As temperatures on land rise, glaciers and frozen land will melt. This will increase the amount of freshwater reaching the oceans – making the water less salty and less dense. This could affect the ocean currents that distribute heat. The result is that ocean temperatures could actually decrease in some places, and increase in others. Such changes could greatly affect the ocean foodwebs – for example, if the UK's waters increase in temperature, the cod might move north to find cooler waters. If temperatures decrease, we might find other species from the Arctic moving south.

## Indirect impacts

These relate to rises in sea level. Warmer water temperatures cause the oceans to expand (thermal expansion). Melting glaciers and ice sheets also add to the volume of water in the oceans. In the 20th century, the oceans rose by an average of 15 cm. By the end of the 21st century, this rise could be between 20 cm and 1 metre.

Many of the world's coral reefs suffer from bleaching. The coral loses colour (bleaches) due to the loss of the algae with which it lives. Changes in ocean temperature by 1 °C-2 °C can stress coral. Eventually, bleached reefs weaken and collapse.

Algae

*On your planet*

✚ Oceans act as a **carbon sink** by storing carbon dioxide. However, increasing windiness is stirring up the oceans and carbon dioxide is being released back into the atmosphere at a greater rate than it is being absorbed.

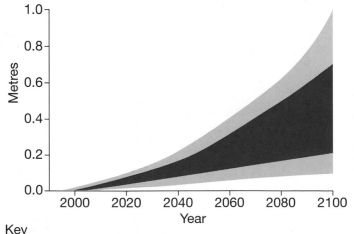

**Key**

⬛ Average of 35 projections of greenhouse gas emissions

⬜ Area of uncertainty

▲ The range of predictions for sea-level rise (1990-2100).

A sea-level rise of 1 metre would have dramatic consequences for coastal communities. Thousands of square kilometres of land would be flooded.

- Coral reef islands, such as the Maldives or South Pacific islands, would be almost completely submerged.
- Heavily populated mega deltas, such as the Nile or Ganges-Brahmaputra, would be at risk from storm surges and floods. Supplies of freshwater would be contaminated.
- Without flood defences, low-lying countries – such as the Netherlands – would be almost completely submerged.

- Already the first **environmental refugees** have been recorded from the Char Islands of Bangladesh and the islands of Kiribati in the South Pacific.
- Mega cities, such as Mumbai or Lagos, would be under threat. Their water and sanitation systems would be strained. Millions of poor people living in shanty towns would be at even greater risk from disease.
- Many ports would cease to operate – affecting the economies of many countries.

**On your planet**
+ If the Greenland ice cap and West Antarctica ice sheet were to completely melt, sea levels would rise by 12 metres!

▶ The effects of projected sea-level rise by 2100.

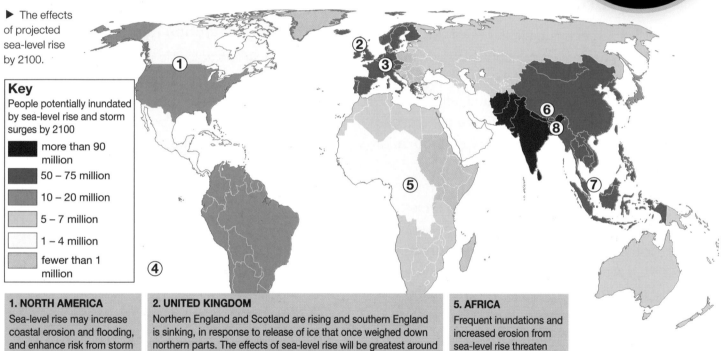

**Key**

People potentially inundated by sea-level rise and storm surges by 2100

- more than 90 million
- 50 – 75 million
- 10 – 20 million
- 5 – 7 million
- 1 – 4 million
- fewer than 1 million

**1. NORTH AMERICA**

Sea-level rise may increase coastal erosion and flooding, and enhance risk from storm surges in Florida and along the Atlantic Coast. Coastal wetlands will be threatened.

**2. UNITED KINGDOM**

Northern England and Scotland are rising and southern England is sinking, in response to release of ice that once weighed down northern parts. The effects of sea-level rise will be greatest around southern coasts. While protecting the most valuable land, the authorities may allow waters to advance in some places. The risk of flooding due to storm surges will increase.

**5. AFRICA**

Frequent inundations and increased erosion from sea-level rise threaten coastal settlements including in Senegal, Gambia and Egypt.

**3. EUROPE**

The greatest losses from sea-level rise will be in the east of England, the Po Delta of northern Italy and along the coast from Belgium, through the Netherlandsand north-west Germany into western and northern Denmark. Some seaside communities in the Netherlands, England, Denmark, Germany, Italy and Poland are already below normal high-tide levels.

**4. SOUTH PACIFIC**

Here there are:
- disappearing islands
- evacuated islands
- abandoned islands
- threatened islands

**6. ASIA**

Sea-level rise and increasing intensity of tropical cyclones could displace people in low-lying coastal areas of temperate and tropical Asia. It would also threaten mangroves and coral reefs.

**7. SINGAPORE**

As a small, low-lying nation,Singapore is vulnerable to sea-level rise. Strengthening sea defences is probably the cheapest solution.

**8. BANGLADESH**

Bangladesh is vulnerable to sea-level rise. Storm surges already make people's lives a misery. Catastrophic floods have resulted in damage 100 kilometres inland. A sea-level rise of 45 centimetres could displace 5.5 million people.

**your questions**

1 Make a table of the possible effects of climate change. Include the areas of the world affected, and the effect on people and the economy.

2 **Exam-style question** Using examples, explain the impacts of climate change on either developed or developing countries. (6 marks)

In this section you will explore how ocean ecosystems can be managed sustainably.

Pressure on marine ecosystems is growing, due to rising populations in coastal areas and increased demands for ocean resources.

**Sustainable management** is a balancing act between ecosystem conservation and helping local people to make a living without overharvesting resources.

For ocean ecosystems, this involves:

- using fishing equipment which doesn't damage delicate habitats, such as coral reefs.
- using marine resources at a rate that won't destroy them for future generations. Overharvesting of fish, shellfish and other marine organisms must be avoided. For example, if net size is too small, many young fish are caught, which reduces the breeding stock. Marine resources should be sustainable – unlike **finite resources**.
- allowing poor people to use resources for subsistence activities. This often conflicts with government or business interests.
- local people (community-based management), so they can decide how fishing and other uses should be managed. Just creating protected areas, which prevent locals from carrying out their livelihoods, is bound to lead to conflicts.

+ Resources which will one day run out are called **finite resources**. Examples are oil and coal.

## Sustainable success story?

St Lucia in the Caribbean pioneered the idea of community-based management of ecosystems. In 1986, 19 areas were declared Marine Reserve Areas (MRAs). These included coral reefs, turtle breeding grounds and mangroves. Fishing Priority Areas were also created (FPAs) (see the map opposite). However, the boundaries were never fully defined (always difficult in the ocean), so conflicts arose.

## Why was protection needed?

- St Lucia is a volcanic island with most of its population concentrated along narrow coastal plains. Land-based damage of the ocean is therefore likely. The population is rising at nearly 2% a year. Densities average 300 per km$^2$.
- The continental shelf is narrow, which leads to overfishing. Most fishermen are subsistence fishermen who don't have boats for deep sea fishing. Their methods – such as placing pots on coral reefs, or chasing fish into nets by throwing rocks into the water – can be very damaging to the reef.
- The tourist industry has developed very rapidly, leading to problems of waste disposal and pollution. Tourism earnings are worth nearly half of St Lucia's annual earnings.
- More tourists have led to more snorkelling, diving and yachting. Soufrière is a focus for reef-based tourism.
- 20% of people live below the poverty line. Many have no jobs. They harvest mangroves for charcoal, hunt wildlife and catch fish. This puts pressure on local resources.
- Forests are cut down for banana plantations. This leads to heavy siltation, especially during storms.

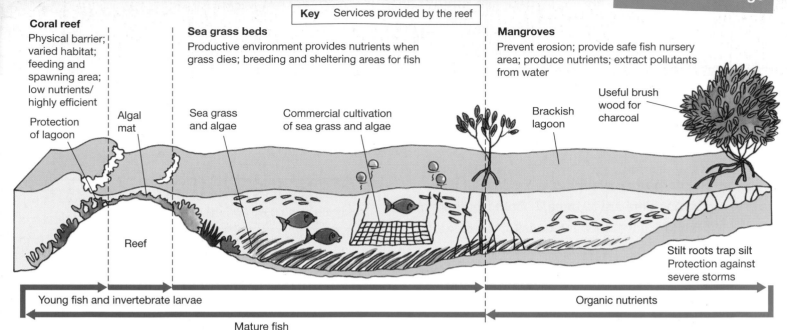

**Key** Services provided by the reef

**Coral reef**
Physical barrier; varied habitat; feeding and spawning area; low nutrients/highly efficient

Protection of lagoon

Algal mat

Reef

**Sea grass beds**
Productive environment provides nutrients when grass dies; breeding and sheltering areas for fish

Sea grass and algae

Commercial cultivation of sea grass and algae

**Mangroves**
Prevent erosion; provide safe fish nursery area; produce nutrients; extract pollutants from water

Useful brush wood for charcoal

Brackish lagoon

Stilt roots trap silt
Protection against severe storms

Young fish and invertebrate larvae

Organic nutrients

Mature fish

▲ How the marine ecosystems of St Lucia (coral reefs, mangroves and sea grass beds) provide important services.

# Managment conflicts in Soufrière and Mankòtè

In St Lucia, the coral reefs of Soufrière and the mangrove forests of Mankòtè were suffering damage. For both areas, any management plan had to allow for community participation.

In Soufrière, the conflicts were between fishermen, divers, snorkellers and yacht owners. They all had the potential to damage the reef.

In Mankòtè, there was a lot of unregulated hunting and fishing in the 1980s. It had become a site for rubbish disposal and a target for mosquito eradication – this involved spraying with insecticide. The Mankòtè mangrove is St Lucia's largest remaining mangrove forest, so it has ecological importance. It also provides social and economic services for local communities (subsistence and commercial fishing, fuelwood/charcoal, honey and salt).

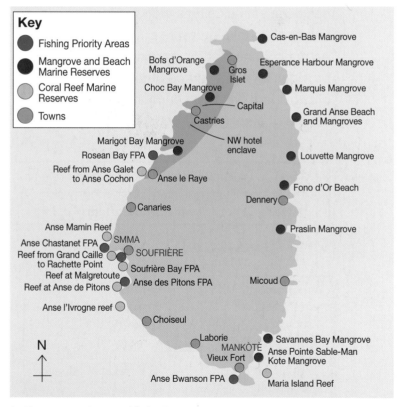

**Key**
- Fishing Priority Areas
- Mangrove and Beach Marine Reserves
- Coral Reef Marine Reserves
- Towns

Cas-en-Bas Mangrove
Bofs d'Orange Mangrove
Gros Islet
Esperance Harbour Mangrove
Choc Bay Mangrove
Capital
Marquis Mangrove
Castries
Grand Anse Beach and Mangroves
Marigot Bay Mangrove
NW hotel enclave
Rosean Bay FPA
Reef from Anse Galet to Anse Cochon
Anse le Raye
Louvette Mangrove
Fono d'Or Beach
Dennery
Canaries
Praslin Mangrove
Anse Mamin Reef
Anse Chastanet FPA
SMMA
Reef from Grand Caille to Rachette Point
SOUFRIÈRE
Reef at Malgretoute
Soufrière Bay FPA
Micoud
Reef at Anse de Pitons
Anse des Pitons FPA
Anse l'Ivrogne reef
Choiseul
N
Laborie
Savannes Bay Mangrove
MANKÒTÈ
Anse Pointe Sable-Man Kote Mangrove
Vieux Fort
Anse Bwanson FPA
Maria Island Reef

▲ The protected areas of St Lucia.

➕ **In this section you will learn about the stages of participatory planning.**

**Participatory planning** is essential if all stakeholders are to benefit. Both Soufrière and Mankòtè have been successful, to an extent, because they have supported the incomes of local people and involved them in managing resources in the marine reserves.

At Soufrière, fishermen have been provided with modern boats and a refrigerated ice house to improve fish processing. In Mankòtè, land has been set aside as a woodlot for fuel and charcoal (to take pressure off the mangroves). An agricultural plot for growing vegetables provides an extra source of food. In both cases, **ecotourism** has been encouraged as an extra source of employment and income.

> ➕ Some people prefer to take a holiday in natural areas so that they gain an understanding of conservation and wildlife. This is known as **ecotourism**.

> ➕ In **participatory planning**, all stakeholders are involved in the development of a scheme.

▲ Fishermen at Soufrière hauling in their catch.

▼ The key stages for successful participatory planning.

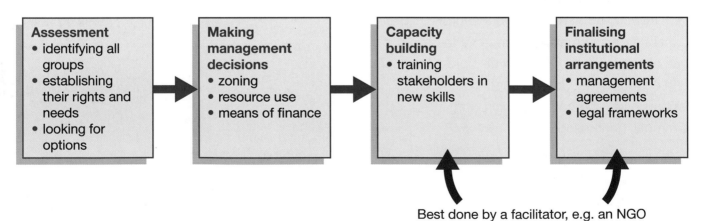

**Assessment**
- identifying all groups
- establishing their rights and needs
- looking for options

➡

**Making management decisions**
- zoning
- resource use
- means of finance

➡

**Capacity building**
- training stakeholders in new skills

➡

**Finalising institutional arrangements**
- management agreements
- legal frameworks

Best done by a facilitator, e.g. an NGO

# Soufrière Marine Management Area (SMMA)

In 1992, the St Lucia Department of Fisheries (in the Caribbean) brought the following people together:

- the local town council
- local hotel owners
- water-taxi owners
- dive businesses
- fisherman
- marine managers.

They went out in boats and cruised along the coast discussing how coastal zoning could work. The SMMA developed from this trip.

Overall, the SMMA has been successful as a model of sustainability. But there have been problems. It's very difficult to get stakeholders with so many different interests to agree. For example, establishing a marine conservation area, means part of it has to be a no-go area for fishermen. The fishing community became angry seeing divers in the conservation area. They didn't believe that a conservation area would help to conserve future fish stocks.

Local people had to be trained and educated to manage the scheme. The rangers who police the area had to be equiped, and this costs money. Fees from divers and yacht owners now make the scheme self-funding.

However, there are some problems:

- The area has become so popular that the marine environment is threatened by mass tourism.
- Rapid development in Soufrière encourages siltation and pollution.

But, on the positive side:

- the numbers, sizes and diversity of fish species have increased
- many stakeholders are now involved in marine conservation.

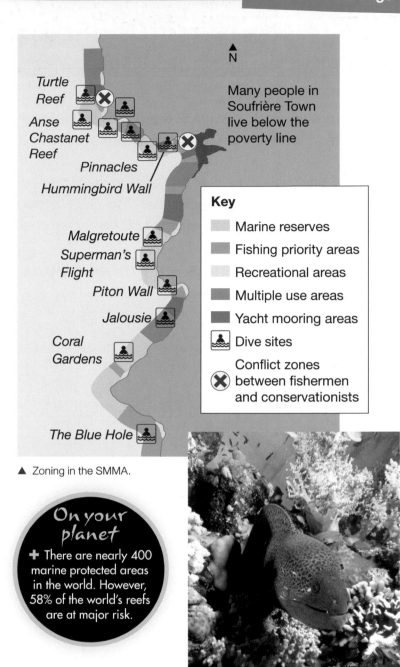

▲ Zoning in the SMMA.

**On your planet**

+ There are nearly 400 marine protected areas in the world. However, 58% of the world's reefs are at major risk.

## your questions

1 In pairs, decide what have been the SMMA's three biggest successes, and why.
2 Why is involving local people essential for schemes like the SMMA to work?
3 **Exam-style question** With reference to an example, explain the problems and successes of sustainable fishing. (4 marks)

➕ In this section you will learn about Marine Protected Areas and also look at management of fish stocks in the North Sea.

Sustainable management needs to take place at local, national, international and global scales. At a global scale, countries can get together to provide legal frameworks for regional and local schemes.

**On your planet**
➕ In 2008, the UK had only one-fiftieth of 1% of our seas as MPAs!

## Marine reserves – a global solution?

▼ The distribution of Marine Protected Areas (MPAs).

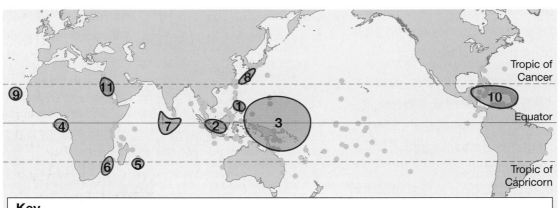

**Key**

| | | | | | | |
|---|---|---|---|---|---|---|
| • | Marine Protected Area | ? | 1 Philippines | × | 7 North Indian Ocean |
| ⬭ | Marine hotspots | ? | 2 Sundaland | ? | 8 Southern Japan/Taiwan |
| × | Not protected | ? | 3 Wallacea | × | 9 Cape Verde |
| ? | Partially protected | ? | 4 Gulf of Guinea | × | 10 Western Caribbean (tourism) |
| ✓ | Good protection | ? | 5 Mascanene | ? | 11 Red Sea (oil spills) |
| | | ✓ | 6 Eastern South Africa | | |

As you can see from the map, there is no truly global network of sites. Indeed, 35% of countries have no Marine Protection Areas at all. The map also only shows the distribution of MPAs – with no indication of their size. Of the 500 marine reserves, most are only 1 km² in size. There are few reserves above 1000 km² – and big is beautiful when it comes to marine conservation. Kiribati in the South Pacific has just created the world's largest reserve (the size of California). In addition, the map tells us nothing about the quality of the reserves. Many are very poorly managed. Another problem is that many of the **marine hotspots**, the areas of highest biodiversity under the greatest threat, aren't currently protected.

### How to set up a successful reef reserve

- Make laws and police them to limit damage.
- Clean up pollution from sewage outfalls. Try to find the polluter. Provide incentives to stop people polluting.
- Find other jobs for unemployed fishermen.
- Help local fishermen to market their goods.
- Educate people about reef management.
- Limit numbers of boats, divers, etc.
- Charge for use of the reef. Use the money to manage the reserve.
- Use the results of scientific research to conserve species.

# Managing fish stocks in the North Sea

This regional-scale case study illustrates the need for international co-operation. The EU Common Fisheries Policy has tried to bring back fish stocks from catastrophically low levels.

Every year, the EU reviews its fisheries policy. The fishermen want to fish as large a quota as possible, but marine scientists argue that only 'no fishing' marine reserves will save species like the cod. These are very expensive to set up, because you have to compensate the fishermen for loss of earnings – you have to pay them not to fish! The fishermen also argue that the calculations about fish stocks aren't correct.

A **whole ecosystem approach**, not a single species approach, might save the North Sea. This would include:

- ensuring that the mesh of nets won't catch undersized young fish
- limiting the number of hours and days that fishing boats can operate each year
- quota management – each year a limit is placed on the number of tonnes of fish from various species which can be caught, based on a 'state of stock' survey
- discard management – fewer unwanted fish are discarded (by catch)
- setting up marine reserves which protect all species – some of which could be temporary to protect spawning and nursery grounds
- further research into how fishing affects the whole ecosystem of the North Sea, e.g. a lack of sand eels results in fewer puffins.

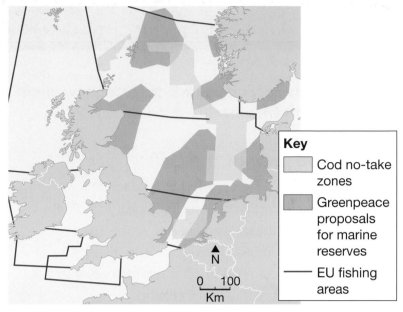

**Key**

- Cod no-take zones
- Greenpeace proposals for marine reserves
- EU fishing areas

N

0    100
Km

▲ European waters are divided into sectors. Each year a limit (**quota**) is set on the number of each species that may be caught in each sector.

## Global warming

It's not just a case of too many cod being caught by efficient factory trawlers. Global warming has also played a part. The cold water species of plankton, which bloomed in the spring at just the right time for baby cod to eat, has been replaced by warm water species. These bloom in the late summer, when baby cod need larger food. This partly explains the drop in numbers of cod surviving to adulthood since 1980. Also, as the North Sea warms, cod will migrate to Arctic areas.

*On your planet*

✚ 'Has our cod had its chips?' It's up to us as consumers, too. Should we stop eating cod and change to less threatened species which are guaranteed to be sustainably produced?

## your questions

1 Why is fishing a difficult industry to manage?
2 Draw a large spider diagram. Label on it as many problems as you can think of facing the North Sea.
3 In a different colour, write as many solutions as you can think of to solve these problems.
4 Write a 300-word letter to the UK Minister for Fisheries, explaining what you think should be done to save fish stocks.

➕ In this section you will consider how global actions could help to improve the health of the oceans.

## The Law of the Sea

The **Law of the Sea** was developed to prevent certain nations from taking an unfair share of the ocean's wealth. At the Third Law of the Sea Conference (UNCLOS), the aim was to develop a treaty which covered a wide range of issues including:

- fisheries
- navigation
- continental shelves
- the deep sea
- scientific research
- pollution of the marine environment.

The treaty was finally ratified in 1994. 40% of the ocean was placed under the law of the adjacent coastal states. These were defined in four zones of increasing sizes.

1 Territorial seas, with a 12-mile limit of sovereign rights
2 Contiguous zones, with a 24-mile limit – controlled for specific purposes.
3 Exclusive Economic Zones, with a 200-mile limit. Rights are guaranteed over economic activity, scientific research and environmental conservation.
4 Continental shelves, where nations may explore and exploit without infringing the legal status of the water and the air above.

The traditional 'freedom of the seas' remains for the other 60% of open oceans. Within this, the deep seabed area is designated 'the common heritage of mankind', and is controlled by the International Seabed Authority.

UNCLOS addresses the main sources of ocean pollution:

- land-based/coastal activities
- continental shelf drilling for gas and oil
- seabed mining
- ocean dumping
- pollution from ships.

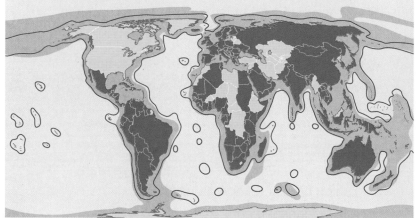

**Key**
- Continental shelf
- Open ocean
- ⎯⎯ EEZ (Exclusive Economic Zones)
- Signed up to UNCLOS, 2002

▲ The impact of UNCLOS – how the Law of the Sea controls oceans.

The state of the world's oceans is a test for the emerging skill of planet management. New challenges are putting even more pressure on the oceans.

- New technologies allow deep sea exploration.
- Increasingly complex pollutants are reaching the oceans from advanced industrial processes.
- And, of course, global warming has introduced new pressures – such as opening up the Arctic.

In 1974, the Helsinki Convention developed programmes to manage marine pollution. This included preventing pollution from ships (see right), and preventing the dumping of radioactive waste into the sea. It led to a series of UN Environment Programmes (UNEP), sponsored by Regional Action Plans, where states work together to clean up seas such as the Mediterranean.

International action has also provided frameworks for marine conservation. One scheme, known as the Global Marine Species Assessment, is trying to find out what species really live in our oceans. After all, how do you protect biodiversity if you don't know what is there? The idea is to develop a red (for danger) list of threatened marine species. Marine 'hotspots' will also be identified.

At the 2002 World Summit on Sustainable Development, a target was set for the development of Marine Protected Areas. It was suggested that these should amount to 10% of the world's oceans. MPAs are the main hope for conserving biodiversity in the oceans.

▲ There are many similarities between conservation of oceans and land-based conservation, but the challenges are even greater.

On your planet
+ Did you know that the Mediterranean Sea is one of the dirtiest in the world?

## your questions

1 In groups of 2-3, draw a labelled spider diagram to show reasons why the world's oceans are threatened. Feed back in class and decide the 3 or 4 biggest problems that all the groups identified.

2 Take two of these problems. For each make a copy of the following table. In it, explain how far the limits shown would help to solve the problem.

3 **Exam-style question** Using an example, explain why marine environments are so difficult to protect. (6 marks)

| Problem: (e.g. pollution) | Whether these will help to solve the problem and why |
|---|---|
| 12 mile exclusion zones | |
| 24 mile exclusion zones | |
| 200 mile exclusion zones | |
| Continental shelf exclusion zones | |

➕ In this section you will learn what desert climates are like in Australia.

### Next stop Cook!

Twice a week, the Indian-Pacific train trundles through the Nullarbor Plain on its way across Australia from Perth to Sydney – and twice back again. The Nullarbor covers an area the size of the UK! And it's just as extreme as Antarctica in terms of challenging people's survival skills. Summer temperatures are above 40 °C, rain is rare, and soils are thin and infertile.

The train is full of tourists – the journey takes 3 days, so it's hardly a commuter route! But for some outback residents, it's a lifeline for on-line shopping and supplies. When the train stops at Cook to refuel, it brings supplies for the two people living there … and others who've driven over 100 km to this, their nearest settlement.

▼ Cook, in South Australia. Once it was a town, with a secondary boarding school for 'local' farm children (up to 500 km away), a swimming pool and shops. Isolation and a harsh environment forced people away.

### The back of Bourke

Bourke is a small town in the outback of New South Wales. Australia's outback – as its semi-deserts are known – is huge! Fly to Sydney from Asia and you're over it for 4-5 hours – making it as big as the Atlantic! The outback is one of the world's most barren and least populated places. Mostly it's desert or semi-desert, with scattered cattle farms (or 'stations'). Mining towns are linked by a few tarmac-covered roads and dirt tracks. Australians rarely travel there, referring to its remoteness as the 'back of Bourke'.

After Antarctica, Australia is the world's driest continent; more than one-third is desert. There is some rainfall, but most evaporates quickly in the heat. The deserts vary in appearance and vegetation, so they are given different names, as shown below.

▼ This shows the extent of Australia's deserts and semi-deserts. The desert isn't true desert like the Sahara, but it's extremely barren.

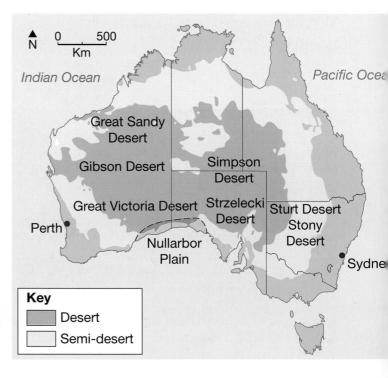

# Why is most of Australia desert?

Normally, rain-bearing winds blow across the Pacific towards Australia. The mountains that border the coast – the Great Dividing Range – cause this air to rise and cool rapidly. This leads to condensation, and then rain (see right).

As the air descends from the mountains, it is drier – and a 'rain shadow' is created. This results in low rainfall in western areas. The further west the winds blow, the drier they are – so the driest areas are in Western Australia.

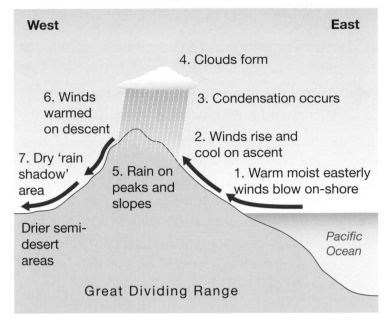

West     East

4. Clouds form

6. Winds warmed on descent

3. Condensation occurs

7. Dry 'rain shadow' area

2. Winds rise and cool on ascent

5. Rain on peaks and slopes

1. Warm moist easterly winds blow on-shore

Drier semi-desert areas

Pacific Ocean

Great Dividing Range

▲ How rain shadow areas are made. They turn much of central and western Australia into desert.

*On your planet*

✚ The longest hot spell in Australia was at Marble Bar in 1923-24. For 160 consecutive days, temperatures rose above 37.8 °C (100 °F) – that's 5 months!

*On your planet*

✚ Australia's highest temperature was recorded at Oodnadatta in South Australia – 50.7 °C on 2 January 1960!

## your questions

1 In pairs, discuss and write down the benefits and problems of living in places like Cook for: **a** young children, **b** families, **c** the elderly.

2 Use Google Images or Maps to obtain images of three different Australian desert landscapes. Locate them on a map of Australia. Then prepare a poster or PowerPoint showing why people might want to see Australia's deserts.

3 Place these phrases in order, to show why Australia's deserts get so little rain: *clouds form, condensation, cooler temperatures, deserts form, little moisture left, on-shore winds from the Pacific, rain falls, winds descend, winds forced to rise.*

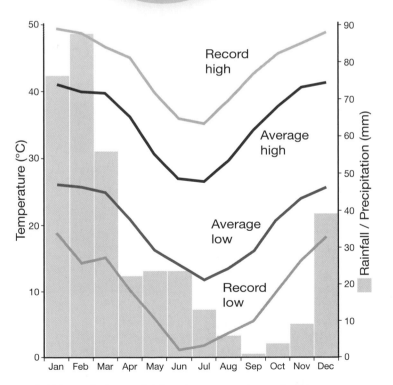

Record high

Average high

Average low

Record low

▲ This graph shows rainfall, and average and record maximum and minimum daytime temperatures for Marble Bar in north-western Australia.

➕ **In this section you will learn how plants (flora) and animals (fauna) survive Australia's desert climate.**

Australia's **biodiversity** (its range of plants and animals) is unique. It has over 1 million species – many of which are found nowhere else, because of Australia's geographical isolation. The unique plants and animals have to be able to cope with the arid (dry) climates of Australia's interior.

## Desert plants

The three main ways in which desert plants have adapted to arid climates are:

- succulence
- drought tolerance
- drought avoidance.

▼ Australia's desert is actually semi-desert, as shown here. Plants have to be capable of resisting extremes of temperature and drought. Often they survive near streams.

### Succulence

Australia has over 400 **succulent** species. They store water in fleshy leaves, stems or roots. Desert rains are infrequent, light, short-lived and evaporate quickly, so water must be captured and stored.

- To survive, succulents can very quickly absorb large amounts of water through extensive, shallow root systems. They can store this water for long periods.
- Their stems and leaves also have waxy cuticles (surface layers) which make them almost waterproof when their stomata close.
- Their metabolism slows down during drought and their stomata remain closed, so that water loss almost stops. Growth, therefore, stops during drought.
- However, their water stores make them attractive to thirsty animals, so most have spines or are toxic; some are camouflaged.

Australian eucalypt trees, like those in the background, have very deep roots, and also grow close to river courses that hold water for long periods. Their leaves have few stomata, to minimise water loss.

Most desert plants resume growth within a few hours of rain.

## Drought tolerance

Drought tolerance means having mechanisms that help to survive drought.

- During drought, plants of this type shed leaves to prevent water loss through transpiration. They become dormant. Others, like eucalypts, remain evergreen but have waxy leaves with few stomata, to minimise water loss.
- These plants have extensive, deep roots which penetrate soil and rock to get at underground water.
- They photosynthesise with low leaf moisture levels, which would be fatal to most plants.

## Drought avoidance

Most drought avoiders are annuals – they survive one season, have a rapid life cycle, and die after seeding.

- Their seeds last for years, and germinate only when soil moisture is high.
- Some germinate during autumn, after rain and before winter cold sets in. The seedlings survive winter frost and flower in spring.

# Australian desert animals

Australia's deserts are not the world's driest, but rain is unpredictable. Several years can pass between showers. Many animals settle near small watercourses – known as billabongs. Desert animals have had to evolve to survive (see right).

### your questions

1 Define *succulence, drought tolerance* and *drought avoidance*.
2 Draw a spider diagram to show the different ways in which different animals, birds and plants survive drought.
3 Select one plant and one animal. Produce an A4 sheet to show how they both survive drought. Include an image, and 150 words of text for each one.

### The Bilby

A small marsupial (mammal with a pouch), bilbies once lived throughout Australia's deserts. Having been hunted by domestic cats, dogs and wild foxes (all imported to Australia), few survive.

How it survives
- It is nocturnal, sheltering from the daytime heat to avoid dehydration.
- It burrows for moister, cooler conditions.
- It has low moisture needs, obtaining enough from its food, such as bulbs, fungi and insects.

### The Perentie

This is a giant lizard. They can grow up to 2.5 metres in length, and weigh up to 15 kg. It's one of the desert's top predators.

How it survives
- To escape the desert heat, it digs burrows or hides in deep rock crevices. It emerges from these to hunt.
- It hibernates from May to August to avoid cold.
- Like the bilby, it has low moisture needs.

### The Red Kangaroo

The Red Kangaroo, the world's largest marsupial, is one of many kangaroo species.

How it survives
- It survives by hopping (a fast, energy-efficient form of travel) to find food in the sparsely vegetated desert.
- It feeds at dawn and dusk when the air is cooler, and sleeps during the heat of the day.
- Dew is an important part of its water intake.
- Rain triggers a hormonal response in females, so that breeding only occurs during rains.

➕ **In this section you will learn how people cope with living in the outback's extreme conditions.**

If you're a white-skinned European, there's one rule for the midday Australian sun – stay out of it. Early European explorers walked across Australia's interior and few survived. They failed to use the lessons of Australia's own aboriginal peoples – to live sustainably with the environment.

However, white settlers do live in Australia's most extreme environments. The problem is making a living. The soils are poor, with little organic material (which retains moisture) and few nutrients. Poor soils mean that plants have low nutrient content. And many plants resist being eaten by animals because they have thorns or toxic sap. If water is available, there is just about enough grass to feed cattle or sheep – but only in very low numbers. So outback farms are huge – some the size of Wales.

## Managing supplies of water

With little rain, farmers have two water sources:
- Most farms have dams and reservoirs to store water which cattle and sheep can drink.
- Farms also use **boreholes** to tap into underground **artesian water**. Rain soaks into the desert soils and **percolates** (trickles) down into the bedrock. Over many years, water gradually collects. If you drill a borehole, it comes up – either under natural pressure or by using **windpumps** (see right). This water can be used for domestic supply or for animals.

▶ Artesian water – how it's formed and extracted.

The system of obtaining and storing water by boreholes and reservoirs is fragile. Many people question now whether water and the landscape are being used sustainably.
- Recent droughts have put pressure on the landscape. Animals still graze, but the grass starts to die in the arid conditions. As the roots die, nothing binds the soil together – so it's eroded by windstorms.
- Underground water is also being over-used, beyond the amount of annual rainfall which recharges it. Water tables are falling each year.

◀ A windpump in the outback – these draw water from underground boreholes, which lead to aquifers of artesian water (see below). In front of the windpump is a reservoir for storing water for sheep.

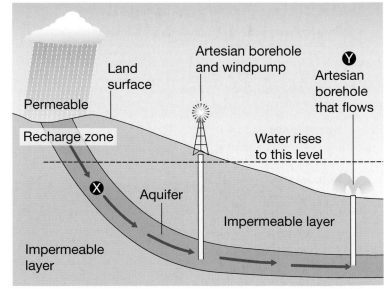

# Going underground

Most people in Australia's outback don't farm – they work in the mines. Australia has some of the world's largest reserves of quality iron ore, silver and gems (such as opals). Wages are high to attract people. But living there means using lots of energy to run the air conditioning needed to make life bearable.

Traditional settlers lived differently. The town of Coober Pedy in South Australia is built almost wholly underground. Houses are cut into solid rock, where daytime heat and night-time cold are evened out.

*On your planet*

+ Most outback farmers have their own planes – as do the Flying Doctors. Many outback children learn over the radio, instead of going to school.

## The Prairie Hotel, Parachilna

There's not much at Parachilna – just the Prairie Hotel and a population of 7! But the hotel's extension won awards in 2000 for its 'green' approaches to energy-saving.

- The building is 1 metre below ground, and is cooled by the surrounding rock. That's important during summer when it's 45 °C.
- Solar panels generate electricity for lighting and fans; no air conditioning is needed.
- Kitchen and bathroom water is recycled as 'grey water' for watering the gardens.

▶ The Prairie Hotel, southern Australia.

## your questions

1 What makes it so difficult to survive in the Australian outback?
2 Copy the diagram showing artesian water. Explain:
   **a** how rainwater gets underground at point X
   **b** why water comes up without any need for a pump at point Y.
3 Why can artesian water only support small numbers of people?

4 In pairs, list reasons why farming in the outback would be difficult: **a** economically, **b** socially, and **c** environmentally.
5 **Exam-style question** Giving examples, explain how people cope with living in an extreme climate.
(4 marks)

➕ **In this section you will learn something about Australia's aboriginal people and their culture.**

Each afternoon, an aboriginal man takes his seat on the ground in Circular Quay in Sydney. It's the place where you can catch ferries for tours around Sydney's harbour. He plays a didgeridoo, a traditional instrument – but he plays to an electronic dance rhythm mixed with sounds of the outback, such as kookaburras. His creative mix earns huge applause and he sells copies of his CD. He probably wonders how many customers have been to the outback, or met aboriginal people.

He is one of the few aboriginals who white Australians are likely to meet. Aboriginal people remain hidden, both within certain parts of cities and also in remote outback camps. White Australians are more likely to watch TV programmes about aboriginal life instead.

## Healthy eating, aboriginal style

The traditional aboriginal diet varies, depending on the area of Australia they are in. For desert groups, there is a huge variety of food available (see the table). Now Australia has a growing 'native food' industry, based on traditional aboriginal knowledge of what's edible in the outback. Several plants have multiple uses as food, medicine, utensils, tools, musical instruments or weapons.

The aboriginal population:
- has only two-thirds of the life expectancy of white Australians – 52 years instead of 78
- has Australia's worst drug and alcohol abuse; and homelessness is also a problem
- sees its traditional lifestyle disappearing – there is a risk that aboriginal customs, knowledge and beliefs will disappear unless action is taken to preserve them.

| Fruits | • Bush tomatoes taste of tomato and caramel, and are used for chutney or as a seasoning for meat.<br>• Desert limes have a strong citrus flavour and are used in jams and sauces.<br>• Quandongs (pictured) are also called 'native peaches'. They are bright red berries, high in Vitamin C, which taste a bit like apricots.<br>• Bush bananas are vines with long fruit that tastes like green peas and avocado. | |
| --- | --- | --- |
| Seeds | • Wattle (acacia) seeds from wattle trees, like the one on the right, are used in biscuits, in drinks or as dressings.<br>• Sandalwood nuts are eaten. | |

| | |
|---|---|
| **Grubs** | • Witchetty grubs are the larvae of moths and beetles, which are eaten raw or cooked. They taste like scrambled eggs and peanut butter with a crispy 'chicken skin' coating. |

| | |
|---|---|
| **Meat** | • Traditional wild animals, such as kangaroo, crocodile or emu (pictured). |

## Aboriginal beliefs and lifestyle

Aboriginal beliefs focus on the land. They see it as sacred and something to be protected. We think of aboriginal people as being nomadic, but in fact their tribal groups followed strict paths according to the seasons.

Traditionally, aboriginal people survived by **hunting and gathering** – finding edible plants and animals.

- They created conditions in which grubs could live and breed.
- They built dams across rivers to catch fish, and to make pools where birds would gather.
- They used fire to drive out animals for hunting, to clear wood and allow grass to grow. As a result, fire-tolerant plants (eucalyptus trees) came to dominate the landscape. These re-grow quickly after fire. The seeds of some species actually need fire to burst and germinate.
- Aboriginal crafts were based on hunting (boomerangs) or music and tribal celebrations (didgeridoos).
- Their customs and stories were spoken, never written. It's only now that some stories are being written down.
- As generations split and younger people move to the towns, stories about care of the land are being lost.

*On your planet*
+ Aboriginal Australians generally don't believe in owning land – they see themselves as caretakers.

### your questions

1 In pairs make two lists – one of ways in which the aboriginal lifestyle is unique, the second about ways in which understanding this lifestyle is valuable to us as a human race.
2 List as many ways as you can to show how well aboriginal people have adapted to Australia's extreme environment.
3 Type 'aboriginal food' into Google. Gather some information and create a pamphlet or poster about how aboriginal people use the resources around them for food and medicine.
4 Is hunting and gathering sustainable? Make a table to show ways in which it is and isn't sustainable.

# Cultural threats

➕ In this section you will learn about the threats to aboriginal culture from tourism.

Watching the sunset at Uluru is on the BBC's list of '50 things to do before you die'. Uluru, also known as Ayers Rock, is the most visited spot in Australia. It's also a sacred site of the aboriginal Anangu people.

## Tourism and aboriginal culture

The aboriginal peoples are some of the world's longest established communities. They first reached Australia at least 43 000 years ago. About 10 000 years ago, the first aboriginal people came to Uluru. You can't begin to appreciate how long ago that is – even when you see their cave paintings or sacred sites. It's for that reason that Uluru is a World Heritage Site.

Uluru is located within the Uluru-Kata Tjuta National Park. The surrounding environment is very sensitive, but the number of visitors is rising rapidly (see the table).

*On your planet*

➕ Australia's oldest aboriginal settlements are 20 000 years old – they were already 15 000 years old when the Egyptian pyramids were built!

| | |
|------|--------|
| 1961 | 5000 |
| 1984 | 100 000 |
| 1987 | 180 000 |
| 2000 | 380 000 |
| 2005 | 400 000 |

▲ Visitors to Uluru: 60% of tourists come from overseas.

▲ Uluru at sunset. Several kilometres from the rock itself, the view from the car parks at sunset or sunrise is startling. So coach tours arrive and offer tourists 'the experience' – champagne at sunset.

▲ Uluru – these cave paintings show how important this site is to aboriginal peoples. Uluru is one of the oldest aboriginal sacred sites. Some paintings at Uluru aren't shown to the public.

# What problems do tourists bring?

Tourists visit Uluru because of the landscape, and because they are interested in aboriginal culture. They bring economic benefits when they buy aboriginal art and crafts. However, there are problems too.

1. Aboriginal culture can be exploited to provide entertainment. Paintings are produced to suit visitors' tastes, rather than express culture.

2. Tourists may leave without learning a thing about aboriginal culture or beliefs. People come for the 'experience' – the sunset. And perhaps to climb the rock, even though Uluru is sacred. It's against Anangu spiritual beliefs to climb Uluru.

3. The Anangu have no part in the management or development of the tourist resort where most visitors stay.

4. Tour guides often ignore 'awkward' aboriginal history. How many mention these facts?

   - 90% of aboriginal people were slaughtered during the first 60 years of British rule in Australia.
   - Aboriginal children were forcibly taken from their parents until the 1970s.
   - Only recently has aboriginal land been handed back – and then only in desert areas, not where there are minerals.

## Tourism is changing

The new Uluru Aboriginal Cultural Centre educates visitors about aboriginal peoples. Its displays include photos, spoken histories, aboriginal language learning, videos and artefacts. There are outdoor walks about bush food ('tucker'), guided walks around Uluru, and cultural sessions – all led by aboriginal people. The aim is to educate as well as entertain.

There are other benefits. Income from admission fees goes to the Anangu community. Today, over 30 aboriginal people work in the park and the park's management is dominated by aboriginal owners.

Now visitors are requested not to climb Uluru, and one of the biggest tourist souvenirs is an 'I Didn't Climb Ayers Rock' t-shirt!

CELEBRATING OUR 43RD MILLENNIUM..

▲ Welcoming in the 43rd Millenniu[m]
*Sydney Morning Herald* on New Y[ear]
the world is a little bit older than [...]

**On your planet**

+ Watch the film *Rabbit-proof fence* – the story of 'stolen generations' of aboriginal people taken from their parents.

132

## your questions

1 Draw a line graph to show the increase in tourist numbers at Uluru since 1961. Why do think the numbers have increased so much?

2 Explain why the culture of aboriginal peoples at Uluru is of interest to tourists.

3 Complete a table to show benefits and problems brought by tourism to Uluru.

4 In pairs, decide whet[...]
Uluru. Feed back yo[...]

5 **Exam-style quest[...]**
value of traditiona[...]
world. (6 marks)

✚ In this section you will learn about the threats to the Australian outback from climate change.

Extreme temperatures are common in Australia's deserts:

- Many desert locations have exceeded 48 °C, and on three occasions reached 50 °C!
- Hot spells can be very long in the outback; most places have had periods of 10 or more consecutive days over 40 °C.

## Climate change and Australia's deserts

Global temperatures are increasing. And it's likely to get hotter during the next 100 years, because of increasing greenhouse gases. Temperatures in the outback may be 1.4–5.8 °C higher by 2100. These would make the outback difficult to live in.

How will the outback change as temperatures increase?

- Droughts will become more frequent.
- Evaporation will increase, particularly in the north and east.
- Bushfire weather will increase across south-eastern, central and north-eastern Australia.

However, rainfall changes are less certain. The Australian Bureau of Meteorology estimates that rainfall will decrease in southern Australia, especially in the south-west. So the desert boundary will move ...th by 100-200 km, while the northern ...ry stays the same.

...hange in Australia, 2001-2005. Central ...ore rainfall – the highest ever! But ...lia had its driest periods ever.

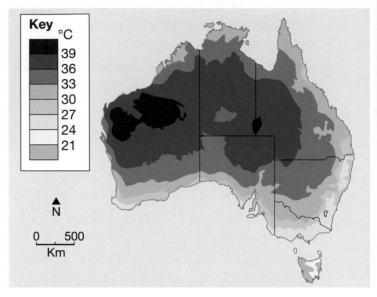

▲ The hottest summer areas in Australia. Data collected 1961-1990. Maximum January temperatures are shown. The highest temperatures occur in the Pilbara region in Western Australia, and between south-west Queensland and Alice Springs.

Rainfall in the outback varies from year to year. Only once has an Australian place been rainless for a whole year, but less than half the average rainfall is common. Occasionally, the entire average annual total falls in just a month – sometimes in a day! Most of central Australia has fewer than 25 'rain days' per year – just 2 days a month.

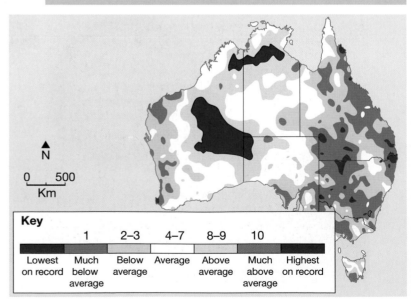

**Key**

| | 1 | 2–3 | 4–7 | 8–9 | 10 | |
|---|---|---|---|---|---|---|
| Lowest on record | Much below average | Below average | Average | Above average | Much above average | Highest on record |

# The threat of El Niño

One problem for farmers in the outback is unreliable rainfall. But in fact there is a pattern. Every 5-7 years, an event called **El Niño** occurs. Instead of winds bringing rain from the Pacific, El Niño reverses everything. Winds blow across Australia from west to east. By the time they reach the east, these winds are dry and bring drought. El Niño lasts for up to 2 years before the winds change back – and in that time many farmers have gone out of business.

> + **El Niño** is a reversal of normal air currents across Australia and the Pacific Ocean. It brings drought to Australia every 5-7 years.

*On your planet*

+ The most severe droughts in Australia have been those in 2002–03 and 2006–07, each being linked to El Niño. Others lasting over a year were in 1901-2, 1982–3, and 1994–5. These were all caused by El Niño.

▼ The effects of El Niño in Australia. Droughts have a severe economic impact.

## The impacts of climate change – long term

Farm incomes could fall so low by 2030 that farms become abandoned, and rural communities are destroyed.

Rainfall totals are estimated to drop by 20-40% across Australia by 2070.

In just one year (2002), one drought dragged Australia's exports down by A$1 billion.

14% of bird species and 25% of all mammals in Australia could become extinct by 2100.

**Key**
Areas affected most by recent El Niño droughts

The Australian landscape could change as native plants like eucalypts and banksia die off with increased drought and higher temperatures – 30 eucalypt species out of 80 could die back.

Winter crops failed in Western Australia – seeds sown but there was no rain to help them germinate.

Irrigation water reduced so that there would be enough water for Adelaide.

Melbourne plans a desalination plant to guarantee water supply.

Severe water shortages in mid-Queensland.

Dust storms in New South Wales.

More bushfires in New South Wales.

Farmers in New South Wales and Victoria demand subsidies to help them through the drought – they've lost 80% of their income.

Water restrictions in Melbourne and Sydney.

## The impacts of the 2006-7 drought – short term

| Year | Annual rainfall in mm |
|------|----------------------|
| 1976 | 272 |
| 1977 | 203 (El Niño) |
| 1978 | 406 |
| 1979 | 432 |
| 1980 | 348 |
| 1981 | 380 |
| 1982 | 110 (El Niño) |
| 1983 | 424 |
| 1984 | 282 |
| 1985 | 268 (rainfall declining and heading for another El Niño in 1987) |

▲ How annual rainfall varies at Walpeup in Victoria.

## your questions

1 Draw a bar graph to show annual rainfall at Walpeup, 1976-1985. Label the El Niño periods. Describe what happens to rainfall in El Niño periods and why.

2 On a blank map of Australia, shade in those areas affected most by drought.

3 Now label the effects that recent droughts have had on your map. Use three coloured pens – one each for environmental, economic, and social effects.

4 In pairs, list 'survival actions' that farmers may have to take if temperatures increase. Consider how they could change: **a** their crops or animals,
**b** their working day, **c** how they'll get water.
**d** how might they survive increased temperatures in their homes?

✚ **In this section you will learn how local actions can help to protect people against climate change.**

## Climate change in Africa

Africa is already 0.5 °C warmer than a century ago, and temperatures have risen even more inland. Droughts are common. Areas that were already arid are now drier. Africa is also less able to deal with climate change than other areas of the world. Of its 55 countries, two-thirds are among the world's poorest 50 countries. These are least able to cope with change.

## The Sahel

The Sahel has been badly affected by climate change. It lies on the southern edge of the Sahara Desert, in a belt extending 3000 km east to west and 700 km north to south. It's a transition zone between the dry Sahara to the north and wetter grasslands further south. It depends on the monsoon rainy season for its rainfall, but this varies hugely from year to year (see below).

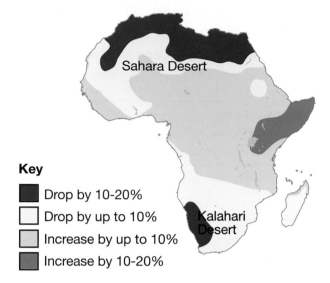

**Key**

■ Drop by 10-20%

□ Drop by up to 10%

▨ Increase by up to 10%

▨ Increase by 10-20%

▲ How rainfall in Africa is predicted to change by 2030. Rainy seasons are likely to be more unreliable. There will be downward trends in rainfall in some areas, and more rain in others.

Several factors have made the Sahel one of the poorest regions in the world. Over 50 years of rapid population growth, deforestation, overgrazing and drought have turned large areas into barren land. The soil is poor and water resources are scarce. Trees which protect the soil from wind erosion are disappearing as people cut them for fuelwood. Grassland is under pressure from grazing too many cattle. Farmers have to grow food for more people, so farming becomes more intensive – squeezing more out of the soil.

▲ The Sahel region.

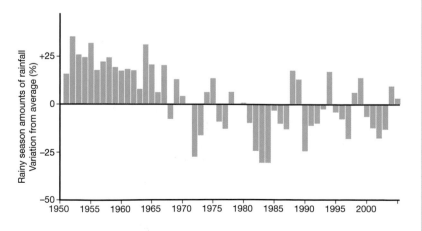

▲ How rainfall varied in the Sahel, 1950-2006. The '0' line is the average for the period, and shows whether rainfall fell above or below that average. Anything above the line shows wet years, and anything below shows dry.

Rainfall variation is a major cause of poverty in the Sahel. For people living there, climate change threatens their survival.

- If the rains don't arrive, the grass dies – exposing the soil to removal by wind erosion.
- When the rains do arrive, heavy rains erode and wash away the soil.

## Diguettes in Siguin Voussé

Siguin Voussé is a small village in Burkina Faso in the Sahel. It was badly affected by drought. Deforestation and over-use of the land left villagers unable to grow enough food to feed themselves. Trees and grass were cleared for farming, so that, whenever it rained, the precious topsoil was washed away.

In 1979, Oxfam started a project called *Projet Agro-Forestier*. It aimed to:
- prevent further soil erosion
- preserve as much rainfall as possible.

Local farmers were encouraged to build **diguettes** to form barriers to erosion. A diguette is a line of stones, laid along the contours of gently sloping farmland. It slows down rainwater and gives it a chance to soak into the hard ground. The diguettes also trap soil, which builds up behind the stones. Soil erosion is therefore reduced.

The diguettes were a success. Almost everyone in the village had improved crop yields. And families now feed themselves. Over 400 villages in Burkina Faso have now built diguettes. In one test done by farmers, soil depth in an area *without* diguettes decreased by 15 cm; while in a field *with* diguettes, it increased by 18 cm.

**On your planet**
+ Diguettes are an example of intermediate technology – little know-how is needed, the materials exist locally, and labour is free. It's a cheap solution to a problem.

▲ Building diguettes in Burkina Faso.

Awa Bundani, from the village, explains why diguettes are important to her community.

*'Last year the rains were good. But in some years they stop, and crops die. The diguettes have made a huge difference. Before, the compost and soil were washed away. And when the rain was poor, the soil would dry out quickly. We knew it was a problem, but we didn't know what to do about it. Since we built the diguettes, the land produces more. We would have had only one bag of groundnuts, where now we get two.'*

### your questions

1 Use an atlas to identify: **a** five countries which will become wetter with climate change, and **b** five countries that will become drier.
2 What problems will each group of countries face? Make a list and explain your ideas.
3 **a** Write a 300-word speech to persuade a village elder in Burkina Faso that diguettes are the way to go.
  **b** Draw a poster to leave in the village, that shows how diguettes work.
4 **Exam-style question** Using examples, explain how people can overcome problems of drought, in one extreme climate you have studied. (6 marks)

✚ In this section you will assess how global action can protect Africa's dry lands from climate change.

### African vulnerability

Climate change is a reality. Of all continents, Africa is most vulnerable to the effects of climate change, even though only 4% of global $CO_2$ emissions come from there.

In 2006, Oxfam published a report explaining how Africa is threatened by climate change.

- Arid or semi-arid areas in northern, western, eastern and parts of southern Africa are becoming drier. This could be disastrous – Africa depends on rain for its farming.

- Food emergencies in Africa have tripled in 20 years. Increased drought would make this worse.

Climate change could reduce African crop yields by 10% or more. Maize production might fall by 33%. In Africa, 70% of the working population relies on farming to make a living. Farming contributes 40% to the Gross Domestic Product of Africa as a whole.

So, how should Africa deal with this?

- Global agreements could **mitigate** (reduce) climate change by cutting emissions of $CO_2$.

- Charities and voluntary organisations, like Oxfam, could help Africa to **adapt** to climate change.

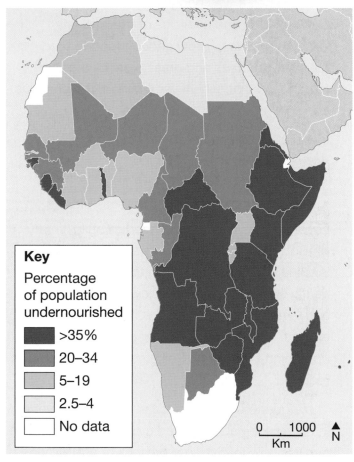

**Key**

Percentage of population undernourished

| | |
|---|---|
| ■ | >35% |
| ■ | 20–34 |
| ■ | 5–19 |
| ■ | 2.5–4 |
| □ | No data |

0    1000
Km

N

▲ 'Hunger hotspots' in Africa. These countries are also the poorest and the ones most vulnerable to climate change.

## Adapting to drought

Charities and voluntary organisations – usually called Non-Governmental Organisations (NGOs) – are working with communities in Africa. One of the biggest is Oxfam. This works with a charity in Zambia, The Evangelical Fellowship of Zambia (EFZ), to help people adapt to drought.

EFZ trains people to use **conservation farming**. It's a method that traps moisture, improves soil quality, minimises soil erosion and resists drought. Crop yields often increase by 10 times. It's a form of **multi-cropping**, where farmers plant several species instead of just one (see the diagram). This calls for periods of moderate but constant farm work, allowing many people to work who would otherwise be too weak. 17% of Zambia's population is HIV-positive so this method is suitable for them.

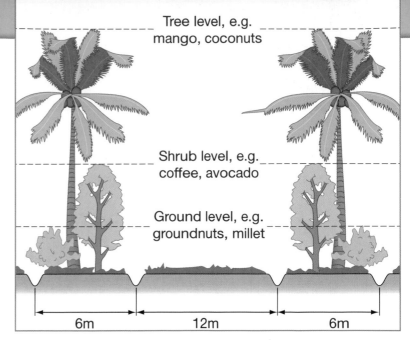

▲ How conservation farming brings many crops together.

Tree level, e.g. mango, coconuts

Shrub level, e.g. coffee, avocado

Ground level, e.g. groundnuts, millet

6m   12m   6m

▨ How conservation farming differs from single crop farming.

| Single crop farming | Conservation farming |
|---|---|
| Plough all the land. | Plough only where you plant crops. |
| Ploughing done all at once – increases soil erosion from rain or wind. | Only small areas are dug or ploughed, so no risk of erosion. |
| Moisture easily evaporates from all of the land. | Moisture evaporates only from parts that have been dug. |
| Planting done all at once. | Planting spread over the whole year. |
| One crop planted. | Several crops planted, mixed together. |
| Land wasted between rows of crops. | All land used – plants use all the space. |
| Harvesting done all at once. | Harvesting done over the whole year. |
| If the crop fails all income is lost. | If one crop fails there are others. |
| Demands huge effort at ploughing, sowing and harvest. | Work is evenly spread over the year. No need for any extra labour. |
| If the main crop does well, prices for everyone will fall. | Prices for all crops remain more stable. |
| If the main crop fails, food aid is needed to feed those who might starve. | If one crop fails, there are others, so no one starves. |

## Cutting emissions – the Kyoto Summit

In 1997, the Kyoto Summit was held to cut greenhouse gas emissions by 5.2% by 2012. By the end, 141 countries had signed up to a protocol. 181 had signed by 2008. These countries fall into four groups:

**Group 1:** Signed and now meeting Kyoto targets – UK, France, Germany, Sweden, Poland.

**Group 2:** Signed but not meeting Kyoto targets – Canada, Denmark, Spain, Portugal, Italy, Ireland, Japan, Norway.

**Group 3:** Signed but weren't set targets – China, India, and other LEDCs. MEDCs produce most emissions, and LEDCs ought to be given time to develop targets.

**Group 4:** Didn't sign – USA (the world's biggest polluter) and Australia. (Following a change of government in 2008, Australia has now signed.)

### your questions

1 In about 100 words, explain why Africa will suffer most from climate change.
2 Why is climate change something that all countries should discuss?
3 Draw up a table of the achievements and failings of the Kyoto Summit.
4 In class, discuss in groups: 'Was Kyoto a waste of time?'
5 Explain how 'conservation farming' is: **a** different from other methods of growing food, **b** well suited to Africa's land, changing climate and people.

➕ In this section you'll learn about the themes in unit 2, and how they link together.

## The human challenge!

How many people can the Earth support? In the late 18th century, Thomas Malthus believed that the world was already way beyond its ability to support the human population. He thought that mass starvation would result, because he believed that the food supply could never keep up with population growth.

Britain's population in 1801 was just over 5 million people, but rising fast. The world's population at the time – though no one is certain about this – was probably about 800 million. Today, Britain's population is about 60 million, and the world's population is nearing 7 billion.

Many people are now beginning to agree with Malthus. The four themes in this Unit are things that he was concerned about over 200 years ago!

- Many people think that the world's population cannot support itself if it continues to grow. But we have survived so far - can't we go on?
- People consume resources – from food and water to oil and metals. The world's oil won't last forever. But aren't we bright enough as a human race to invent other ways of providing what we need?
- Many of the world's living spaces are crowded. Many billions of people live in slums and squalor.
- People around the world make a living in different ways. But what happens to people, and the environment, when employment changes?

▲ Children in Rwanda – how many people will there be in the world when these children are 64?

> ### On your planet
> ➕ By the year 2050, will there be 9 billion people or 15 billion people in the world? Even the experts can't agree!

▲ How much longer will the world's oil last?

▲ Many people moved here from rural areas – is their life here in Shanghai, China better?

▲ A Chinese factory providing goods for Western countries – but what are working conditions like here?

| Theme | What's it about, and what are the challenges? |
|---|---|
| **Population dynamics** | • How the world's population is changing – how big will it get?<br>• How some countries control their population. |
| **Consuming resources** | • Resources around the world – inequalities between rich and poor.<br>• Growing demand for resources.<br>• Sustainable use of resources. |
| **Living spaces** | • What makes good living spaces?<br>• Pressures on living spaces. |
| **Making a living** | • Employment change in different parts of the world.<br>• The environmental impacts of employment change.<br>• Is there such a thing as 'sustainable employment'? |

## your questions

1 List any resources that you've consumed already today. Did you turn on a tap? How did you get to school? What did you have for breakfast? Think of as many as you can.

2 Write down ways in which you have put pressure on the Earth's resources already in your lifetime.

3 Draw a large version of this diagram. It will need to be a whole page.

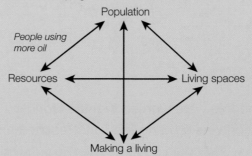

Label any ways you can think of to show how the four themes are linked. One has been done for you.

4 Which do you think might be the bigger problem – if the world's population keeps increasing as it is now, or if we keep using up its resources? Explain your reasons.

In this section you'll learn that the world's population was growing very quickly, but is now beginning to slow down.

## Billions of people!

In 2008, there were just over 6.7 billion (that's 6 700 000 000) people living on planet Earth. The six billionth person was born in 1999, and the seven billionth will probably arrive sometime in 2012. That means that, in the space of 13 years, a further one billion people will have been added to the world. These people will all require food, water and shelter. They will all be born with ambitions, and they will want access to technology, education and opportunities.

In 1804, it had taken the human race just over 300 years to double in number from half a billion to 1 billion. Yet, in 1999, the population was 6 billion and the doubling time had fallen to 39 years (see below). In other words, the bigger the population, the faster it has grown. This is called **exponential** growth.

| Year | World population |
|------|------------------|
| 1500 | 500 million (half a billion) |
| 1804 | 1 billion |
| 1927 | 2 billion |
| 1960 | 3 billion |
| 1974 | 4 billion |
| 1987 | 5 billion |
| 1999 | 6 billion |

Many people were worried that the world's population was growing too fast. People feared that the world could become **overpopulated**. There were estimates that the number of people could reach 10 billion, 20 billion or even 50 billion by the year 2050! How would so many people be fed, or find shelter, or find a job?

## World population reaches six billion

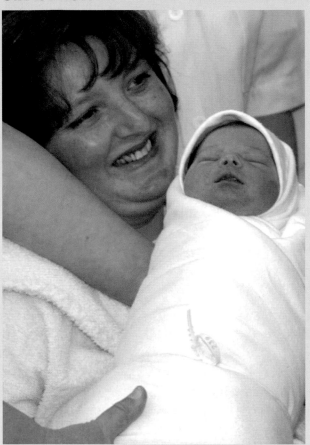

Adnan Nevic was the symbolic six billionth person to join the world's population when he was born on Tuesday 12 October 1999 in Bosnia. His mother was greeted by the UN Secretary-General Kofi Annan. The celebration was purely symbolic – with three babies being born every second, it would have been impossible to know exactly when the six billion figure was passed.

# Something changed

In the 1960s, nearly every country on the planet had an expanding population. However, from the 1970s, the number of babies being born in developed countries began to drop. Some of these countries (such as Sweden and Italy) are now seeing their populations begin to fall. Some developing countries (such as China and India) also introduced controls to limit their populations. The overall number of babies being born in developing countries is also starting to level off. With fewer babies being born, the United Nations now expects world population to peak at 10 billion (see below). It may slowly begin to drop after 2200.

| Year | World population |
|------|------------------|
| 2013 | 7 billion |
| 2028 | 8 billion |
| 2054 | 9 billion |
| 2183 | 10 billion |

▲ No country in the European Union is now producing enough babies to stop their population declining.

### What do you think?
✚ Why do you think people living in the European Union are having fewer children?

# Population Growth Uncertain

The United Nations has changed its forecast for world population growth. Two years ago, the UN Population Division expected the world population to grow from 6 billion to 9.3 billion by 2050. It is now expected to reach 8.9 billion.

This is because people are having fewer babies than expected, especially in developing countries. Indian women are now having fewer babies than American women were in the 1950s.

▲ Adapted from *The Times*, 28 February 2003

### On your planet
✚ Poorer countries now experience nearly all of the world's population growth.

## your questions

1 Use the data in the two tables to draw a line graph to show world population growth 1500–2183.
  a Mark data for 1500–1999 as a solid line.
  b Then use the estimates for 2013–2183 and draw a dotted line for these. Link 1999 to 2013 with a dotted line also.
  c What does your graph show you about global population growth?
2 Look at the table showing world population 1500–1999.
  a Starting with 1804, calculate how long it has taken to add a further billion people each time.
  b What do you notice about how long it takes each time?
  c What word describes this growth?
3 Now look at the estimates for global population 2013–2183.
  a Starting with 2013, calculate how long it takes to add each additional billion people.
  b What do you notice?
  c How does this compare with previous population growth?

✚ In this section you'll learn how to compare population change between countries.

### It's all about babies

The population of a country is constantly changing. In some countries the population will be growing, in others it may stay level, or may even decline. The change of population in a country is the difference between the number of babies being born and the number of people who die. This difference is called the **natural increase**. If this is a positive number, there are more births than deaths (**population increase**). If it is a negative number, there are more deaths than births (**population decline**). If birth and death rates are almost equal, the country will have a **population balance**.

**Population balance: Poland**
**Total population: 38 500 696**

Birth rate: 9.99 per 1000
Death rate: 10.01 per 1000
Natural increase: -0.02 per 1000
This decrease is so small that it means that the population in Poland remained level in 2008.

**Population increase: Uganda**
**Total population: 31 367 972**

Birth rate: 48.15 per 1000
Death rate: 12.32 per 1000
Natural increase: 35.83 per 1000
For every 1000 people living in Uganda, nearly 36 more people were added to the population in 2008.

**Population decline: Japan**
**Total population: 127 288 419**

Birth rate: 7.37 per 1000
Death rate: 9.26 per 1000
Natural increase: -1.89 per 1000
This means that for every 1000 people living in Japan, nearly 2 people were lost from the population in 2008.

✚ **Birth rate** is the number of babies born alive for every 1000 people in one year.

✚ **Death rate** is the number of people who die for every 1000 people in one year.

✚ **Natural increase** is the number of people added to, or lost from the population for every 1000 people in one year.

**Note:** It is easier to compare countries by measuring birth rate, death rate and natural increase per 1000 of the population, because the total population of each country is different.

# Why so different?

Birth rates and death rates vary greatly between countries. Some of the many factors that can affect differences between birth rates and death rates are the:

- level of **development** of a country
- religious views of people in a country
- policies of the Government.

For example, birth rates can be affected by the availability of, and attitudes towards, contraception, while death rates can be affected by the level of access to health care and medicines.

## How to work out natural increase

Natural increase = birth rate – death rate

**United States of America, 2008**
**Total population: 303 824 646**
Birth rate: 14.2 per 1000
Death rate: 8.3 per 1000

The natural increase for the USA in 2008 is:
14.2 – 8.3 = **5.9** per 1000 = **0.59%**
(Note that natural increase is often given as a percentage.)

| Country | Total population | Birth rate (per 1000) | Death rate (per 1000) |
|---|---|---|---|
| Bangladesh | 153 546 901 | 28.68 | 8.00 |
| Burundi | 8 691 005 | 41.61 | 12.91 |
| Denmark | 5 484 723 | 10.71 | 10.25 |
| Germany | 82 369 548 | 8.18 | 10.80 |
| Haiti | 8 924 553 | 35.69 | 10.15 |
| Iraq | 28 221 181 | 30.77 | 5.14 |
| Niger | 13 272 679 | 49.62 | 20.26 |
| Turkey | 71 892 807 | 16.15 | 6.02 |
| Uzbekistan | 28 268 440 | 26.45 | 7.62 |
| Venezuela | 26 414 815 | 20.92 | 5.10 |

*On your planet*

+ There are on average 4 births and nearly 2 deaths somewhere in the world every second!

◀ Population data for selected countries, 2008.

## your questions

1 Write down what each of the following terms means:
   a population increase
   b population balance
   c population decline
2 Use the population data table for 2008 to:
   a work out the natural increase for each of the 10 countries
   b list the countries in rank order by their natural increase
   c describe any patterns that you may notice in the rank order
3 Choose one of the countries from the population data table for 2008. Calculate how many people will be added to the total population for that country in one year.
4 Copy and complete the following table:

| Country | Birth rate (per 1000) | Death rate (per 1000) | Natural increase (per 1000) |
|---|---|---|---|
| Austria | 8.7 | 9.9 | |
| Bolivia | | 7.4 | 15.0 |
| Canada | 10.3 | | 2.7 |
| Thailand | 13.6 | 7.2 | |
| Zambia | 40.5 | | 19.2 |

In this section you'll learn that population change is occurring at different rates in different countries.

## What's happening where?

Although the population of the world is expected to continue to grow until about 2200, this growth will not be even. The map below shows that:

- **population increase** is mainly happening in Africa, the Middle East, and parts of South America and South Asia
- **population balance** is mainly found in North America and Europe
- **population decline** is happening in Russia and parts of Central and Eastern Europe.

It is important to understand that:

- higher levels of population increase are occurring in developing (**low-income** and **middle-income**) countries
- lower levels of population increase, population balance or even population decline mainly occur in developed (**high-income**) countries. See the table.

The **World Bank** groups countries by their **Gross National Income (GNI)** per person. This is the average amount of money earned by a country in one year, divided by all the people who live there.

In 2007, the groups were:

**high-income countries:** $11 456 or more
**middle-income countries:** $936–$11 455
**low-income countries:** $935 or less

| | Average GNI per person in US$ (2007) | Average % natural increase of population (1990–2005) | Average estimated % natural increase of population (2005–2020) |
|---|---|---|---|
| Low-income countries | 578 | 2.0 | 1.7 |
| Middle-income countries | 2872 | 1.1 | 1.2 |
| High-income countries | 37 566 | 0.7 | 0.4 |

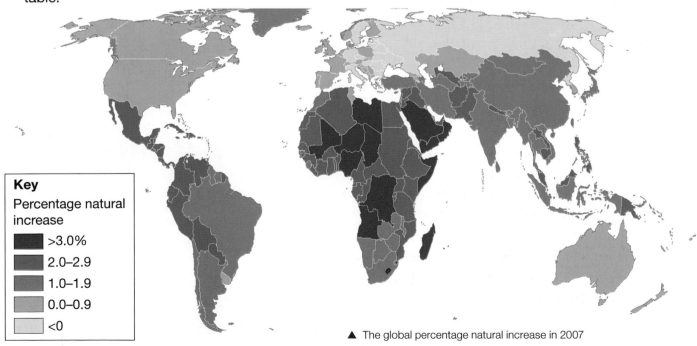

**Key**

Percentage natural increase

- >3.0%
- 2.0–2.9
- 1.0–1.9
- 0.0–0.9
- <0

▲ The global percentage natural increase in 2007

# Let's compare two countries

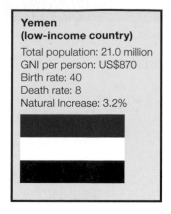

**Russia
(middle-income country)**

Total population: 143.1 million
GNI per person: US$7560
Birth rate: 10
Death rate: 16
Natural Increase: -0.6%

**Yemen
(low-income country)**

Total population: 21.0 million
GNI per person: US$870
Birth rate: 40
Death rate: 8
Natural Increase: 3.2%

| Country | GNI per person, in US$ (2007) | % Natural increase of population (1990–2005) |
|---|---|---|
| Brazil | 5910 | 1.5 |
| Cambodia | 540 | 2.5 |
| Chile | 8350 | 1.4 |
| D. R. Congo | 140 | 2.8 |
| Greece | 29 360 | 0.6 |
| Mongolia | 1290 | 1.3 |
| Nicaragua | 980 | 1.8 |
| Norway | 76 450 | 0.6 |
| Portugal | 19 950 | 0.4 |
| Senegal | 820 | 2.5 |
| Sierra Leone | 260 | 2.0 |
| UK | 42 740 | 0.3 |

- In 1950, the total Russian population was 103 million. In Yemen there were just 4.3 million. This meant that there were 24 Russians for every Yemeni.
- In 2008, Russia's population had grown to 143 million and Yemen's population to 21 million.
- By 2057, Russia's population is expected to fall to about 104 million, while Yemen's population is expected to rise to 105 million. Between now and 2057, the population of Yemen will equal that of Russia – and then overtake it!

Russia's population will decline because of:
- falling life expectancy for men (60 years) caused by industrial disease and alcoholism
- **outward migration** of young men and women
- a low **fertility rate** of 1.2 children per woman.

Yemen's population will grow quickly because of:
- early age of marriage: 48% of women are married by the age of 18
- low literacy rates among women: as most girls marry early they rarely complete secondary school
- a high fertility rate of 6.7 children per woman increasing life expectancy due to improved child vaccinations.

> **+ Fertility rate** is the average number of children born to a woman in her lifetime.
>
> **+ Replacement level** is the average number of children required to be born to a woman to ensure that the population remains stable (it is 2.1).

## your questions

1. Look at the GNI per person table above. Draw a scatter graph for the 12 countries, showing their GNI and percentage rates of natural increase:
   a. Plot the GNI per person on the X axis.
   b. Plot the % natural increase on the Y axis.
   c. Draw a line of best fit.
   d. Divide your graph into three sections: high-income, middle-income and low-income countries.
   e. Describe the pattern that your graph shows.
   f. Try to explain what the pattern means.
2. In a table compare the reasons for Russia's falling population, with Yemen's increasing population.
3. Why do you think low literacy rates among women can cause a high fertility rate?
4. Exam-style question Referring to examples, explain the factors that can lead to *either* population increase or population decrease.

In this section you'll learn that Japan has a population which is both declining and ageing.

## The silvering of the yen

Japan has the oldest population in the world:

- Over-65s make up 20.8% of the population.
- There are 26.8 million pensioners, including 25 606 people over the age of 100.
- The average age is 44 – the highest in the world (compared to the mid-30s for the USA and China).
- The birth rate remains below **replacement level**.
- Under-15s make up just 13.6% of the population.

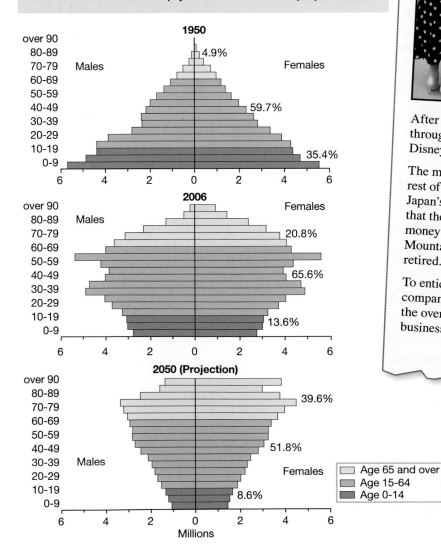

▲ The changing structure of population in Japan.

## Japan runs out of children for Disneyland

After 25 years of successfully luring children through the gates of the Magic Kingdom, Tokyo Disneyland has decided to chase the 'silver yen'.

The move comes because Disney, along with the rest of Japan, is running out of children. With Japan's birth rate in decline, Disney has accepted that the largest group of customers with the money and the time to spend a day on Splash Mountain or Pooh's Hunny Hunt is mostly retired.

To entice these people on to the rides, the company is offering a cut-price season ticket for the over-60s and has made them a target for new business.

▲ Adapted from: *The Times*, 22 February 2008

# Why is Japan's population structure changing?

There are two main reasons:

- People in Japan are living longer. The average life expectancy is 79 for men and 85 for women. This is due to a healthy diet (low in fat and salt) and a good quality of life. Japan is one of the richest countries in the world and it has good health and welfare systems – Japan spends 8.2% of its Gross Domestic Product (GDP) on healthcare, and there are 210 doctors for every 100 000 people (compared to 190 in the UK).

- The birth rate in Japan has been declining since 1975, as the graph shows. This is partly due to the rise in the average age at which women have their first child. The average age rose from 25.6 years in 1970 to 29.2 in 2006. Throughout this period, the number of Japanese couples getting married has fallen, and the age at which people get married has risen (see the table).

## What does this mean for Japan?

An increasing elderly population, together with a shrinking younger population, means that there will be a number of challenges in the future:

- An increase in the cost of pensions. More elderly people, living longer, will require pensions for longer. With the falling birth rate, there will be fewer workers in the economy, so higher taxes will be needed to fund those pensions. The Government has already raised the pension age from 60 to 65.

- A rising number of elderly people living in nursing homes. Since 2000, everyone over 40 has had to contribute £20 a month to pay for care for the elderly.

- An increase in the cost of healthcare, as more elderly people require medical treatment. The Government has already raised patient contributions for medical expenses from 10% to 20%.

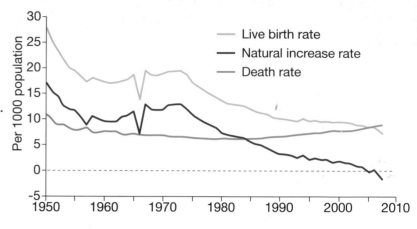

▲ The natural increase of population in Japan.

| Year | Men | Women |
|------|------|-------|
| 1950 | 25.9 | 23.0 |
| 1960 | 27.2 | 24.4 |
| 1970 | 26.9 | 24.2 |
| 1980 | 27.8 | 25.2 |
| 1990 | 28.4 | 25.9 |
| 2000 | 28.8 | 27.0 |
| 2006 | 30.0 | 28.2 |

▲ The average age of first marriage in Japan.

▶ A Japanese bride and groom in their traditional wedding clothes.

## your questions

1 Look at the three population pyramids for Japan.
   a Describe the shape of each pyramid.
   b Explain the changes between 1950 and 2050.
2 Look at the graph showing the natural increase in population. Describe the changes that it shows.
3 Explain why the population of Japan is:
   a getting older
   b decreasing
4 In pairs, discuss the benefits and problems if the Japanese Government were to a increase retirement age to 70 b increase the cost of health care to £60 a month.

In this section you'll learn about Mexico's increasing population and the country's high percentage of young people.

▼ Adapted from: www.imediaconnection.com, 2007.

## What's happening to Mexico's population?

**Mexico (middle-income country)**

Total population: 108.3 million
GNI per person: US$8340
Birth rate: 20.04 per 1000
Death rate: 4.78 per 1000
Natural increase: 15.26 per 1000

- Mexico has a large youthful population. Under-15s make up 31% of the population. Just over 5% of the population are over 65 (see the population pyramids below).
- The population grew from just 20 million in 1940 to 70 million by 1980, and is fast approaching 110 million.
- The fertility rate in 1970 was as high as 7.1 but is much lower now. In 2008, it was still above the replacement level, at 2.4.
- The average age in Mexico is just 26.
- 47% of Mexico's young people are now entering the childbearing age.

## Mexico online: Targeting the young population

Mexico has the second largest online market in Latin America (behind Brazil), with more than 20 million Internet users, and it ranks 13th out of the top 20 Internet-user countries. The online population in Mexico grew from 2.7% in 2000 to 19.2% in 2006.

The main reason for this massive growth is that 53% of Mexico's population is under 24 years old, and Internet users in Latin America tend to be younger than the population as a whole. The growing number of young Mexicans are reading books, newspapers and journals, surfing the Internet and using multimedia education in order to improve their chances of success in the over-crowded job market.

Mexico is proving to be one of the key markets for online advertisers to break into, due to the growing population of young people.

▼ The changing structure of population in Mexico.

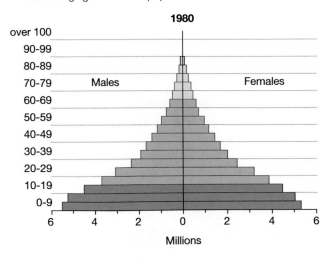

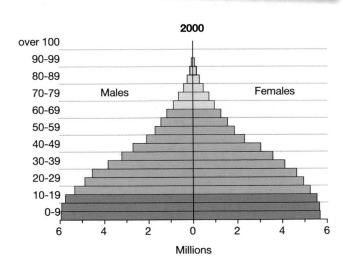

# Why is Mexico's population changing?

There are several reasons why Mexico's population structure, with its current pyramid shape, is changing:

- A low death rate – just 4.78 deaths per 1000. Not only are more babies being added to the population, but people are living longer as well! This is due to more childhood vaccinations being introduced, an increase in doctors, and efforts to reduce infant mortality.

- Although the birth rate is falling, there is still a large percentage of young people. Even if these young people have fewer children than their parents did, the population of Mexico will still continue to increase. Today's children are tomorrow's parents.

- It is expected to take at least 50 years before the population structure of Mexico loses its pyramid shape, and the population levels. Today's young people in Mexico will then be moving into old age.

# What does this mean for Mexico?

A growing population, coupled with an increasing percentage of young people, means that there are a number of challenges and opportunities facing Mexico:

- A large youthful population requires an increase in school places.

- Large numbers of young people are unable to find work, so some migrate to the USA in order to find employment.

- There is a growing manufacturing industry. The Mexican economy is expected to grow to overtake the UK's and become the seventh largest economy in the world by 2050.

- Although Mexico is a strongly Catholic country, abortion has been legalised in Mexico City in an attempt to reduce the number of abandoned children.

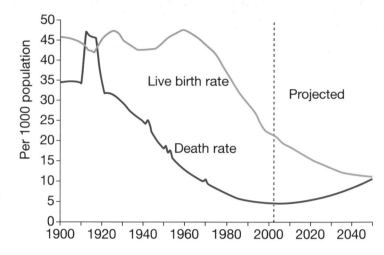

▲ Birth and death rates in Mexico.

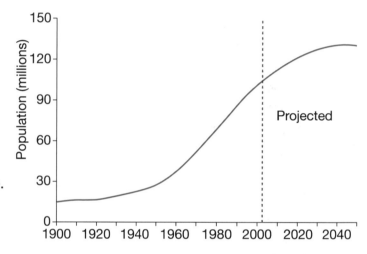

▲ Population growth in Mexico, 1900–2050.

✚ In this section you'll find out how governments try to influence population growth in their countries.

> ✚ **Population policies** are measures taken by a government to influence population size, growth, distribution, or composition.

## Population control

Many countries around the world have introduced **population policies** to encourage people to have either more or fewer children.

## The one-child policy

The best-known population policy is China's one-child policy. China's population grew exponentially throughout the 1950s and 1960s. The birth rate reached 5.8 per 1000, which was unsustainable given China's natural resources of food, water and energy. In 1979, the Government introduced rules to limit population growth. Those couples who only have one child receive financial rewards and welfare benefits. Heavy fines are imposed on those who have more than one child.

### China faces population imbalance crisis

China will be short of 30 million brides within 15 years, according to a report from the State Population and Family Planning Commission in China. About one in every 10 men aged between 20 and 45 (equivalent to almost the population of Canada) will be unable to find a wife, it is predicted. By 2020, the report says, the number of men of marriageable age will outnumber women by 30 million.

▲ Adapted from *The Times*, 12 January 2007

**On your planet**
✚ Even with the one child policy, there are still another 16 500 babies born in China every day.

## Impacts of the policy

The policy has been successful in preventing 300 million births. However it has has some more negative effects.

Some people now believe that China's rapidly growing economy will not have enough workers to keep it going. If this happens, the Chinese Government may be forced to relax the one-child policy.

There is now a serious imbalance of men to women (see extract above right). Couples have often used illegal methods to ensure that their one child is a boy, as boys are traditionally more able to care for their parents when they are elderly.

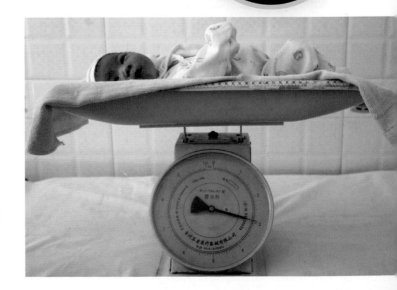

**Anti-natalist policy: India**
**Total population: 1.15 billion**
**Population continues to grow**

India was the first country in the world to introduce a population policy, in 1951. The Government aims to limit the population to 1.11 billion by 2020 and to reduce the fertility rate to 2.1 (replacement level). The Government began by offering contraceptive advice and sterilisation. Throughout the 1970s, forced sterilisation was used on men. This policy has now ended and the Government now uses incentives to encourage couples to have only two children.

**Pro-natalist policy: Sweden**
**Total population: 9 million**
**Population is ageing and has been decreasing**

Sweden's fertility rate was at a peak of 2.1 in 1989, but fell throughout the 1990s to 1.5 in 1999. Since then, the Government has introduced a range of benefits to encourage couples to have more children:

*Paternal leave:* Available for 13 months at 80% of earned income

*Speed premium:* An extra payment if there are less than 30 months between children

*Cash child benefit:* Currently €900 per child per year

*Sick child care:* 120 paid days per child per year

All-day child care and all-day schools: high-quality, low-cost services for all

**Immigration policy: Russia**
**Total population: 143.1 million**
**Population is decreasing**

Russia introduced its population policy in January 2007. The aim is to stabilise the population and ensure that there are enough workers to maintain economic growth (page 145). The Government has used two strategies:

- A cash incentive to encourage Russians who have moved abroad to come back and live in Russia.
- Encouragement for migrant workers to come and work in highly-skilled professions in Russia. 6.5 million immigrants were welcomed in 2007.

**Immigration policy: UK**
**Total population: 60.2 million**
**Population remains balanced**

The population policy of the UK Government has been one of 'no intervention'. However in 2008 a points system was introduced to reduce, but not stop, the flow of migrant workers into the UK. To gain a visa a person must be English-speaking and well-qualified, raising the skill level of the immigrant workforce.

+ **Pro-natalist policies** include incentives, such as financial payments, to encourage people to have more children.

+ **Anti-natalist policies** are policies to encourage people to have dewer children, for eample by providing free State education for only the first child in a family.

## your questions

1 In pairs, discuss why some countries need to use pro-natalist population policies, while others need to use anti-natalist population policies. Then copy and complete the following table:

| Reasons why countries need to use pro-natalist population policies | Reasons why countries need to use anti-natalist population policies |
|---|---|
| | |
| | |
| | |

2 What other population control policies might countries use, and why?

3 Think about how population policies affect the lives of people living in a country where the policy operates. Write down how you think people might react to these policies.

4 As people in China become wealthier, they are expected to live longer. Why should this be?

5 Every baby born under the one-child policy in China is likely to have two parents and four grandparents. This is known as the 4-2-1 problem. What problems will this create for people living in China?

6 Compare the migration policies of Russia and the UK.
   a What is each government 's policy on immigration?
   b Why might each want migrants to come to their country?
   c What kind of migrants do they want?

In this section you'll learn about population policies in two very different countries, Iran and Estonia.

## Anti-natalist population policy: Iran

**Iran (middle-income country)**
Total population: 71.2 million
GNI per person: US$3470
Birth rate: 16.89 per 1000
Death rate: 5.69 per 1000
Natural increase: 11.2 per 1000

### The history of Iran's population policy

**1967–78:** The population policy was introduced in order to reduce population growth (see the box on the right). The aim was to increase economic growth, improve the status of women, and make family planning a human right. It had no effect at all.

**1979–85:** After the Islamic Revolution, the policy was reversed. Family planning was banned as a 'Western influence'. During the war with Iraq (1980–88) couples were encouraged to have as many children as possible, with the aim of producing an army of 20 million soldiers. The population increased to 55 million by 1988 (it had been 27 million in 1968).

**1986 onwards:** A depressed economy caused another change of policy. Rapid population growth was seen as a problem. The family planning policy was restarted and, as a result, Iran has achieved one of the fastest drops in population growth ever recorded. The growth rate dropped from 3.2% in 1976 to 1.0% in 2008, and the fertility rate dropped from 7 in 1986 to 2 in 2007 – below **replacement level**.

### What is the policy?

- Women are encouraged to wait 3–4 years between pregnancies, and not to have children before they are 18 or if they are older than 35.
- Families should be limited to 3 children. Maternity leave benefits are restricted if couples have more than 3 children.
- The State media was told to raise awareness of population issues and family planning programmes.
- The Government covers 80% of family planning costs. A health network of mobile clinics and 'health houses' provides family planning services to Iran's rural population. Birth control, including the provision of condoms, pills, and sterilisation, is free.
- Religious leaders have encouraged family planning as a social responsibility. Iran is the first Muslim country where religious laws have been agreed to allow all types of contraception.

| Year | Growth rate (%) |
|------|------|
| 1926 | 0.60 |
| 1936 | 0.70 |
| 1946 | 1.80 |
| 1956 | 2.71 |
| 1966 | 2.71 |
| 1976 | 3.20 |
| 1986 | 2.50 |
| 1992 | 2.30 |
| 1993 | 1.80 |
| 1994 | 1.75 |
| 1995 | 1.56 |
| 1996 | 1.41 |
| 2000 | 1.20 |

▲ Population growth in Iran. ▶

# Pro-natalist population policy: Estonia

**Estonia (high-income country)**
Total population: 1.3 million
GNI per person: US$13 200
Birth rate: 10.28 per 1000
Death rate: 13.35 per 1000
Natural increase: −3.07 per 1000

## Why does Estonia need a population policy?

Estonia became independent from Russia in 1992. Falling population (see the graph below) was one of the most important issues that the new government had to deal with. The total fertility rate dropped from 2.2 in 1988 to 1.4 in 1998. The United Nations has projected that, by 2050, the Estonian population could halve. Surveys found that Estonians were planning fewer children because:

- poverty was on the increase
- they were unsure how job prospects in the country might develop
- there was a lack of childcare facilities for women who wished to work
- more women were following 'single' lifestyles
- many young people were migrating overseas
- there were competing values, such as family, career, or social life.

The Estonian Government saw the declining birth rate as a danger to the survival of the country, and launched a pro-natalist population policy in 2004.

## What is the policy?

The Estonian Government introduced the 'mother's salary'. Women are paid to have children. Working women receive 15 months' fully paid maternity leave, and non-working women receive $200 per month. By 2006, the fertility rate had risen to 1.5. This is still below replacement level but is an improvement. The Government is now looking at other ways to encourage people to have more children.

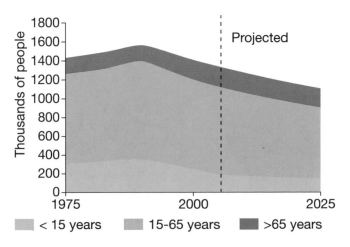

▲ Population by age group in Estonia, 1975-2025.

## your questions

1 Look at the table showing Iran's population growth rate for 1926 to 2000.
   a Draw a line graph of the population growth.
   b Mark on it the dates when the three population policies began in Iran: 1967, 1979 and 1986.
   c Copy and complete the table below, explaining the policies and noting the changes that occurred to the growth rate:

| Year | Aim of policy | Changes in growth rate |
|------|---------------|------------------------|
| 1967 |               |                        |
| 1979 |               |                        |
| 1986 |               |                        |

   d Which policy achieved the best results? Why might this be?
2 Look at the graph that shows the population by age group in Estonia, 1975 - 2025.
   a Describe the changes that the graph shows.
   b How might the 'mother's salary' scheme change the projected growth to 2025?
   c What other problem might Estonia face in the future if not enough children are born?

➕ **In this section you'll learn how the USA tries to control its number of immigrants.**

'America has constantly drawn strength and spirit from wave after wave of immigrants...They have proved to be the most restless, the most adventurous, the most innovative, the most industrious of people.'

*Former US President Bill Clinton*

## An immigration policy in action: the USA

The USA has had some form of immigration policy ever since its founding in 1783. Since 2006, more legal immigrants have settled in the USA than in any other country in the world. There are now some 37 million foreign-born US residents: 12.4% of the total population. However, at least a third of these are thought to have entered the country illegally (see photo).

**USA (high-income country)**
Total population: 303.8 million
GNI per person: US$46 040
Birth rate: 14.2 per 1000
Death rate: 8.3 per 1000
Natural increase: 5.9 per 1000

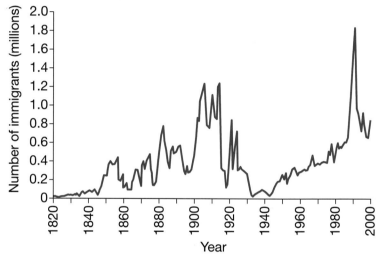

▲ Immigration to the USA.

**Key**
Immigration rate relative to total population of sending country, 2001–2005 (rates per thousand) based on 2000 population US census

- 10.00–53.89
- 3.00–9.99
- 1.00–2.99
- 0.30–0.99
- 0–0.29

▲ Source countries of legal migrants to the USA, 2001–2005.

## Migrating to the USA

In 1990, the US Government passed the Immigration Act. The aim of this was to help businesses attract skilled foreign workers. An annual limit of 675 000 permanent immigrant visas (permits to enter the USA) was set. It was also agreed that 125 000 refugees would be admitted each year.

Anyone who wishes to migrate to the USA has to obtain a visa. There are two types of visa:

- an *immigrant* visa for people who intend to live and work permanently in the USA
- a *non-immigrant* visa for people who live in other countries and wish to stay temporarily in the USA, e.g. tourists, students, or diplomats.

The immigration policy of the USA aims to:

- admit workers with specific skills for jobs where workers are in short supply.
- reunite families by admitting immigrants who already have family members living in the USA.
- provide refuge for people who face racial, political or religious persecution in their country of origin.
- increase ethnic diversity by admitting people from countries with previously low rates of immigration to the USA.

## Illegal migrants

The USA's immigration laws are applied strictly. The long land border with Mexico is difficult to police, and there are many people who try to enter the country illegally (see opposite). There are an estimated 12 million illegal immigrants living in the USA, of which 57% are from Mexico, 24% from Latin America and 19% from other countries.

## Advantages and disadvantages of immigration

In a survey in 2004, 55% of Americans said that legal immigration should remain at current levels or be increased. 41% said it should be decreased.

Some of the advantages are that:

- immigrants add more than $30 billion to the US economy
- foreign-born workers fill unskilled, low-wage jobs
- 40% of US PhD scientists were born abroad
- immigrants are more likely to start up businesses and provide further employment – immigrants start up 40% more new businesses than non-immigrant Americans
- over their working lifetime, immigrants contribute nearly $80 000 more tax revenue than US-born Americans.

Some of the disadvantages are that:

- wages are forced down
- health and welfare systems are strained
- immigrants often live in their own communities and do not integrate into broader American society and culture.

## your questions

1 List reasons why the USA would want to control the number of immigrants entering the country each year.
2 Take each of the four aims of US immigration policy and explain why you think these might be suitable for the USA.
3 Look at the graph showing immigration to the USA.
   a When were the peak periods for immigration?
   b What is the current trend for immigration?
4 Look at the map showing the source countries of legal migrants to the USA, 2001-2005. Describe where the highest number of migrants come from.
5 Outline the advantages and disadvantages that immigration has for the USA.
6 How is US migration policy a similar to b different from that of Russia and the UK? (page 151)

✚ **In this section you'll investigate the impacts that the world's growing population might have on resources.**

## More people = more demands

In 2007 and 2008, there were street protests and riots in West Africa, and outbreaks of violence in Mexico, Morocco, Yemen, Haiti, Egypt, India, Indonesia and other places (see the map). The reason: **food insecurity**. The cause: global population growth means that there are more mouths to feed. At the same time, there is a world shortage of wheat, rising milk and rice prices, and droughts in Australia. The result: people across the world are struggling to afford basic foodstuffs. With more than 6.5 billion people on the planet, there is a delicate balance between the **resources** that these people need and the resources that are available.

+ **Food security** is the ability to obtain sufficient food on a day-to-day basis. People are considered to be 'food secure' when they do not live in fear of hunger.

+ **Food insecurity** is when it is difficult to obtain sufficient food. This can range from hunger through to full-scale famine.

+ A **resource** is a naturally occurring substance (e.g. water, minerals) which can be used in its own right, or made into something else.

## Food crisis threatens security

Ban Ki-Moon, the UN General Secretary, issued a gloomy warning yesterday. He said that the deepening global food crisis, which has triggered riots and threatened hunger in dozens of countries, could have serious implications for international security, economic growth and social progress.

▲ Adapted from *The Guardian*, 21 April 2008.

▼ Global price protests and their main causes.

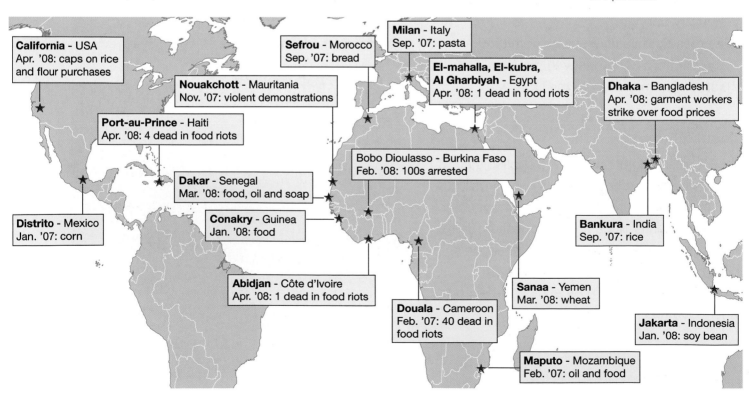

**California** - USA
Apr. '08: caps on rice and flour purchases

**Sefrou** - Morocco
Sep. '07: bread

**Milan** - Italy
Sep. '07: pasta

**El-mahalla, El-kubra, Al Gharbiyah** - Egypt
Apr. '08: 1 dead in food riots

**Dhaka** - Bangladesh
Apr. '08: garment workers strike over food prices

**Nouakchott** - Mauritania
Nov. '07: violent demonstrations

**Port-au-Prince** - Haiti
Apr. '08: 4 dead in food riots

**Bobo Dioulasso** - Burkina Faso
Feb. '08: 100s arrested

**Dakar** - Senegal
Mar. '08: food, oil and soap

**Conakry** - Guinea
Jan. '08: food

**Bankura** - India
Sep. '07: rice

**Distrito** - Mexico
Jan. '07: corn

**Abidjan** - Côte d'Ivoire
Apr. '08: 1 dead in food riots

**Douala** - Cameroon
Feb. '07: 40 dead in food riots

**Sanaa** - Yemen
Mar. '08: wheat

**Jakarta** - Indonesia
Jan. '08: soy bean

**Maputo** - Mozambique
Feb. '07: oil and food

The population of the world will continue to grow. The United Nations predicts that it will peak at around 10 billion by 2183. That's another 3.5 billion people in the world compared to now. They are likely to face challenges of:

- more expensive food
- more expensive fuel – more people will mean greater demand for oil, etc.
- climate change – more people will mean a greater release of $CO_2$ into the atmosphere
- water shortages – already many people in the world lack access to safe water
- more migration – many people will be born in some of the world's poorest countries, and they will want to move to where they can achieve a better quality of life
- political instability and war – with more people competing for fewer resources, individuals and governments may resort to desperate means.

## Consuming resources

There are two possible outcomes:

- A future where there are not enough resources for the global population. This will lead to mass starvation, and ultimately a fall in population. This is called the Malthusian view (see page 158).
- A future in which people successfully use technology in order to provide resources for the growing population.

## China's growth could spark political tensions

China's booming economy is expected to consume more than half of the world's key resources within a decade.

The rapid development of China's economy means that it is likely to consume the majority of the world's supply of all the major metals and minerals, potentially leading to clashes with other countries over controlling access to resources. China already accounts for 47% of all iron ore consumption, 32% of aluminium and 25% of copper.

▲ Adapted from *The Times*, 28 January 2008.

▲ Police quelling food riots in Haiti in April 2008.

### your questions

1 Look at the world map showing the location of food riots in 2007-2008.
   a Describe the location of the countries where the riots took place. Do you notice a general pattern?
   b Are these wealthy or poor countries?
   c What conclusions can you come to about which countries may be most affected by resource shortages?
2 Ban Ki-Moon, the UN General Secretary, stated that he was concerned that food shortages could have severe global impacts. Do you agree with him? Use evidence from the information on these two pages.
3 As you have seen, China's population is still growing. What impact could this have on its future use of resources?
4 What challenges does a bigger world population create?

✚ In this section you'll learn that there are different views about the relationship between population and resources.

## Malthus – 'We're all doomed!'

Thomas Malthus (1766-1834) was born into a wealthy family. He was educated privately and studied at Cambridge University. He later married and had three children and became Professor of Political Economy at the East India Company's college in Hertfordshire. His students affectionately referred to him as 'Pop' or 'Population' Malthus.

In 1798, Malthus wrote an influential essay about population. He believed that population grew **exponentially** (doubling at each stage – 1:2:4:8:16, etc.), but that food production grew **arithmetically** (adding one unit at each stage – 1:2:3:4:5, etc.). This meant that population would eventually outstrip food supply (see the graph). At this point, the population would decrease through starvation. Malthus called this a 'natural check' on population growth. Other 'natural checks' were war, disease and morality.

### Where does morality come into it?
Malthus had deep religious beliefs that affected his work. He believed that people had a moral duty to keep the population low. This could be achieved by restraint from marriage and sexual relations. He called this 'delaying the gratification of passion from a sense of duty'.

According to Malthus, whenever population outstripped resource supply, 'natural checks' would come into play. The population would be reduced to a more manageable level and would then continue to grow again until the next 'natural check'.

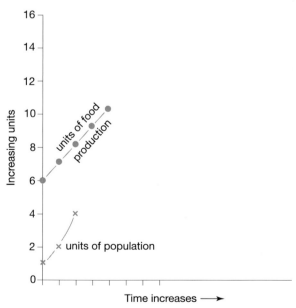

2000 AD – 25% world population is starving
25% more is hungry
Conditions are better than they were in the 1970s

'I think I may make two points: first, that food is necessary for the existence of man. Secondly, that passion between the sexes is necessary and will always remain. Assuming this then, I say that the power of population is greater than the power in the earth to produce food for man'.
Thomas Malthus, 'An essay on the principle of population', 1798.

## The Honourable East India Company

The East India Company was founded in 1600 by Elizabeth I, and traded cotton, silk, tea, spices and opium that it sourced from Asia. For many years, the company effectively governed India. This came to an end in 1858, when the British government took direct control of India after the Indian Mutiny.

Malthus's views often influenced the policies of the East India Company. When famines struck in India, the British authorities would do nothing to relieve them. Famine was seen as a 'natural check' on population growth. This became known as the 'Malthusian' policy of the East India Company.

### your questions

1 Look at the graph that shows Malthus's prediction.
   a Copy the graph on page 158. Extend the lines. Remember that population doubles at each stage. Food production only gets bigger by one unit at each stage.
   b Shade in the area where there would be too many people and not enough food. Where this happens, people would die of starvation. Label this area on your graph.
   c Do you think that this is the situation in the UK today? Why?
   d Is it true of other parts of the world?
2 Look at the graph that shows Böserup's view.
   a Describe what the graph shows.
   b Explain why there is no point on the graph where there are too many people and not enough food.
3 Exam-style question Explain why some people believe that the world's resources will run out soon, while others think that will not happen. (6 marks)

## Böserup – 'Necessity is the mother of invention'

Ester Böserup (1910-1999) was a Danish economist who worked for the United Nations. In 1965 she published *The Conditions of Agricultural Growth*. This book opposed the ideas of Malthus. Böserup did not like the idea of 'natural checks'. She argued that food production does not limit or control population growth. Instead, she said that population growth controls farming methods. She believed that people would try not to give in to disease or famine. Instead, they would invent solutions to the problem. She used the term 'agricultural intensification' to explain how farmers can grow more food from the same piece of land using better farming techniques and chemical fertilisers. The graph below shows Böserup's view.

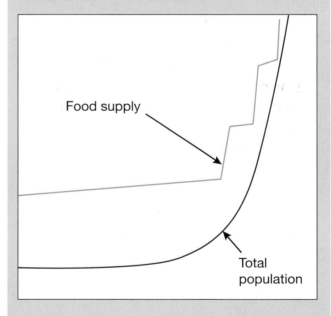

➕ In this section you'll learn how resources can be classified into different categories.

## How can we classify resources?

Resources are all of the things we need to live and work. They can be divided into three main categories:

- **Renewable:** These will never run out and can be used over and over again, e.g. wind and solar power. They are infinite resources.
- **Sustainable:** These are meeting the needs of people now, without preventing future generations from meeting their needs, e.g. bio-fuels and hydrogen-powered vehicles.
- **Non-renewable:** These are being used up and cannot be replaced, such as coal and oil. They are sometimes known as finite resources.

## World energy resource use

Energy supplies are a good example of where people use a variety of different resources. Global energy use has continued to grow. That means that people are using more resources. Different types of resources are being developed to meet rising energy demands (see the graph).

## Classifying energy resources

**Renewable energy: wind power in the USA**
Wind turbines convert the power of the wind into electricity. There are now more than 13 000 large wind turbines in California, and hundreds of homes and farms across the state are also using smaller wind turbines. In 2007, wind energy provided 2.3% of California's total energy requirements.

▲ A wide variety of resources are extracted from the ground every day – from diamonds to copper, from sand to uranium. Is this sustainable?

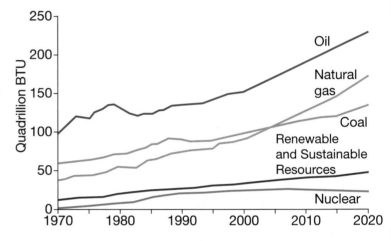

▲ Global energy use by fuel type, 1970 – 2020 (projected).
*Note:* Energy use is recorded in quadrillion (million billion) BTUs. One

## Sustainable energy: bio-gas in India

Bio-gas plants convert organic matter, such as wood chips and animal dung. This ferments, releasing methane gas. This is collected in a tank and can be burnt to provide electricity or gas for cooking. There are over 2.5 million bio-gas plants across India, providing 57% of the country's energy.

## Non-renewable energy: natural gas supplies in Europe

Natural gas is used for electricity production, heating and cooking in Europe. In Britain, gas is collected from underneath the North Sea (see the photo). However, much of this gas has now been used up. In 2006, the UK imported 50% of its gas supplies. This is expected to increase to 80% by 2020. Like many other countries in Europe, Britain is now dependent on gas supplied from Eastern Europe and Russia. Large pipelines carry the gas across the continent to where it is needed. In 2004, the Government produced a report identifying that global gas supplies will fall after 2030. Other energy sources will need to be found after this date.

## your questions

1 What is meant by the term **resource**?
2 Make a copy of the table below in Word:

|  | *Renewable* | *Sustainable* | *Non-renewable* |
|---|---|---|---|
| Definition: |  |  |  |
| Examples: |  |  |  |
| Locations (where found): |  |  |  |
| Pictures: |  |  |  |

a Complete a definition of the three categories of resources.
b Classify the following resources into the correct columns:
solar power, oil, apples, wave power, vegetable oil, iron ore, limestone, grass, milk, sand, wind power, water, chicken, coal, natural gas, biogas.
c Complete the locations row. Use Google images to find one image for each column.

3 Look at the graph showing global energy use by type of fuel from 1970 to 2020.
a Describe what is happening to world energy use.
b Is world energy use sustainable? Explain your answer.
4 Look at the three examples of energy resources. Research one more type of energy resource, describe where it's found and explain whether it is renewable, sustainable or non-renewable.

In this section you'll learn that resource supply and consumption are not evenly spread.

## The trade in metals

It's summer 2008 and the price of metal on the world market just keeps rising. Anglo-American, the world's fourth largest mining company, announces record profits. The price of platinum has risen 57% in one year. So what is driving the rise in metal prices? Which countries import these metals, and where do they all come from?

The two maps on the right have been drawn to show which countries export and import metals. Metals include nickel, zinc, copper and aluminium. They also include metal items such as tools and cutlery. Of all the money spent on trade worldwide, 3.8% is spent on metal exports. On each map, the size of each country has been adjusted to show how much metal it either exports or imports.

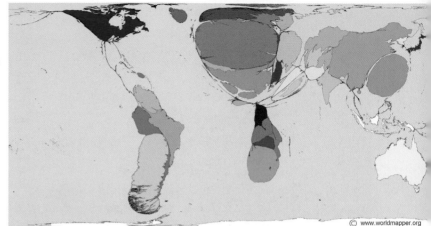

© www.worldmapper.org

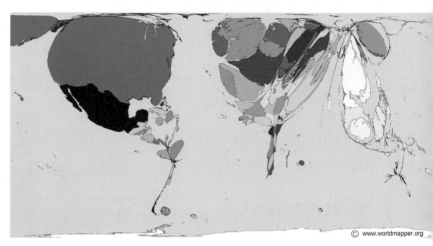

© www.worldmapper.org

▲ Countries that export metal and metal products (top) and countries that import metal and metal products (bottom)

## A new enemy for 4x4 owners

Owners of 4x4s now have a new enemy – thieves who slip under their car and saw off its catalytic converter. The police believe that thousands of the exhaust parts have been stolen from 4x4s in the past six months by criminal gangs who want the precious metals inside. The prices of these metals have soared over the past two years.

The thieves, who sell the catalytic converters on to metal traders for between £100 and £200, are the first stage of an international trade in scrap metal that takes the stolen car parts to India, China and Eastern Europe to be recycled and sold on.

▲ Adapted from *The Times*, 31 August 2008

# Does every cloud have a silver lining?

In June 2008, the *New Scientist* magazine reported in detail about why the price of metals on the world market was increasing. Metals are being used up (see right). The report claimed that if metals continue to be used at the current rate, there will only be 45 years' supply of gold left, 46 of zinc, 29 of silver and 59 of uranium. Worse still, if the rest of the world were to use metals at the rate of the USA, things could get even worse – there would only be 36 years' supply of gold, 34 of zinc, 10 of silver and 19 of uranium!

Stocks of silver have declined worldwide to about 300 million ounces. That's a seventh of what they were 10 years ago. Recycling of silver is difficult. Some can be recycled, but most of it is lost forever once it has been used. The result is that the price of silver continues to rise (see right).

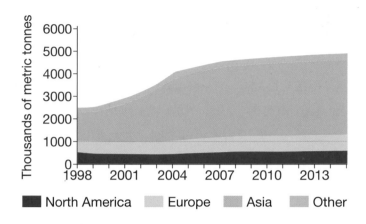

▲ Worldwide metal consumption by continent 1998–2015 (estimated)

Legend: North America — Europe — Asia — Other

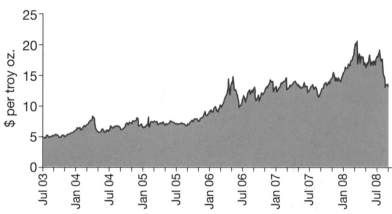

▲ The price of silver on the world market, July 2003 to September 2008.

▲ Silver has the highest thermal conductivity of all metals. It is used in computer keyboards, mobile phones, washing machines, batteries, TVs, and cameras.

## your questions

1 Look at the maps showing metal exports and imports.
  a Describe the distribution of countries that export metals.
  b Explain why the distribution of countries importing metals is different from that of countries exporting metals.

2 Look at the graph showing worldwide metal consumption by continent.
  a Which continent uses the largest quantity of metals?
  b Using the export map, explain where most of these metals come from.

3 Explain why the price of silver continues to rise on the world market.

4 **Exam-style question** In the future, it may be more difficult for countries to obtain metals needed to make the everyday products. Explain:
  a how countries that export metals may benefit from this.
  b what problems countries that import metals may face.
  (6 marks)

➕ **In this section you'll find out about oil production and consumption, and what the future holds for this 'black gold'.**

## Black gold!

Oil is used in a great many ways in modern society. It fuels cars, heats buildings, provides electricity, and makes the plastics that we use in everything from milk containers to computers. Oil is a **finite resource** – in other words, it will run out someday. There is only so much oil under the ground, and, once it is all used up, it will be gone for ever. The question is, how much oil is left?

Oil consumption has risen from less than a million barrels a day in 1900 to over 85 million barrels today (see the graph). The International Energy Agency predicts that demand will rise to 116 million barrels a day by 2030. Unfortunately, the oil industry believes that it is impossible to produce more than 100 million barrels a day. This is the problem that the world faces — oil is needed for energy and transport, but there is not going to be enough of it to go round!

> ➕ **Black gold** is another name for oil, because it is such a valuable commodity.
>
> ➕ A **finite resource** is one that is limited or restricted.

The map and table on the next page show which countries have the largest known oil reserves.

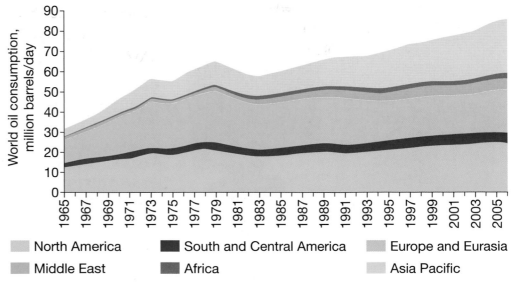

▶ World oil consumption, 1965–2005. Note the increase in consumption by the Asia Pacific region.

World oil consumption, million barrels/day

| North America | South and Central America | Europe and Eurasia |
| Middle East | Africa | Asia Pacific |

| World reserves of oil | | |
| --- | --- | --- |
| | Billions of barrels | Percentage of world reserves |
| 1  Saudi Arabia | 262.73 | 22.3 |
| 2  Iran | 132.46 | 11.2 |
| 3  Iraq | 115.00 | 9.7 |
| 4  Kuwait | 99.00 | 8.4 |
| 5  UAE | 97.80 | 8.3 |
| 6  Venezuela | 77.22 | 6.5 |
| 7  Russia | 72.27 | 6.1 |
| 8  Kazakhstan | 39.62 | 3.4 |
| 9  Libya | 39.12 | 3.3 |
| 10 Nigeria | 35.25 | 3.0 |
| 11 USA | 21.37 | 1.8 |
| 12 China | 17.07 | 1.4 |
| 13 Canada | 16.80 | 1.4 |
| 14 Qatar | 15.20 | 1.3 |

> **+ Peak oil** is the point at which oil production reaches its maximum level and then declines.

## Drilled out?

It is difficult to tell how much oil remains under the ground, or how much demand there may be in the future. However, what is certain, is that once **peak oil** is reached, it will become much more difficult and expensive to extract what is left.

Oil pessimists believe that the world has already reached – or is close to reaching – peak oil. They say high oil prices are evidence of this. Oil optimists believe that this point is still decades away, because there is so much oil yet to be discovered – including huge reserves of tar that can be refined in Canada. The chief economist at BP predicts: 'People will run out of demand before they run out of oil.'

### A post-oil future?

Oil pessimists argue that we are at a turning point in oil production. From now on, they argue, there will be less oil to extract. This could lead to recession and possibly war, as countries that import oil try to get access to oil reserves. In the 1970s, an earlier time of high oil prices, the USA considered military action to seize Middle-Eastern oil fields. The USA decided against such action then, but the future could be different. Many people argued that the recent war in Iraq was directly linked to oil.

### your questions

1  Look at the graph that shows world oil consumption.
   a  Give reasons why world oil consumption has continued to rise.
   b  Explain why you think Europe and North America account for such a large proportion of oil consumption.
2  Look at the map and table showing who has the largest oil reserves. Describe the distribution of these countries.

3  In the future, it may be more difficult for countries to import oil. Explain the problems that countries may face as follows:
   a  Countries that rely on oil for energy and transport.
   b  Countries that were major exporters of oil.

+ In this section you'll learn about sustainable solutions to the world's dependence on oil.

## Plastic, plastic everywhere ...

In Britain we drink more than 2 billion litres of bottled water each year. This is despite clean tap water being available in every office, factory and house. Many people believe that bottled water is in some way cleaner, fresher, or more pure. However, the plastic bottles that hold the water are made from oil, and oil is used to transport the bottled water across great distances.

'It is the great irony of the 21st century that the most basic things in the supermarket, such as water and bread, cost the most. Getting water from the other side of the Earth to sell here is ridiculous.'
Sir Bob Geldof

Of course, in many high- and middle-income countries around the world, water is a readily available resource. The issue of importing bottled water has now been highlighted, and recently an 'eco-trend' has been growing. More people are carrying refillable bottles of tap water to the office, gym or school. In terms of the use of oil as a resource, people need to reuse water bottles, and ensure that plastic bottles are recycled.

**On your planet**
+ Plastic bottles of water are responsible for 33 200 tonnes of $CO_2$ emissions each year – the equivalent of providing electricity to 20 000 houses!

Most plastic bottles are made from PET (polyethylene terephthalate), extracted from crude oil. Plastic water bottles account for about 0.25% of the world's annual oil consumption. To make a one-litre plastic bottle requires 162g of oil. The manufacturing process releases 10 balloons-full of $CO_2$. Most bottles end up being buried in landfill sites, where they will take about 450 years to break down. In 2007, only 3 billion of 13 billion bottles sold in Britain were recycled.

In the USA, the world's biggest consumer of bottled water, 29 billion plastic bottles are used each year, requiring more than 17 million barrels of oil. That's enough oil to fuel more than a million cars for a year!

▲ Adapted from: *The Observer*, 10 February, 2008

# ... and cars that produce water!

Hydrogen is the most abundant element in the universe, but it is usually found in combination with other elements such as carbon (oil, natural gas) or oxygen (water). Once separated, hydrogen has the potential to provide an alternative to oil – to power vehicles that release no harmful emissions. The only thing to come out of the exhaust is water – so pure that you can drink it! Separating hydrogen requires energy, but this can be provided from a renewable resource, such as solar or wind power. Hydrogen offers the potential of reducing dependence on oil, while also emitting zero $CO_2$.

## CaH2Net: California's hydrogen highway

In 2005, the world's first hydrogen highway, the CaH2Net, was launched in California. The plan is to ensure that California has at least 50 hydrogen filling stations by 2010. This will encourage Californians to adopt the new technology as hydrogen fuel cell vehicles become more affordable. This will reduce the state's dependence on oil for transport, and reduce harmful emissions.

## The Ford Edge

The Ford Edge is a huge leap forward for hydrogen fuel cell technology. It is the first alternative fuel car to have a similar travelling distance to that of a traditional petrol or diesel fuelled car. A special 'fuel cell' converts hydrogen into electricity to power the car's four motors (one for each wheel). The Edge has a range of up to 225 miles before refuelling with hydrogen is required.

## your questions

1 Describe the ways in which oil is used as a key resource by the bottled water industry.

2 **a** How might the bottled water industry become more sustainable?

  **b** What sustainable alternatives to bottled water are there to ensure that you drink enough water each day?

3 'Hydrogen fuel cell cars are the only sustainable future for transport.' How far do you agree with this view?

4 Do you think that we are working towards a sustainable future in which the world depends less on oil as a resource?

5 **Exam-style question** Using examples, explain how renewable energies could replace the world's dependency on fossil fuels. (4 marks)

In this section you'll look at the sustainability of food supplies.

# Lots of oil, but no food

Abu Dhabi is the largest of the seven emirates that make up the United Arab Emirates (UAE). It is a major producer of oil, producing over 2.7 million barrels a day. However, it is a desert country. Abu Dhabi has very little rainfall or land that is suitable for growing crops. The emirate relies on the profits made from selling oil to import food resources for its people. There was a global food shortage in 2008, coupled with price rises in wheat and grain. This led the Abu Dhabi Fund for Development (ADFD) to search for a sustainable solution to its food security requirements.

The answer has been to develop 30 000 hectares of farmland in the country of Sudan. The government of Sudan has agreed to lease the land free of charge, in exchange for technological expertise in improving farming techniques. Sudan lies close to the Equator, but has a plentiful water supply from the River Nile. It also has large areas of land that are suitable for agricultural development. Crops, such as wheat, potatoes, beans and alfalfa, will be grown and exported to Abu Dhabi.

The ADFD is worried about global warming and continued population growth. The time may come when, even if you have money, buying resources such as food may not be easy. Abu Dhabi is committed to developing guaranteed food supplies, with similar projects in Uzbekistan and Senegal.

▲ Adapted from *The Guardian*, 2 July 2008

# Can the planet feed us?

The world's population reached 6 billion in 1999, doubling from 3 billion in less than 40 years. Even so, our food supply has kept pace with this growth. Not only has it kept pace, but more people are consuming more calories than 40 years ago (see the graph).

- World cereal production has doubled since 1970.
- Meat production has tripled since 1961.
- The number of fish caught grew more than six times between 1950 and 1997.

Malthus (in Section 10.2) predicted that population would grow faster than the production of food. This would provide a 'natural check' on population growth. In recent times, this has not been the case. Böserup's view, that 'Necessity is the mother of invention', would seem more accurate. But the huge growth in food output was only possible by giving nature a 'helping hand' through the use of fertilisers, irrigation systems and pesticides.

▲ Genetically modified corn has been developed to be resistant to pests.

## Can the rate of food production continue to grow?

The world's population is expected to peak at around 10 million people by 2183. Most of the world's best farmland is already being used, or even overused. Pesticides are beginning to have less effect than in the past, as pests develop resistance. Perhaps the future will depend on bio-technology? Genetically modified (GM) crops are being developed to be drought- or pest-resistant. The concern is whether such foods will be safe for people to eat, or have other unforeseen consequences.

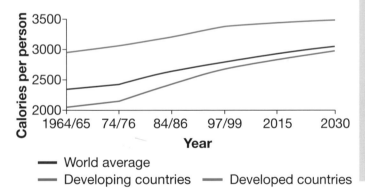

- —— World average
- —— Developing countries —— Developed countries

▲ Calories consumed per person per day.

## your questions

1 Read the extract about Abu Dhabi.
  a Give reasons why Abu Dhabi is looking to lease land from other countries.
  b Do you think that this is a sustainable solution for Abu Dhabi? Give reasons why you think this.
2 Using the United Arab Emirates as an example, explain how some countries can be rich in one resource, yet poor in another.

3 Look at the graph showing calorie consumption levels per day.
  a Describe what the graph shows.
  b Give reasons why the number of calories consumed has continued to increase, despite the world's population continuing to grow.
4 Using the Internet, research GM foods and produce a table showing arguments for and against their use. Do you think that GM crops are the answer to a sustainable food supply?

## 10.8 Achieving sustainability

**On your planet**

+ You can calculate the size of your own ecological footprint on the Internet by going to: http://footprint.wwf.org.uk/

+ In this section you'll investigate the impact that resource use has on our planet, and whether sustainable solutions can be found.

## Measuring resource use

### What is an 'ecological footprint'?

The term ecological footprint was first used in a book called *Our Ecological Footprint: Reducing Human Impact on the Earth* written by two Canadian University researchers in 1996. In the book, they used a calculation to estimate the area of the Earth's land and water that was needed to supply resources to an individual, or a group of people. They stressed that an ecological footprint should also measure the amount of land required to absorb the waste produced by the individual or group.

+ The Brundtland Commission defined **sustainable development** as that which 'meets the needs of the present without compromising the ability of future generations to meet their own needs'.

### How big is your ecological footprint?

In 2007, WWF-UK produced a report on the ecological footprint of 60 cities in Britain. They analysed the 'footprint' of the average resident of each city. They found that if everyone in the world lived the same way as an average person in Britain, we would need three planets to support us. The tables below show the 10 best- and the 10 worst-performing cities. The 'planets consumed' columns show how many planets would be needed to support us if everyone in the world lived the same way as the people in the cities listed.

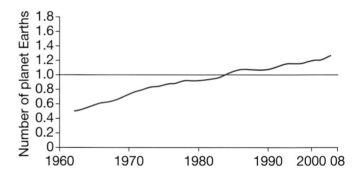

▲ Humanity's Ecological Footprint. The graph shows a comparison between the resources people use and the Earth's ability to provide them. In 2008, we needed around 1.25 planets to supply the resources used.

| Rank | City | Planets consumed |
|------|------|------------------|
| 1= | Newport | 2.78 |
| 1= | Plymouth | 2.78 |
| 3= | Salisbury | 2.79 |
| 3= | Kingston upon Hull | 2.79 |
| 3= | Stoke on Trent | 2.79 |
| 6= | Gloucester | 2.81 |
| 6= | Wakefield | 2.81 |
| 8 | Sunderland | 2.83 |
| 9= | Truro | 2.84 |
| 9= | Wolverhampton | 2.84 |

| Rank | City | Planets consumed |
|------|------|------------------|
| 51 | Portsmouth | 3.21 |
| 52 | Cambridge | 3.22 |
| 53 | Durham | 3.24 |
| 54 | Southampton | 3.27 |
| 55= | Oxford | 3.40 |
| 55= | Canterbury | 3.40 |
| 57 | Brighton and Hove | 3.47 |
| 58 | Chichester | 3.49 |
| 59 | St. Albans | 3.51 |
| 60 | Winchester | 3.52 |

◄ The best … and the worst. The 20 cities in the UK with the smallest (left) and the largest (right) ecological footprints.

# What does the future hold?

People are using up natural resources at least 25% faster than the planet can renew them. It takes just 9 months for people to use up the resources that the planet requires one year to replenish. According to the Global Footprint Network, the world has been in 'ecological debt' since 1987. The United Nations estimates that with population and economic growth continuing, it is likely that many of the Earth's ecosystems will collapse. This would cause untold damage to our planet. So what are the possible solutions? Three possible ecological footprint futures are shown below. The graph shows that current resource consumption is unsustainable and that alternatives are needed.

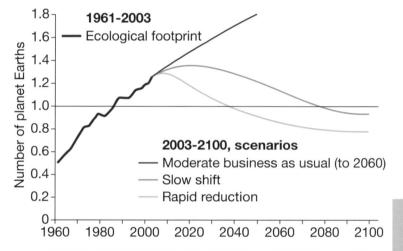

| | Total ecological footprint (million gha) | Per person ecological footprint (gha/person) |
|---|---|---|
| **World** | **14 073** | **2.2** |
| USA | 2819 | 9.6 |
| China | 2152 | 1.6 |
| India | 802 | 0.8 |
| Russia | 631 | 4.4 |
| Japan | 556 | 4.4 |
| Brazil | 383 | 2.1 |
| Germany | 375 | 4.5 |
| France | 339 | 5.6 |
| UK | 333 | 5.6 |
| Mexico | 265 | 2.6 |
| Canada | 240 | 7.6 |
| Italy | 239 | 4.2 |

▲ Ecological demand in selected countries, 2003. *Note:* gha = global hectares: the amount of productive land and water required to produce the resources used and the waste generated by each country.

## One Planet Future

To achieve sustainability, and ensure that our ecological footprint is reduced to just one planet, action will need to be taken. WWF have called for radical changes to achieve sustainability by 2080. These include reducing $CO_2$ emissions and fish catches by 50%. The 'One Planet Future' campaign encourages people to:

- measure their own ecological footprint
- 'green' their lifestyle – by saving energy and travelling less
- support sustainable energy solutions
- consider their diet – think about the distance food has travelled
- lobby their MP and MEP to take action
- support the UK Climate Change bill.

## your questions

1 What is meant by the terms:
   a ecological footprint
   b sustainable development?
2 On a blank outline map of the UK, use an atlas to mark on:
   a in green, the top 10 cities of the UK with the smallest ecological footprints
   b in red, the bottom 10 cities of the UK with the largest ecological footprints.
   c Describe any patterns that you notice.
3 Which city on your map is nearest to where you live? Work with a partner to list the ways in which you contribute to the ecological footprint of the place where you live.
4 Look at the table showing the ecological footprint of selected countries.
   a Using graph paper, produce a scatter graph of the two data sets.
   b Describe any patterns that you notice.
   c Explain what the graph tells you about resource use.
5 Look at the graph which shows the three possible future ecological footprints. Describe how you could contribute towards achieving a more sustainable future.

In this section you will learn about living spaces and what makes a good living space.

## Ingredients of a living space

What makes a living space good or bad? People live in widely differing geographical areas of the world. Some areas might be better places to live than others. People vary in their ideas about what makes a good living space.

## An attachment to place

Place attachment is the 'bonding of people to places'. This can take two forms:

- Practical or **functional attachment**. This gives you the things you need – somewhere to live, water, food, work, etc.
- **Emotional attachment**. This involves the feelings, moods and emotions that people have about certain places.

You form a stronger bond to a living space if it meets your needs, and if the living space matches what you like doing. The strongest influence on place attachment is how long you have lived there. The longer you have lived in an area, the stronger your sense of place identity. People also feel more attached when they have many local friends and relatives, and when there is a strong community 'feel'.

In **deprived areas**, there are higher crime rates and a greater fear of crime. These factors reduce place attachment so that people become 'detached' from a living space. Similarly, older people may feel 'out of place' in a town centre at night when it is filled with young people.

A 2001 Mori questionnaire survey investigated the most important factors in making somewhere a good place to live (see right). Interestingly, there were differences in the perceptions of good living spaces between:

- men and women – low crime vs. affordable housing
- young and old people – jobs vs. good health care
- people in rural areas and city dwellers – affordable housing vs. low crime.

▲ 10 Radio, Somerset: community radio strengthens community togetherness. It also helps local people to connect with their place identity.

Are there jobs? (economic health)

Is this where I want to live? (personal preference)

Is it pleasant to live here? (physical environment)

Is there food and water? (resources)

Is the political system and culture OK for me?

**Results of 2001 MORI survey (%)**

| | |
|---|---|
| Low level of crime | 56 |
| Health services | 39 |
| Affordable decent housing | 37 |
| Shopping facilities | 28 |
| Public transport | 27 |
| Education provision | 25 |
| Job prospects | 25 |
| Clean streets | 24 |
| Activities for teenagers | 23 |
| Facilities for young children | 22 |

▲ What makes somewhere a good place to live?

# The nicest place to live in Britain?

Bournville, near Birmingham, was a village created in the last years of the 19th century. George Cadbury, a Victorian businessman, wanted homes for the workers next to his chocolate factory.

Cadbury said that the estate should 'ameliorate the condition of the working-class and labouring population ... by the provision of improved dwellings, with gardens and open space to be enjoyed therewith'. A century on, Cadbury's model village is now a large garden suburb of 400 hectares. There are 7800 homes and 50 hectares of open space.

Research found that, across a range of national indicators, Bournville residents were simply happier. A researcher noted: *People just tend to be generally friendly in Bournville. There are newsletters and notice boards about what's going on. I hadn't seen that anywhere before.*

However, other surveys contradict these findings – so the best living spaces are all really a matter of opinion!

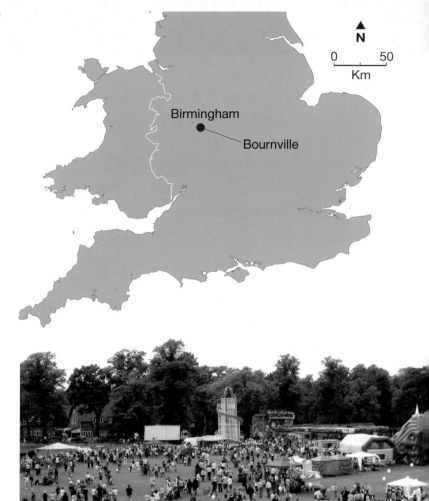

▲ A funfair in a Bournville park supported by the community.

## 10 Best places to live in the UK
### (according to a BBC survey)

1 - Epsom and Ewell
2 - City of Westminster
3 - Harrogate
4 - Ashford, Kent
5 - Stratford-upon-Avon
6 - East Hertfordshire
7 - South Cambridgeshire
8 - Mole Valley
9 - Guildford
10 - West Oxfordshire

## your questions

1 Draw a mental map of your favourite place, including as many details as you can.
2 Identify one feature on your mental map that is particularly special. Describe why it is special to you.
3 Write down what you don't like about your favourite place (it could be noise, traffic, people, etc.).
4 How might you go about improving the quality of the living space in this area?
5 How might people of different ages feel about this place?

✚ In this section you will learn about factors which affect the quality of living spaces.

## What factors affect living spaces?

At a global scale, factors can be broadly divided into social, economic, political and environmental/physical (see right).

At a smaller scale, other factors change the quality of living spaces. **Gentrification**, for example, is where wealthier people move into the area and carry out house improvements. This improves the quality and condition of local housing.

Alternatively, processes may cause the quality of living spaces to decline. An example is the **spiral of multiple deprivation** (see right), which affects poor estates. Richer people can afford to live elsewhere – so in many urban areas this causes **social segregation**. In cities, areas of poor housing show certain patterns, in both developed and developing countries.

+ **Social segregation** is where richer people live in certain areas and less well off families in others.

**Social factors:**
crime rates, healthcare, education availability, energy sources (electricity), risk of disease, communications, e.g. broadband

**Economic factors:**
access to services (range, choice, quality), job opportunities, interest rates, transport links, business infrastructure

**Political factors:**
political stability, opportunity to participate in a democracy, freedom of speech and movement

**Environmental factors:**
available water, steepness of ground, soil fertility, vandalism and litter, local levels of pollution, possibility of hazards like flooding or earthquakes

▼ Patterns of poor quality housing in cities.

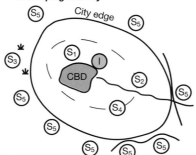

**Developed country**

*City edge*

① Inner-city areas, a mixture of redeveloped nineteenth-century slums and modern high-rise flats.

② Inner-city transitional areas; former villas have become bedsits and multi-let homes

③ Outer-city council estates, many very deprived

**Developing country**

*City edge*

I   Inner slum zone
S   Squatter settlement:
($S_1$) on temporary site
($S_2$) on a river bed
($S_3$) on a marsh
($S_4$) on a hillside, subject to landslides
($S_5$) new shanty towns on edge of city — septic fringe'

▼ The spiral of multiple deprivation. One problem leads to another and the estate goes downhill.

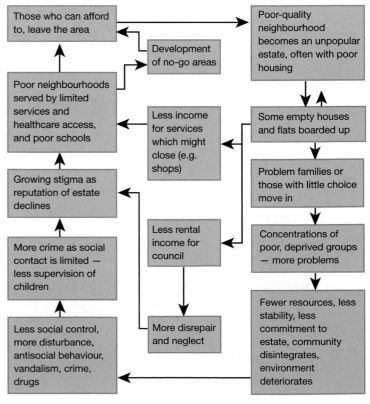

Those who can afford to, leave the area

Poor-quality neighbourhood becomes an unpopular estate, often with poor housing

Development of no-go areas

Poor neighbourhoods served by limited services and healthcare access, and poor schools

Less income for services which might close (e.g. shops)

Some empty houses and flats boarded up

Growing stigma as reputation of estate declines

Problem families or those with little choice move in

Less rental income for council

Concentrations of poor, deprived groups — more problems

More crime as social contact is limited — less supervision of children

More disrepair and neglect

Fewer resources, less stability, less commitment to estate, community disintegrates, environment deteriorates

Less social control, more disturbance, antisocial behaviour, vandalism, crime, drugs

# The worst living spaces

The worst living spaces result from both human and environmental factors – with a bit of chance thrown in. The table shows where many people think the worst places to live are. This includes cities and countries. The key processes are events which have led to a dramatic decline in the quality of the living space.

| | Location and description | Key processes |
|---|---|---|
| 1 | **Chernobyl, Ukraine.** Since the devastating meltdown and explosion of a nuclear power plant at Chernobyl in 1986, over 5 million people have been living in contaminated areas surrounding the blast zone. A 19-mile exclusion zone around the plant is totally uninhabitable. Around this zone, thyroid cancers in children have risen by 90%. | Environmental accident – radiation |
| 2 | **Baghdad, Iraq.** A war-torn city with political unrest and instability. Political insurgents cause problems; the city is still very dangerous. | Geo-political tensions and war |
| 3 | **Detroit, USA.** More than 30% of its residents live below the poverty line. There are abandoned buildings and rubbish in the streets. Detroit has lost 50% of its population since the 1950s. The declining car industry has led to many job losses. | De-industrialisation, unemployment, out-migration |
| 4 | **Mexico City.** The developing world's third largest metropolis. Very overcrowded and noisy. There are problems with air, land and water pollution. Organised crime is out of control in Mexico City, as is drug abuse and alcoholism. Slum areas are growing. More than 50 000 children live on the streets. Water and electricity are unreliable. Over 3000 kidnappings were reported in 2006, mostly by cab drivers. | Overcrowding, rural-urban migration, pollution, homelessness, crime, poverty |
| 5 | **Zimbabwe.** Zimbabwe has a hard currency shortage. This has led to hyperinflation and chronic shortages in imported fuel and consumer goods. The economic meltdown and food crisis have been attributed to government economic mismanagement, the government refusing relief efforts from other countries, a drought affecting the entire region, and the HIV/AIDS epidemic. A major cholera epidemic broke out there in 2008. | Poverty, hyperinflation, famine, disease, drought |

## your questions

1 Pick out two social, economic, political and environmental factors, that are most important to you.

2 Rank these in importance from 1 (most) to 8 (least). Write out and explain your rankings.

3 In pairs, design two diagrams to show the effects of
   a gentrification in a developed city
   b poor slum dwellers on a developing city.

4 Look at the table above. Pick out the two key processes that have you think have had the worst impact. Explain your choices.

+ In this section you will learn about the rural idyll and rural-urban migration

This table compares rural and urban living spaces.

|  | **Urban** | **Rural** |
|---|---|---|
| **Developed countries** | • High densities of people.<br>• Good public transport systems.<br>• High numbers of shops and services, e.g. healthcare and education.<br>• High-speed broadband access and cable. | • Fewer shops and services.<br>• Lower density of transport facilities.<br>• Higher levels of car ownership.<br>• Quieter.<br>• Less transport infrastructure.<br>• Lower-speed Internet access. No cable. |
| **Developing countries** | • Very high densities of people.<br>• Significant proportion of urban poor.<br>• May live in illegal, spontaneous settlements.<br>• Some range of services in central areas.<br>• Range of employment, both 'formal' and 'informal'.<br>• Some public transport systems. | • Very poor public transport systems.<br>• Many people employed in agriculture.<br>• Poverty common.<br>• Disease and poor educational facilities.<br>• Limited or no services, e.g. electricity, telephone (land line). |

## The rural idyll

The rural idyll is an image that people have of rural life in developed countries. They imagine close communities with less traffic and crime, near the countryside, with pubs and village shops.

But there's one problem. Demand means that housing there has become expensive, beyond most people's reach. In 2007, poverty in UK rural areas rose by 3%, compared to only 1% in urban areas.

At weekends the population of many villages increases, like Dent in North Yorkshire. Dent's population is under 200, but at weekends and in summer it swells as the number of holiday homeowners return. In winter, they stay away. Now many rural villages are losing post offices and shops, and in 2008 five rural pubs closed every day in the UK. There simply isn't the permanent population to keep them open.

> *Our garden is much bigger than the one we had in Notting Hill, and we are in the Chiltern Hills. We haven't really missed London and, if we want to, we can get there in 40 minutes on the train. We have all the amenities, including two pubs, a post office, newsagent and butcher. And, if we want a bit of theatre, we have the village hall next door.*

# The rush for towns and cities in developing countries

Rural villages struggle in developing countries too. Young people leave in search of jobs, and families leave to find schooling for their children. This is known as **rural-urban migration**. In most developing countries, health care is also better in cities, with more hospitals and doctors. These cost more than families can ever afford – but people still arrive in increasing numbers. 'Bright lights syndrome' draws more people in.

For the first time in 2005, half of the world's population lived in cities. Most of the increase came from countries like Brazil and India, whose cities are enormous. Most people there end up living in slums like in the photos, without water or sanitation. Over time, as the children grow up and work to bring money into the family, life for many can begin to improve.

▲ Life for some migrants to Mexico City.

▶These migrants from rural India had hoped for something better when they arrived in the city.

## your questions

1 Explain what the cartoon is saying about village life in the UK.

2 Draw a list of advantages and disadvantages of living in villages like Dent for **a** families **b** young people.

3 What reasons might make people in rural villages in developing countries **a** want to leave for the city **b** prefer to stay put?

4 In pairs, find one picture of a shanty town (or favela) on the internet. Write 150-200 words on **a** where it is **b** what life is like **c** the services available. The best cities to try are Rio de Janeiro, Sao Paulo, Mumbai or Cape Town.

✚ In this section you will learn that personal choices about living spaces depend on the ages and needs of people.

When people think about moving from one living space to another, there are factors which act as 'strains' (pulls) or 'stresses' (pushes). These factors can be driven by political, economic or personal circumstances. Often personal circumstances are strongest in determining where people might live. These can override the economic and political factors.

## Getting out of here – retiring to Spain

One of the 'hottest' places to retire to for people in the UK is the coastal strip in eastern and southern Spain. There are a number of reasons why Spain is particularly attractive as a retirement location.

- It has a warmer climate than the UK, which has health benefits for those with arthritis and rheumatism.
- Modern and efficient health facilities.
- It has excellent transport links within the country, which are government subsidised.
- House prices are lower than in the UK.
- It has lower heating costs and household bills compared to the UK.
- Private and state pensions can be transferred directly into Spanish bank accounts.
- It is relatively cheap to fly home to see friends and family.
- As part of an 'expat' community, you don't feel culturally isolated.

**STRESSES**

| |
|---|
| Few employment prospects |
| High cost of living |
| Too far from family/friends |
| Too distant from services and leisure facilities |
| Problems with transport |
| Too noisy/quiet |
| Too polluted |
| Climate unpleasant |

**Shall we move or not?**

**STRAINS**

| |
|---|
| Better employment prospects |
| Cheaper housing/cost of living |
| New job |
| Nearer to family/friends |
| Better road/rail links |
| Better leisure facilities |
| Cleaner environment |
| Quieter |
| Better weather |

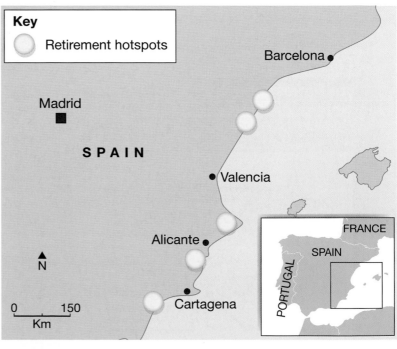

▲ Retirement hotspots in Spain.

# Moving back to the city – 're-urbanisation'

In the past, people have left cities to either live in the countryside (**counter-urbanisation**) or to live in the suburbs (**suburbanisation**). But there is now a slowing and even reversal of these processes in some places. This new internal migration pattern is called **re-urbanisation** and it literally means 'moving back to the city'.

People moving back to the cities tend to be either relatively young (25-35) or retired (60+). Both groups are relatively well off. They are attracted back to the centres of cities for some reasons which are similar – and some which are different.

The table shows some reasons for re-urbanisation.

|  25-35 Young and upwardly mobile |  60+ Recently retired |
|---|---|
| • Good nightlife.<br>• Around similarly aged people.<br>• Close to work.<br>• Lots of high-quality, newly furnished flats.<br>• Close to high-quality shopping. | • May be close to friends/family who live in the city.<br>• Flats are likely to have no garden so there is minimal maintenance.<br>• Accommodation is semi-communal so it is less isolated. |
| • Lots of good places to eat out.<br>• Good public transport infrastructure (can avoid having a car).<br>• Culture and leisure facilities very close by. | |

▲ These city centre flats are in demand.

## your questions

1 Work in groups to list factors that are 'strains' and 'stresses' for where you live. If you had the choice, where would you like to live and why?

2 List the problems and benefits of the migration of retired people to Spain for **a** the UK **b** Spain.

3 Use books/the Internet to research the following terms. Then write a definition for each one.
   • re-urbanisation   • counter-urbanisation
   • suburbanisation   • international migration

4 Make a list of some disadvantages of town centre living – both for the 'twenty something' group and the '60+' group.

✚ **In this section you will look at the pressures put on rural areas by housing and transport.**

The rural idyll is being crowded out by demands for more living space and transport infrastructure. Such pressures are reducing the quality and peacefulness of the countryside.

## Green belt – does it work?

**Green belts** were set up over 50 years ago, and now cover 13% of England. They have three main functions:

- To check the sprawl of cities like London and Birmingham.
- To protect the surrounding countryside from further development.
- To prevent neighbouring cities (like Leeds and Bradford) from merging into one another.

> ✚ A **green belt** is an area of open land around a city, which is protected from development.

Have green belts been a success? In some respects, the answer is a definite yes. Green belts have checked the urban sprawl from cities such as Manchester, Birmingham and London. For the most part, green belt has remained intact for the last 50 years. However, over 1100 hectares of green belt have been lost each year since 1997. And nearly 50 000 homes (equivalent to a city the size of Bath) have been built on green belts since 1997. Green belts may have also caused environmentally damaging 'leap-frogging', where commuter towns grow outside the green belt, like Chelmsford in Essex. Here, people have to commute to work elsewhere – usually London.

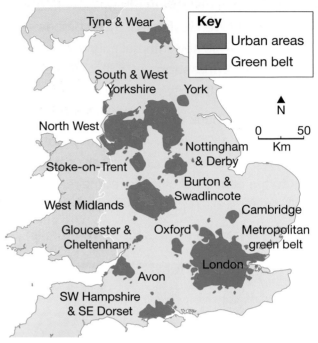

**Key**
- Urban areas
- Green belt

▲ The biggest green belt in the UK is known as the Metropolitan green belt, around London. There are other major green belts in the West Midlands, around Manchester and Liverpool, and in South and West Yorkshire.

**New buildings and infrastructure**
New housing consumes more countryside than any other kind of development. Government figures show a greenfield area nearly the size of Leicester vanishes under bricks, mortar and concrete in the UK each year. Pressures for new housing result from an increasingly elderly population and more people choosing to live on their own.

**New roads**
The noise from a busy road can extend over miles of countryside. Traffic levels are likely to increase by 30% by 2015. The government has allocated billions of pounds to widen motorways, dual single-carriageway roads and build bypasses over the next decade – including in protected rural areas.

**PRESSURES ON LIVING SPACES**

**More planes and runways**
There is likely to be massive increase in air travel and the expansion of airports and associated development. Air traffic in the UK has trebled over the past twenty years. It is forecast to grow at 4-5% each year.

**Increased light pollution**
Use of outdoor lights is blotting out our view of the skies. Between 1993 and 2000, light pollution increased 24% nationally. The amount of very dark night sky fell from 15% to 11%.

# Airport expansion

A massive increase in air traffic is predicted by 2030, especially in Britain's heavily populated South East. Business travellers use London airports for easy access to the City. Heathrow is an important airport for passengers transferring from long-haul international flights to short-hop European ones. At the same time, low-cost carriers such as Easy Jet are increasing their routes and passenger numbers (see below). This puts pressure on smaller regional airports to expand.

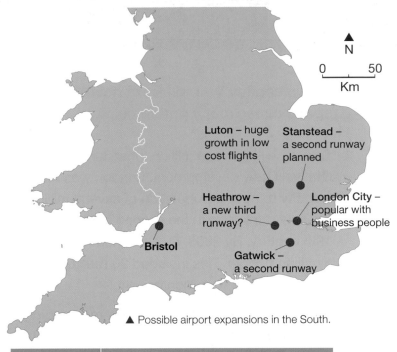
▲ Possible airport expansions in the South.

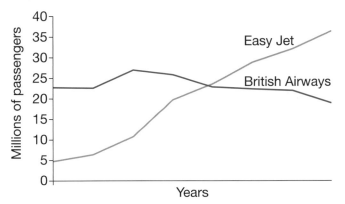

A recent environmental impact assessment (EIA) looked at the possible expansion of Bristol airport. Its findings are shown in the table.

The arguments for and against the airport expansion are controversial and complex. People who support the development of Bristol airport insist that it will help the economic growth of the region. Those against the development raise concerns about air and noise pollution, as well as the loss of valuable countryside.

| Concern | Possible effects associated with expansion |
| --- | --- |
| Air | Increased air pollution (especially $CO_2$ and particulates) from road traffic and aircraft. |
| Archaeology | Loss of special archaeological sites, e.g. historic airfield. |
| Biodiversity | Possible damage to plants and animals, including badgers; increased risk of groundwater contamination. |
| Community | Possible health impacts, but increased employment opportunities. More traffic on local roads. |
| Landscape | Long-term changes in landscape character; loss of mature vegetation. |
| Visual | Visual intrusion of new terminal building; increased lighting – especially at night. |
| Noise | Construction and aircraft noise. |
| Transport | More traffic – both during and after construction |
| Water | More chemicals used to de-ice aircraft; more car parks which cause more surface runoff of rainwater. |

## your questions

1. Draw two spider diagrams to show the reasons why
   **a** green belts are needed **b** they cause problems.
2. In pairs, discuss and list reasons to show whether you think green belts are still appropriate.
3. Write two paragraphs – one to explain why Britain's south east should have more airport expansion, and the second to explain the problems this would cause.
4. List the arguments **a** in favour of Bristol Airport expansion **b** against it.

✚ **In this section you will learn how the demands for urban living space cause many problems.**

The city with several million inhabitants is a relatively new development. London was the first city to have a population of several million. It reached this size in the second half of the 19th century. By 2005, 50 cities had more than 5 million people. This included 20 **mega-cities** with more than 10 million people.

People often think that the world's most rapidly growing cities are in Latin America, Asia and Africa. But this isn't always the case. Several cities in the USA and Japan rank among the most rapidly growing cities of the 20th century.

▲ A lot of central Tokyo is built upwards – due to lack of space. Most shops and places to hang out are located high up.

Tokyo is a city which has experienced phenomenal population growth. In 2007, the city centre population was about 12.8 million, or 10% of Japan's total population. During the daytime, the population swells by over 2.5 million as workers and students commute in from adjacent areas.

One of the big challenges of living in a city the size of Tokyo is making the most of your living space. Japanese homes are small compared with other parts of the world – especially in Tokyo, where house prices are very high.
A typical Japanese home is less than 100 m². It usually combines living and dining areas into one. A flat in Tokyo is typically 40 m², so people must make the most of limited space.

*Living space in Tokyo is definitely a luxury. Most 'flats' are ridiculously small by European standards. But we Japanese have developed a way to make everything fit. In my flat the kitchen is fitted into 1 m² of space. So you start to develop an ability to juggle while cooking. You cut food and do other cooking activities in the most random of places (on the floor for example). It's frustrating at first, but you do get used to it. I now manage to cut food, wash up, dry things and cook in the little space I have*

▲ A tiny living space in Tokyo, Japan.

# Overcrowding and the Alice City

Land prices have soared in Tokyo. In 2008, city land values were £1000 per m². To overcome this, some companies are considering building underground cities. The Taisei Corporation is planning a network of Alice Cities (named after Alice in Wonderland). The idea is to build airy underground spaces connected by underground trains and subterranean roads. Although some buildings and roads would remain above ground, much surface space would be freed up for trees and parks.

# The urban geo-grid

An alternative idea is the Shimizu Corporation's urban geo-grid, a vast network of subterranean city spaces linked by tunnels. The £40 billion project would accommodate 0.5 million people. Each 'grid station' – a complex of underground offices, shopping malls, and hotels – would be connected to several smaller 'grid points', which would provide local services such as public baths and convenience stores. There would be a network for transportation, communication and energy supply, both within a city and between cities.

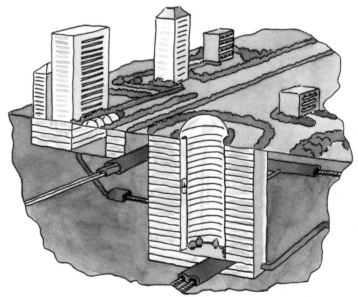

Each Alice City would be divided into three sectors:

- Town space with underground avenues and open-air plazas with no cars. Here are shopping centres, entertainment and fitness centres.
- Office space for businesses, more shops, hotels and parking spaces. Lifts would run between levels. Residential sections will permit some workers to commute vertically, while others will ride in from the suburbs.
- Infrastructure space will contain facilities for power generation, heating and air conditioning, waste recycling, and sewage treatment.

## your questions

1 Use the photos to describe a what living space is like in Japan and how people cope b what being in Tokyo city centre must be like
2 What might Japanese cities of the future look like?
3 Discuss in pairs what it would be like to live in an Alice City. Write a short paragraph to describe a typical day.
4 Exam-style question Describe the advantages and disadvantages for people of living in the urban spaces of one city you have studied.

**On your planet**

+ London intends to be nothing less than 'the most sustainable city in the world'. But in 2007 it was ranked tenth.

+ In this section you will learn how living spaces, both urban and rural, can be made more sustainable.

All living spaces experience some environmental problems. But cities (especially mega-cities) are unsustainable for two main reasons:

- They 'suck in' and consume enormous quantities of resources, e.g. energy, water, food, raw materials (often from around the world). Therefore they have big **eco-footprints**.

- They produce enormous amounts of waste. This is usually got rid of in the surrounding land, rivers, sea and air.

Urban living spaces have problems with:
- waste disposal
- air and water pollution
- unreliable water supplies
- congestion and traffic queues.

Sustainable urban strategies include both **economic sustainability** and **social sustainability**.

> + **Economic sustainability** means allowing people to have access to a reliable income.
>
> + **Social sustainability** means allowing people to have a reasonable quality of life with opportunites to achieve their potential.

But we must also think about the future when planning cities. **Environmental sustainability** means meeting the present needs of people while not putting future generations at risk. This can only be achieved by organising and managing living spaces to:
- minimise damage to the environment
- prevent natural resources being used up.

**Model A – an unsustainable city**

Inputs: Food, Coal, nuclear and oil energy, Goods → City

Outputs → Organic wastes dumped in rivers/coasts; Emissions $CO_2$, $SO_2$, nitrous oxides; Inorganic wastes dumped as landfill

**Model B – a sustainable city**

Inputs: Food, Renewable energy, Goods → City

Outputs → Organic waste recycled; Reduced pollution and wastes; Inorganic waste recycled

**SUSTAINABLE URBAN STRATEGIES**

- Reducing reliance on fossil fuels — and rethinking transport
- Keeping city wastes within the capacity of local rivers and oceans to absorb them, and making 'sinks' for disposal of toxic chemicals
- Recycling water to conserve groundwater supplies
- Providing green spaces
- Conserving cultural, historical and environmental sites and buildings
- Involve local communities and provide a range of employment
- Minimising use of greenfield sites by reusing or using brownfield sites

Curitiba (Brazil) is a good model of sustainability and a 'green city' (see page 203). But London or New York have much work to do. A recent study ranks Britain's 20 largest cities according to social, economic and environmental performance. This goes against the sustainability claims of certain cities. Leicester was the first official 'environment city' – but it was ranked only 14th out of 20.

## Greening rural living spaces

The villagers of Martin in Hampshire are showing how a village can control much of its food system. In a bid to become less dependent on supermarkets, residents are working together to become as sustainable as possible. Many households have joined the Future Farms cooperative. This community allotment on 3 hectares of land grows 45 types of vegetables, in addition to raising pigs and chickens. Produce is priced on a production cost plus 20%. It is sold to all villagers, not just members, at a weekly market in the village hall. This system uses few chemical inputs, which makes it more sustainable.

## Compact communities

A **compact community** is another attempt to reduce the environmental impact of our living spaces. This centres round the idea of 'spatial efficiency' or 'making the best use of space'. Transport infrastructure is integrated with housing and other development – so there's less need to travel. Also, the threat of towns and cities gobbling-up countryside is reduced.

Under the 2003 Sustainable Communities Plan, the government plans for 120 000 additional homes in the Thames Gateway area (North Kent and South Essex) by 2016. This development could form a number of compact communities. Along with the Thames Estuary Parklands initiative, this should lead to landscape enhancement and habitat management throughout the Thames Gateway. If successful, it will attract people and businesses to the area. It will also bring leisure and tourism opportunities.

Linked to the public transport network to allow access to hospitals, workplaces, etc.

Mixed use: they provide everyday social, recreational and shopping facilities (available on foot)

**COMPACT COMMUNITIES**

A mix of housing types/ ownership ensures a range of income groups. Local jobs ensure there aren't pockets of deprivation.

Housing is dense so provides enough people to support local services and use public transport

### On your planet

+ There are approximately 30 000 allotment plots in the UK today, giving about 215 000 tons of fresh produce every year.

### your questions

1 What does 'eco-footprint' mean?
2 In pairs, decide six ways in which a city's eco-footprint could be calculated.
3 Now calculate your own eco-footprint. Type 'calculate my eco-footprint' into an Internet search engine, and carry out one survey. Compare your results with three other people and explain any differences.
4 Go through each of the 'Sustainable urban strategies' and list why each one helps to reduce people's eco-footprints.
5 What are the benefits of **a** the Future Farms cooperative **b** compact communities? Can you think of any problems with them?

In this section you will learn about assessing the potential for sustainable living spaces.

### Eco-villages in Brazil

Ecoovila is a small eco-village in the heart of the Brazilian city of Porto Alegre. It was established in 2001 by 8 farmers. They wanted to develop affordable housing for everyone – and they wanted to use eco-friendly building methods. A village was constructed for 28 families on 2.6 hectares of land.

All the houses face the sun, so they soak up its energy. Solar panels provide hot water. Inside, a central fireplace acts as a thermal heat store – when the fire has gone out it radiates heat during the night. Houses are naturally cooled by air flows using underground chambers; grass roofs also reduce inside temperatures and help to insulate.

Many local materials are used, including bamboo and clay bricks. Sewage is treated in a biological reed-bed system, and waste water irrigates vegetables and flowers. Ecoovila is a community- and eco-friendly model for new settlements. There are plans to build three more villages, on a larger scale, using the same ideas.

### Creating energy – urban algae farming

**Biofuels** are a big hope for the future. The idea is to make fuels from the energy made by plants during photosynthesis. PetroSun Biofuels (an Arizona energy company) has opened a commercial algae-to-biofuels farm on the Texas Gulf Coast, USA. The farm is a network of saltwater ponds, and research will be carried out to develop an environmental jet fuel. PetroSun wants to extract algal oil on-site at the farms and transport it to company biodiesel refineries via barge, rail or lorry.

PetroSun is also working on an algae-based fuel for military use. One day algae technology could also be used to provide fuel for home heating, replacing fossil fuel diesel.

### On your planet

+ Of all the options for jet biofuel production, algae farms are one of the most viable. They yield 30 times more energy per hectare than their closest competitors, and requires neither fresh water nor arable land for cultivation. This gives them an advantage over ethanol.

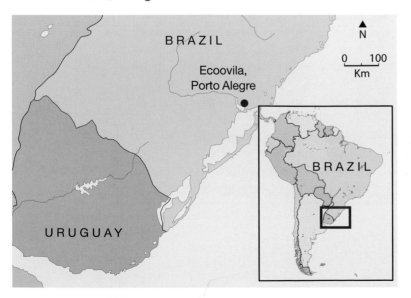

▲ What an algae farm could look like.

# Britain's new eco-towns

Thirteen areas are being considered for new eco-towns, providing up to 15 000 homes in each location (see the map). The key features for the eco-towns include:

- the environment. You need to consider energy use, $CO_2$ emissions, water, materials, waste, transport and access, health, 'green issues', and food production.
- economic issues. Jobs should be within, or accessible from, the community.
- social issues. Sustainable communities should be of a scale to at least support a secondary school (4000 to 5000 homes). They need to be designed with a mixture of people in mind.

But not everyone is in favour of the new proposals. Critics say that the schemes risk being 'car-dependent housing estates'. They also suggest that most towns are planned for **greenfield sites**, and two are in the green belt. Some people say that the sites have been chosen by developers with their own interests at heart.

▲ Britain's proposed eco-towns.

+ **Greenfield sites** are areas currently used for agriculture or left to nature.

▲ Protests against eco-town developments.

## your questions

1 Draw up a table evaluating the advantages and disadvantages of the proposed eco-towns. Use the Internet to get people's opinions.
2 Apart from urban farming, what strategies exist for the production of food and other resources (including energy) in local communities?
3 Research an example of eco-friendly housing. Has it been a success? If not, why not?

✚ In this section you'll find out how employment changes, and how the Clark Fisher model can explain this.

## Employment and industry

Industry provides employment. Employment and industry can be classified into four main groups or **sectors**:

- **Primary industry** – people extract raw materials from the land or sea. Farming, fishing and mining are examples.
- **Secondary industry** – people are involved in manufacturing, where raw materials are converted into a finished product, for example house building, car making or steel processing.
- **Tertiary industries** – provide a service. There is a wide range of service industries, including distribution, retailing, financial services and nursing.
- **Quaternary industries** – provide information and expert help. They are often associated with creative or knowledge-based industries, especially IT, biosciences, media, etc.

A company can employ people in different ways. Someone whose job is manufacturing cars is part of the secondary sector. But the sales people for the same company are in the tertiary sector.

In the UK, there has been a major change in the types of jobs that people have been doing over the past 40 years, as the graph shows. There has been a drop in primary and secondary employment, and an increase in the tertiary and quaternary sector. The tertiary and quaternary sectors now account for about 76% of all employment in the UK. So the UK's employment structure has changed – but why?

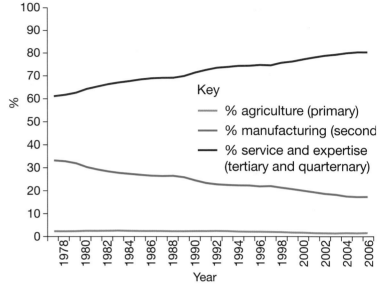

**Key**
— % agriculture (primary)
— % manufacturing (second
— % service and expertise (tertiary and quaternary)

▲ Changing employment in the UK, 1978-2006.

## Spatial employment change

Not only does employment change over time, but it also varies **spatially** – from one place to another. It may vary over a small area, for example within contrasting areas of a town or city, or on a much bigger scale – such as a region or country.  Variations in employment sectors can be complex and are affected by a number of factors, including:

- the **socio-economic** groupings of people
- how wealthy a country is
- investment in manufacturing companies, wage costs and available infrastructure.

## The Clark Fisher model

Two economists, Clark and Fisher, produced a model which can help to explain changes in employment structure. As countries develop, Clark and Fisher said that they go through three stages:

1 Low-income countries are dominated by primary production (**pre-industrial**).

2 Middle-income countries are dominated by the secondary sector (**industrial**).

3 In high-income countries, the tertiary sector is dominant (**post-industrial**).

- As economies develop and incomes rise, the demand for agricultural and manufactured goods will increase. This means secondary industry will grow.
- As incomes continue to rise, people start to consume more services. This means the tertiary sector will grow and develop.
- Finally, tertiary services will support and promote quaternary industries.

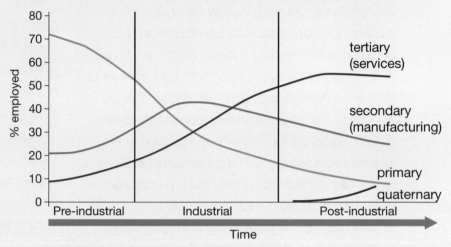

▲ The Clark Fisher model.

## How useful is the Clarke Fisher model?

The model tells us something about how employment changes over time, and how the balance of employment changes as countries develop. But there are some problems with it:

- It assumes that there is a simple straight development path from less-developed to developed. In fact, countries have very different levels of income and there are different economic groupings – as the diagram on the right shows.
- The model also tends to ignore the international context and does not take into account imports of manufactured goods, or the relocation of manufacturing to cheaper, less-developed locations.
- Some developing countries may have a large tertiary sector (linked to a large tourist industry), without having developed a secondary industry first.

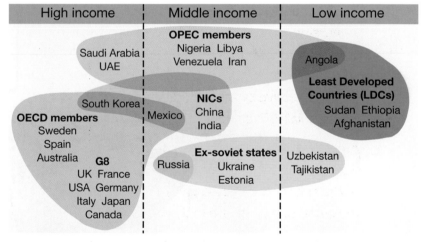

▲ Examples of different economic groupings.

### your questions

1 Describe a changes to the UK's employment shown in the graph on page 188 b where these changes fit the Clark-Fisher model above.

2 a Find a copy of your local free paper, and survey 20 available jobs. Classify them as primary, secondary, tertiary and quaternary. Draw a graph to present this information, and describe what the graph shows.

   b Do the same for a national daily paper. What differences are there between jobs available in your area and nationally?

✚ In this section you'll learn about contrasting employment structures in two different countries.

## Industrialisation

Employment can vary widely between different countries – but why? **Industrialisation** is a social and economic process which changes pre-industrial societies into industrial ones. It is part of a wider process, where social change and economic development are closely related to technological innovation. In the past in the UK, this was linked to the development of large-scale energy and metal processing (for example the iron and steel industries). As a result, a high proportion of people were employed in secondary industries. Industrial output is a good way of measuring how industrialised a country is. The map shows industrial output in 2005.

## Deindustrialisation

Since the 1970s, many cities in developed countries have experienced **deindustrialisation**. Job losses in secondary industries have been replaced with job increases in tertiary and quaternary industries.

These changes can create very different patterns of employment in industrialising and deindustrialising countries. The next page looks at two different countries – Mexico and Germany.

> ✚ **Deindustrialisation** refers to the decline in manufacturing (secondary) industry and the corresponding growth in tertiary and quaternary industries. It can also refer to the decline in the proportion of a country's wealth (or Gross Domestic Product) created by manufacturing industry.

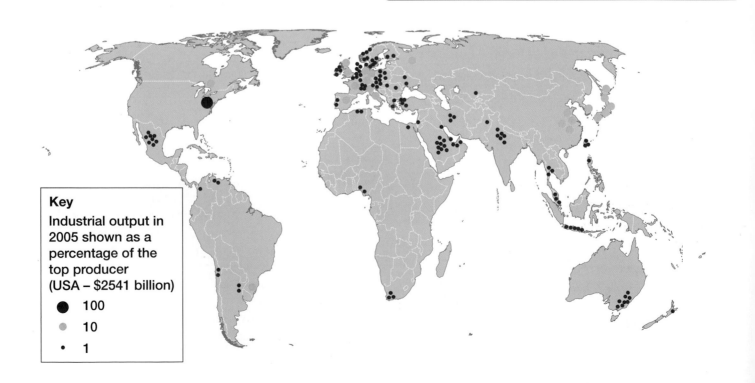

**Key**

Industrial output in 2005 shown as a percentage of the top producer (USA – $2541 billion)

● 100
● 10
• 1

# Mexico – an industrialising country

Mexico has a mixture of modern and traditional industry, including agriculture, but is increasingly dominated by private sector tertiary and quaternary industries. The Government has encouraged competition in seaports, railways, telecommunications, electricity generation, natural gas distribution, and airports.

|  | Primary | Secondary | Tertiary and quaternary |
|---|---|---|---|
| % employed | 15 | 35 | 50 |
| % income generation | 4 | 26 | 70 |

In the early 1950s, manufacturing overtook agriculture as the largest contributor to Mexico's wealth. Mexico's manufactured goods include vehicles, chemicals, machinery and equipment. They are exported and create more than 70% of Mexico's foreign earnings. Manufacturers have been drawn to the capital, Mexico City, for a number of reasons. It has:

- a large and highly skilled workforce
- a large consumer market
- low distribution costs and is close to Government decision-makers.

In recent years, Mexico's manufacturing has been increasingly based on chemical production and food processing. In addition, employment has grown in the re-export processing industry. Re-export businesses are usually located near the border with the USA, and are owned by foreign corporations. They assemble or process goods imported from the USA and then re-export them duty free.

Now there is competition from countries such as China. The Mexican Government has therefore encouraged tertiary activities including tourism and retailing.

# Germany – a deindustrialised country

Today Germany has the fourth largest economy in the world. But it hasn't always been like that. Deindustrialisation during the 1970s and 1980s forced economic change in Germany. Manufacturing moved to lower-cost sites (including overseas) and, in its place, service sector and financial services jobs grew in urban areas. As industry changed, there was a corresponding shift in employment patterns.

|  | Primary | Secondary | Tertiary and quaternary |
|---|---|---|---|
| % employed | 3 | 31 | 66 |
| % income generation | 1 | 29 | 70 |

Although Germany suffered deindustrialisation, it acted to save its economy by setting up many small and medium sized manufacturing businesses. In the UK only 22% of people are employed in manufacturing, but in Germany the figure is 31%. Germany has also refocused activities so that many more people are involved in knowledge-based industries and research and development associated with manufacturing high-end goods, such as cars.

## your questions

1 Using the map opposite, draw up three lists.
- The name of the world's biggest industrial producer.
- The names of the next biggest producers.
- Other important industrial countries.
2 Draw a line graph to show changing employment in Mexico and Germany. Use three colours, one for each type of employment.
3 Explain how the loss of industry in one country can be to the advantage of a developing country.
4 Research one other country to find **a** its employment in each sector **b** how this is changing **c** reasons for the change.

# 12.3 Economic change and urbanisation

+ In this section you'll identify the links between economic change and urbanisation.

## The Asian Tigers

The Asian Tigers aren't a football team but a group of **Newly Industrialised Countries (NICs)**. NICs are countries that developed large manufacturing industries very quickly, and which have seen their exports and GDPs grow rapidly. The Asian Tigers are Hong Kong, Singapore, Thailand, South Korea and Taiwan – but Brazil, Mexico and Argentina are also NICs. And they have been joined recently by the rising economies of India and China.

The economies of the NICs may not match those of developed countries, but they are way ahead of developing nations.

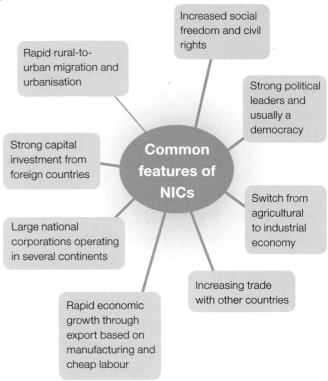

## Moving away from the informal economy

On the streets of any city in a developing country you would be likely to see people working in the **informal economy** – shoe-shiners, food sellers and so on. Although the informal economy is most often associated with developing countries (up to 60% of the labour force works in the informal economy), all economic systems contain an informal economy of some sort. If you've ever heard anyone say that they will do something for 'cash' that's part of an informal economy. The table shows the characteristics of formal and informal sectors of employment.

However, as countries begin to develop, there are accompanying shifts in employment patterns and types, and one of the most significant is a move away from the informal economy.

+ The **informal economy** refers to all economic activities that fall outside the **formal economy**. It is economic activity that is neither taxed nor monitored by a government, and is not included in the country's Gross National Product (GNP).

| Formal | Informal |
|---|---|
| • It may be difficult to get a job (for example for new migrants) | • Easy to start working – no documentation is needed |
| • Employees are taxed | • Do not pay tax |
| • Employee of large company | • May be carried out at home |
| • Reasonable wage for country / location | • Poor pay |
| • Formal skills or qualifications are needed | • No formal skills or education are required |
| • Tends to be large scale | • Tends to be small scale |
| • Health and safety laws in place | • No safety protection at work |
| • No child labour | • Might invovle child labour |

# Urbanisation

As countries develop and their economies grow, **urbanisation** increases. In 2005, the world reached a significant milestone. For the first time in history, more than half its population, 3.3 billion people, were living in urban areas.

People move to cities for better jobs and to improve their lives. However, as people head for the cities, the jobs they do change. For example, in China in 1970, 35% of people were employed in agriculture – by 2002 that figure had dropped to 15% (see the pie charts). The proportion of people employed in Chinese industry increased by 10% in the same period, and the service sector also grew by a similar amount. Other NICs, such as India, have seen similar changes to their employment structure.

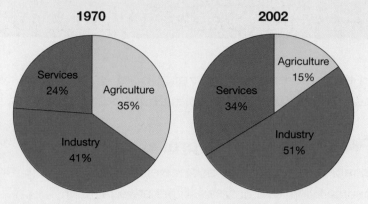

▲ China's employment structure, 1970 and 2002.

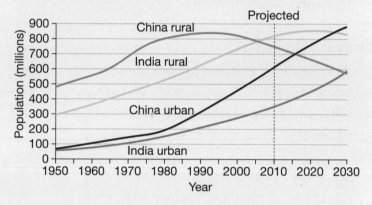

▲ China and India's urban populations are growing.

+ **Urbanisation** is the increase in the percentage of people living in towns and cities.

## your questions

1 Give examples of six informal jobs done in a the UK and b a city in a developing country such as Bangkok.
2 What are the benefits and problems of working in the a formal and b informal economy?
3 Using the graphs describe China's change in employment. What benefits do you think this has brought to the people? What problems?
4 In pairs discuss whether the air pollution brought by industrial growth in countries like China is a price worth paying for more jobs and greater wealth.

## The urban heat island

One significant environmental impact associated with the increasing size of towns and cities is the **urban heat island** effect. Urban areas become warmer because buildings store heat, and car fumes, factories, heating systems and people all add to the heat. This causes the city to become 1-4°C warmer than surrounding landscapes. The diagram shows how temperature rises towards the centre of a city.

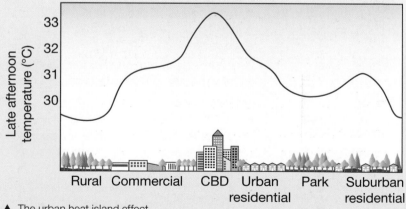

▲ The urban heat island effect.

✚ In this section you'll find out about why and how the rural economy has changed.

## Problems for the rural economy

In recent years, life has become difficult for people living in rural areas. There are a number of challenges that are unique to the countryside, as the spider diagram shows.

## Diversification

To overcome its problems, the rural economy is having to change. As farming declines, people have to plan for a **post-production** countryside and **diversify**. There are three main strands to this:

- **Economic** – using the 'consumption' industries, for example leisure and tourism.
- **Social** – for example the use of 'knowledge industries', i.e. technology, bio-sciences, digital industries, etc.
- **Environmental** – for example developing the use of historic buildings, artefacts and architecture in the countryside.

**Changing rural economy** – the shift from farming towards tourist, which often offers low paid, part-time and seasonal jobs.

**Disappearance of rural services** – the reduction of bus services and health services in rural areas has had a major impact on rural people, especially the elderly.

**Lack of affordable homes** – the purchase of second homes in rural areas by the better-off increases house prices and excludes local people on low incomes.

**Local depopulation** – many communities have lost younger residents, forced out by a lack of job opportunities.

**Challenges for the countryside**

**Lack of transport infrastructure** – public transport may be infrequent and expensive. Car ownership is expensive.

**Agricultural change**
- Low wages, mechanisation, cheaper imports and diseases such as foot and mouth.
- Supermarkets have driven down the prices of farm produce.

+ **Diversify** means to create more variety in jobs and industry so that people are not dependent on one activity.

+ **Post-production** countryside means how the countryside should be used as farming declines. It refers to the diversification of economic activities in rural areas.

| Diversification strategy | What it is | Example |
|---|---|---|
| Food Festival | A celebration of local food. Attracts lots of visitors to an area during festival time. | Ludlow Food Festival |
| Rural sports | A range of ideas, e.g. quad biking, shooting, fishing, paintballing, etc. All attract visitors and money to an area. | Trout fishing at Farletonview Fishery, Cumbria |
| Food trail | A free map detailing good places to eat. Visitors are attracted at all times of the year. | The Ribble Valley Food Trail. |
| Media and film making | Film and TV trails are increasingly popular ways of getting people to explore remoter rural landscapes that have been used for film and TV productions. | Movie Map North Wales |
| Farm diversification | New agricultural enterprises to increase farm incomes. 40% of UK farm income comes from diversification. | Runnage Farm Dartmoor provides bunk/barn accommodation |
| Rural industries | A range of 'knowledge-based' industries and specialist services, e.g. furniture making, web-hosting, Ebay shops, etc. | Jim Laurence traditional ironwork in rural Suffolk |

## The broadband revolution

Broadband allows people and businesses in remote locations to share ideas and reach customers. Near Newquay, Cornwall, Jamie Oliver's 15 restaurant takes most of its bookings over the Internet. 16 miles away in St Austell, Hidden House Bulbs is able to market and sell a range of unusual flowering bulbs across Europe, also using the Internet. Now, about 80% of Cornish holiday cottage rentals are booked through websites or by e-mail. The £12.5 million actnow scheme (www.actnowcornwall.co.uk), led by Cornwall Enterprise, has supported such business innovation in the use of broadband and IT.

## Farm diversification

As incomes from farming have declined, farmers have been encouraged to diversify into other ways of making a living. These include accommodation (B&Bs, campsites, holiday cottages), farm shops and leisure activities, such as clay pigeon shooting (pictured above).

▲ Denbies Vineyard, Surrey. The winery and visitor centre are in the background.

An important aspect of diversification is to celebrate local distinctiveness. Increasingly, local producers and farmers are marketing something that sets them apart from the local competition. One example of a successful farm diversification enterprise is Denbies Vineyard in Dorking, Surrey. Denbies is the largest vineyard in the UK, with 265 acres. It was planted in 1986, with the first wine produced in 1989 – before that it was a pig farm! Denbies is the largest investment the English wine industry has seen so far. Although there are nearly 400 vineyards, most in the South East of England, the average size is around 2 acres.

'Approximately 65% of our wine is sold through the visitor centre – where we have around 300 000 visitors a year. The remaining wine is sold either through mail order or to trade customers, including a few of the larger supermarkets, such as Sainsburys and Waitrose.'

*Denbies Vineyard*

### your questions

1 Why is living in the countryside sometimes a problem for a the young b families c the elderly d those on low incomes?
2 Go through the list of diversification types opposite. Draw a table to show how much they help a the young b families c the elderly d those on low incomes. Give your reasons.
3 Research examples of Cornish businesses on the internet – type in 'Cornwall Holiday cottage', 'farm shops in Cornwall' and one of your choice. How do businesses promote themselves on the internet to attract people?
4 Go through the 'challenges for the countryside' spider diagram opposite. Draw a new spider diagram, showing in one colour the problems which rural diversification will help to solve, and in another colour the ones it won't. Explain your ideas.

# The impacts of changing employment - 1

✚ In this section you'll learn about the impacts of deindustrialisation in developed countries.

## Economic impacts - UK

The shipyard in the photo lies empty – a victim of deindustrialisation. Deindustrialisation has had enormous social and economic impacts, particularly on cities which depended on traditional industries – such as steel making, ship-building and textiles. The problems were made worse by the rapid rate of change.

- **Economic impacts** include: loss of personal income; loss of tax income to national and local governments; rising demand for income-support services; loss of income in the local area due to people's lack of spending power.

## Social impacts - Glasgow

The shipbuilding industry in Glasgow created close communities that were devastated by deindustrialisation in the 1980s. The decline of shipbuilding also caused a decline in steel making (to make the ships) and coal mining (to make the steel).

Generations of well-qualified workers were left on the dole during the 1980s. Glasgow contained some of the most deprived areas in the UK. Those who could, left to get jobs elsewhere, so Glasgow's population fell. There were widespread problems of family breakdown, alcoholism and crime. Without investment, there was a danger of permanently unemployed families.

There are many reasons for deindustrialisation, but they focus on three key factors, as the spider diagram shows. A combination of these factors can create a cycle of decline (see right).

▲ A derelict shipyard - a product of deindustrialisation.

Automation and mechanisation increased productivity and reduced the number of workers required to work in a particular industry.

Greater competition from countries such as India, China, Taiwan and South Korea.

**Reasons for deindustrialisation**

Reduced demand for traditional products, as new materials and technologies are developed.

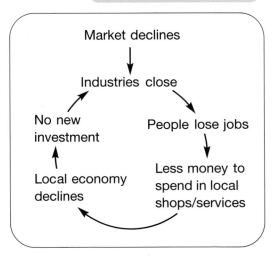

Market declines → Industries close → People lose jobs → Less money to spend in local shops/services → Local economy declines → No new investment → (back to Industries close)

▲ Cycle of decline.

## Environmental impacts

There are a range of environmental impacts associated with deindustrialisation. Some may be positive, others may be negative (see below).

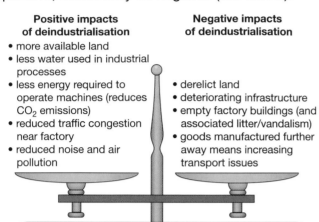

| Positive impacts of deindustrialisation | Negative impacts of deindustrialisation |
|---|---|
| • more available land <br> • less water used in industrial processes <br> • less energy required to operate machines (reduces $CO_2$ emissions) <br> • reduced traffic congestion near factory <br> • reduced noise and air pollution | • derelict land <br> • deteriorating infrastructure <br> • empty factory buildings (and associated litter/vandalism) <br> • goods manufactured further away means increasing transport issues |

When you weigh it up, deindustrialisation has probably had a positive overall effect on the local and regional environment in developed countries like the UK. But if you consider the global scale, the picture is different. As manufacturing has moved overseas, the finished goods have to be transported around the world to different countries. Ships and planes criss-cross our oceans and skies transporting raw materials and goods from place to place. In doing so, they are pumping out $CO_2$, increasing air pollution, and are likely to be contributing to an increase in global warming.

### your questions

1 Explain why the effects of deindustrialisation usually include increased family breakdown, alcoholism, and crime.

2 Make a copy of the scales diagram, but rewrite it for the economic effects of deindustrialisation and investment.

3 Create a poster showing the before and after appearance of Glasgow's docks and riverside. Use an Internet search engine to find some images to use.

4 On balance, do you think Glasgow's suffering in the 1980s has been worth the improved environment and employment opportunities there now?

## New employment in Glasgow

In areas such as Glasgow which have suffered major decline, there has been a lot of government help to kick-start growth. Private industries by themselves rarely invest without a return, so the government invests first to try and create a cycle of growth (see diagram below).

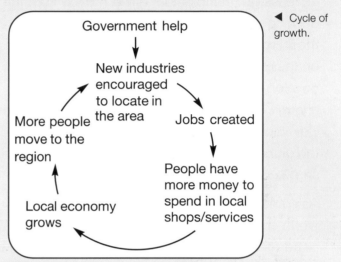

◀ Cycle of growth.

Investment took place in the following areas:

• Arts and culture. Glasgow has a wealth of architecture from famous architects such as Charles Rennie Mackintosh. The government invested in the new Burrell Collection to create an internationally famous art museum that attracts tourists. Increased tourism brings employment.

• The old docks. Here, land lay derelict. Investment took place through private property developers, to build shops, restaurants and riverside flats along the Clyde. This created jobs in building, services and retail.

As manufacturing has declined, tertiary and quaternary industries have grown in importance.

• Tertiary activity includes financial services (for example banking and insurance), retailing, leisure industries, transport, education and health.

• Quaternary activity includes 'knowledge-based' industries where ideas are the main output, including software design and advertising.

In this section you'll investigate the impacts of employment change in India, with a focus on Mumbai.

## Economic growth

Increasingly, manufacturing has moved from the developed world to developing countries. Look at the increase in wealth of India and China shown on the graph. As economies such as India's develop, employment structures and patterns change – and people's quality of life can improve. Developing countries also contribute more to the global economy, as the diagram on the right shows.

As some countries and cities in the developing world continue to prosper, their employment patterns are changing again. Economies driven by steel and textile manufacturing are now beginning to diversify into the service and knowledge-based industries.

## India

India's economy is now developing at its fastest rate since the country became independent in 1947. The 'drivers' of change in India have been the four metropolitan regions of Mumbai, Delhi, Kolkata and Chennai. Since independence, these cities have been the employment magnets, education centres, industrial hubs and trade cores of India – making them the most developed cities in the country. They have developed diverse economies and attracted migrants from varying backgrounds.

But it hasn't all been good news. Attracting migrants to the big cities has led to overpopulation and extreme poverty for some. Overcrowding and unplanned development have also led to environmental problems, such as water and air pollution, and a lack of sanitation.

**World GDP**

Key
- Rest of world
- India
- China
- South America
- Asia Pacific developed
- Eastern Europe
- Western Europe
- North America

Developing world
Developed world

y-axis: Trillion $ (PPP basis) — 0, 50, 100, 150, 200, 250, 300, 350
x-axis: Year — 2005, 2025, 2050

▲ In future the developing world will take a bigger share of the global economy.

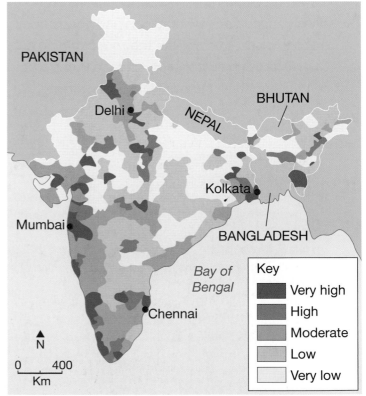

PAKISTAN
Delhi
NEPAL
BHUTAN
Kolkata
Mumbai
BANGLADESH
Bay of Bengal
Chennai

N
0    400
Km

Key
- Very high
- High
- Moderate
- Low
- Very low

▲ Economic growth in India.

## Mumbai – India's economic engine

Mumbai is the commercial capital of India. Home to nearly 13 million people, it contributes 4% of the country's GDP. It acts as an important economic hub, providing:

- 10% of all factory employment
- 40% of all income tax collections
- 40% of India's foreign trade.

The average yearly income in Mumbai is approximately £500 per person, which is almost three times the national average. Many of India's big companies, including State Bank Of India, Tata Group, Godrej and Reliance are located in Mumbai.

Mumbai's economy has diversified:

- It now includes engineering, diamond-polishing, healthcare and information technology.
- Mumbai is home to the Bhabha Atomic Research Centre, and most of India's specialized, technical industries are based there, because Mumbai has a modern industrial infrastructure and large numbers of skilled workers.
- Aerospace, optical engineering, medical research, computers and electronic industries, shipbuilding and salvaging, and renewable energy and power are all represented in Mumbai.

Most of India's major television and satellite networks, as well as its major publishing houses, have their headquarters in Mumbai – and it is the centre of the Bollywood movie industry.

◀ Bollywood produces more films a year than anywhere else.

## Environmental impacts

Although Mumbai has seen the closure of textile mills and the relocation of some of its engineering, chemical and pharmaceutical industries (due to rising land prices and high wage costs), it still suffers from air pollution. The pollution is caused by industries in the eastern suburbs and Navi Mumbai, and also by burning rubbish and increasing emissions from cars and lorries. As a result, Mumbai still has a very high incidence of chronic respiratory problems.

As the tertiary and quaternary sector industries increase in number, there is a corresponding increase in demand for electricity. At present much of Mumbai's (and India's) electricity is generated from fossil fuels, which pump out vast quantities of $CO_2$ – a greenhouse gas.

|  | 1997 | 2002 | 2003 | 2004 | 2005 |
|---|---|---|---|---|---|
| Hydroelectric | 21.7 | 26.3 | 26.8 | 29.5 | 30.9 |
| Nuclear | 2.2 | 2.7 | 2.7 | 2.7 | 2.8 |
| Geothermal/ solar/wind/ biomass | 1.3 | 1.5 | 1.7 | 1.9 | 3.8 |
| Coal | 59.6 | 74.6 | 76.7 | 78.0 | 80.9 |
| Total capacity | 84.8 | 105.1 | 107.9 | 112.1 | 118.4 |

▲ Source of India's electricity generation (in thousands of megawatts).

### your questions

1 Using the graph on the opposite page, calculate the total trillions of dollars earned by India and China together in 2005, 2025, and 2050. Compare their joint earnings to the other economies listed. Which will be the world's biggest economy on each of these dates?

2 Draw a chart listing all of Mumbai's industries classified into primary, secondary, tertiary and quaternary.

3 What are the benefits and problems that Mumbai's industries bring to the city and to India? Create a display with images and text, to illustrate these. Include at least two benefits and two problems.

+ In this section you'll look at how employment change can be managed by regenerating brownfield sites.

## Brownfield sites

As you have already seen, one of the social and economic impacts of deindustrialisation is the loss of employment in an area. But another, equally important, impact is the fact that the closure of industry can lead to dereliction of land and **brownfield sites**.

Derelict brownfield sites like this one in Chesterfield were once used in the city's coal and engineering industries. Like Glasgow, the city is sat in the centre of a large coalfield, creating a demand for mining equipment. When the mines closed in the 1980s and 1990s they left a legacy of derelict land contaminated with pollution from metals, tar and other coal products.

Developing brownfield sites does have disadvantages:

- Brownfield land is often more expensive to develop than **greenfield sites** (because of clean-up costs).
- There is a mismatch between the location of large amounts of brownfield land (in the former industrial heartlands) and the areas of highest housing demand (in the south and south-east).
- Regulations involving reclaiming some brownfield land may be a barrier to new development.
- Some brownfield and derelict land can be important wildlife habitats, public green space, or a core part of urban green networks.

## Birmingham

Birmingham's stock of brownfield sites is an important resource. It can help to meet the demands of both present and future generations in terms of their land requirements. Brownfield land is available for a full range of regeneration options (see right).

> **+ Brownfield site** – an area of land that has been built on before and is suitable for redevelopment.
>
> **+ Greenfield site** – an area of land that has not previously been built on.

Regeneration options for brownfield sites in Birmingham

- Retailing, for example out-of-town shopping centres and shopping complexes
- New commercial sites for office space
- Flood control and management
- Mixed used developments (see Fort Dunlop opposite)
- Residential use, especially flats converted from old warehouses
- Recreational sites such as parks, cycleways and bridleways
- Canalside walks, and riverside cafes and restaurants
- Wildlife habitats, for example woodlands and wildflower meadows

In Birmingham, Government agencies such as the West Midlands Regional Development Agency, helped to secure and support the regeneration of the city areas hit by industrial decline and dereliction.

## Fort Dunlop

At its height, over 12 000 people worked at Fort Dunlop – a tyre storage facility on the outskirts of Birmingham. Built in 1916, an entire village – locally known as 'Tyretown' – developed around the site to meet the workers' needs, including shops, a cinema, concert hall and a sports club. The factory closed in the 1980s, when Dunlop's manufacturing and storage facilities were moved overseas.

'Fort Dunlop's tenants include architects, designers, accountants and regeneration companies. It's unusual to see such a wide range of sectors in one location and it creates a genuine diversity. A number of independent retailers including a coffee company, children's activity centre and homewares store are already in place. Our 2000 plus future office tenants have somewhere not just to work, but a place to shop, eat, drink and unwind.'

*Urban Splash*

▲ An empty Fort Dunlop before regeneration.

▲ The revamped Fort Dunlop.

After being empty and derelict for 20 years, in 2002 Fort Dunlop received planning permission for a mixed-use development to create a sustainable 24-hour community. The phased opening of Fort Dunlop began in Spring 2006. The development includes a 100-bed hotel, a business park with commercial office and retailing space, as well as places to eat and drink.

Fort Dunlop is a good example of how a classic brownfield site has been regenerated to provide employment, and in so doing has improved the local environment.

### your questions

1 Based on what you have seen in Glasgow, summarise the problems faced by Birmingham after the collapse of its traditional industries.

2 In pairs, read the spider diagram of possible uses for brownfield sites. Select three that you think would bring greatest benefits, and present your ideas to the class.

3 Make a list of the options from the spider diagram, and tick which ones the Fort Dunlop scheme has done well. How good do you think the scheme has been?

4 How well would Fort Dunlop cope with a massive cut in consumer spending? Do you think the collapse of the 1980s could ever be repeated here again? Explain.

In this section you'll learn about the potential for growth in the 'green' employment sector.

## The 'green' employment sector

What is 'green' employment? It consists of attempts to improve air and water quality, to recycle and reduce waste, to promote conservation and green tourism, and to improve the environment. Green employment includes the following:

- Making 'green' products – from natural renewable materials, or from recycled goods e.g. making fleeces from plastic bottles.
- Constructing green buildings that use less energy, recycle water, and are built from natural materials.
- Offering 'green' services e.g. eco-tourism.
- Quaternary services e.g. architects (designing 'green buildings').

The UK government estimates that the world market for environmental goods and services will increase to £350 billion by 2010. This section looks at three examples of green employment –

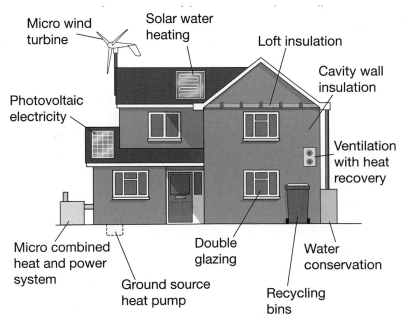

Micro wind turbine

Solar water heating

Loft insulation

Cavity wall insulation

Photovoltaic electricity

Ventilation with heat recovery

Micro combined heat and power system

Ground source heat pump

Double glazing

Recycling bins

Water conservation

## Ecotourism

Ecotourism tries to respect the environment and local people, but reduce the impact of tourists. It is usually small-scale, to lessen the impact on the environment. It is the fastest growing part of the tourist industry, growing by 5% each year.

At Uluru (formerly Ayers Rock) in Australia, ecotourism is offered by the local Anangu peoples on whose land Uluru is located. To them, Uluru is sacred. They offer guided walking tours to cultural sites around the rock and the Uluru Cultural Centre. In this way, they hope that people will both appreciate the desert environment, and learn about traditional aboriginal ways of life.

However, there are problems with this kind of tourism.

- It is 'high cost–low volume', which makes it expensive, attracting mostly high-income tourists. It costs up to A$600 (£280) a night per room at some of the hotels.
- Most tourists arrive by air – which defeats the environmental aims of eco-tourism.
- In the remote desert, everything has to be brought in by road – from building materials to food. Uluru is 2000km from Adelaide, the nearest city!

+ **Ecotourism** is a form of tourism designed to reduce the negative aspects of tourism on the environment.

◀ Building greener homes in the future will provide a range of associated employment opportunities.

# Renewable energy

Renewable, or green, energy offers a range of employment opportunities. These include:

- work on renewable energy sources such as biofuels, wind turbines, tidal and solar power
- jobs in the construction industry, building environmentally friendly buildings
- some commercial finance services.

# Recycling

Curitiba is a city of 1.6 million people, in south-east Brazil. In 1940, the population was 120 000; it quadrupled by 1965, and has trebled again to its current size. Industrialisation and enormous growth meant Curitiba faced the same problems as other developing world cities – unemployment, poor housing, poor nutrition, pollution and congestion.

In 1989, Curitiba became the first city in Brazil to introduce separation and recycling of domestic waste. Since then, recyclers have separated 419 000 tons of waste. Recycling in Curitiba is now so advanced that two-thirds of the city's daily waste is processed. This creates employment, with people sorting organic and inorganic waste. Recycling is done at a plant (itself made from recycled materials) by previously unemployed people – including the homeless. Recovered materials are sold to local industries, and the proceeds are used to fund social programmes (see the photo).

**The photovoltaic generating capacity being manufactured is growing exponentially**

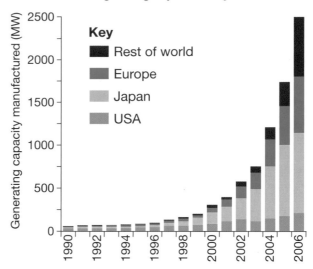

▲ The more photovoltaic, or solar, panels that are manufactured, the more people are needed to produce them.

▲ A green exchange of fresh fruit and vegetables in return for sorted waste in Curitiba - improving both recycling and nutrition.

## your questions

1 Define 'green employment'.
2 In a table, list ten jobs that would count as 'green employment' and beside each explain what makes it 'green'.
3 Copy the table on the right to show the opportunities and threats to jobs in ecotourism, renewable energy, and recycling.
4 Exam-style question Using examples, explain how employment opportunities in the future could be more sustainable. (6 marks)

|  | *Oportunities* | *Threats* |
|---|---|---|
| Ecotourism |  |  |
| Renewable energy |  |  |
| Recycling |  |  |

✚ In this section you'll see how urban lifestyles threaten the natural environment.

## Crisis – what crisis?

Over half of the world's population live in towns or cities. They take up a tiny 2% of the Earth's surface but consume 75% of the world's resources – and produce 75% of its waste. In developed countries, food, water and energy are readily available – but where do they come from? How are they produced?

Did you know that tropical rainforests in places like Sumatra (Indonesia) have been destroyed, and palm oil plantations now grow in their place? And why? To provide us with things like cheap shampoo and margarine (using the palm oil).

- But what happens when that land can no longer provide for the people who live there?
- What happens if that land can no longer support people living in the world's cities?
- What happens when there is not enough land to meet everyone's needs?

▲ Urban lifestyles destroy natural environments around the world.

## Sources, sinks and footprints

We use the natural environment as a **source** – it provides us with our daily needs. We also use it as a **sink** for all of our waste products. The combination of sources and sinks represents the amount of land, water and air (known as **bioproductive areas**) needed to support our lifestyles – and this can be measured in global hectares (gha). This is known as our **eco-footprint**.

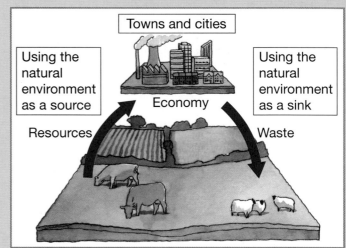

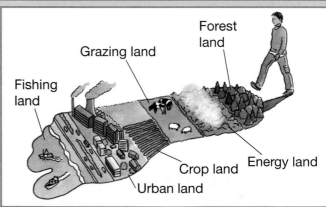

▲ Our eco-footprints extend far beyond the places where we live, because we no longer produce everything we consume for ourselves.

### What do you think?
✚ Does it matter if Sumatra's orang-utans die out because tropical rainforests are being cleared to provide us with margarine and shampoo?

## What's in a footprint?

Everything we buy and everything that everyone does – from Government bodies and businesses to you and me – contributes to our footprint. So it includes everything we consume, our homes, jobs and services, and the journeys that we make. The illustration below shows the components of an eco-footprint. **Carbon footprints** are a measure of our carbon emissions and are included in our eco-footprints. $CO_2$ emissions make up around 70% of the eco-footprint of developed countries like the UK.

**On your planet**

+ Consuming more resources than the environment can provide is not sustainable.

▼ The components of an eco-footprint.

**Food** – land used for crops and animals as well as that needed to absorb $CO_2$ emissions associated with production of farm machines, pesticides, fertilizers and food processing.
**Transport** – land used for transport routes, and for waste and $CO_2$ absorption.
**Housing** – land used for buildings, for waste and $CO_2$ absorption.
**Private services** – resources used by banking, leisure and entertainment, retail and businesses.
**Consumer goods** – all resources used to make everything we buy other than food; to transport it and absorb any waste and $CO_2$ emissions used in its manufacture
**Infrastructure** – resources needed to maintain and renew transport networks, buildings, power supplies, healthcare, education and water supplies.
**Public services** – resources used by governments and councils to support our lifestyles.

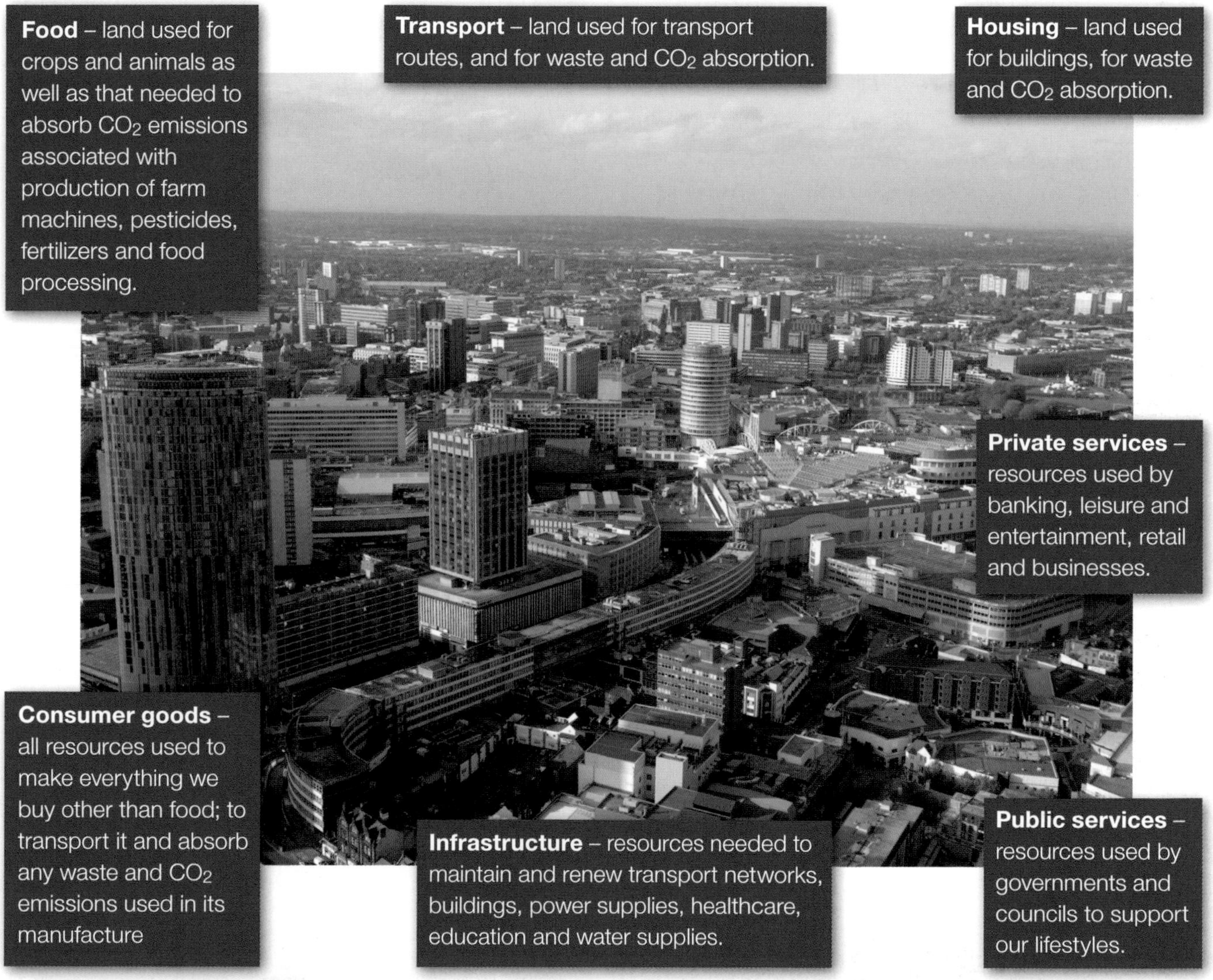

### your questions

1 a Make a list of all the ways in which urban lifestyles put pressure on the environment.
  b Divide your list into those that put pressure on areas close to the consuming city and those that have consequences further away.
2 What factors have allowed cities to develop eco-footprints that are unsustainable?
3 Why are people becoming more concerned about the size of their eco-footprints?
4 Measure your eco-footprint using the websites: http://www.ecologicalfootprint.com/ or http://footprint.wwf.org.uk/

# Different places - different footprints

**On your planet**

+ Every decision we make – where to go, how to get there, what to buy – has an environmental consequence.

+ In this section, you'll learn how eco-footprints vary between different places.

## UK's largest footprints revealed!

A WWF report in 2007 showed that Winchester was the city with the largest eco-footprint in the UK – needing 6.52 global hectares per person to satisfy its needs – equivalent to 3.62 planets! The WWF report concluded that: *'if everyone consumed natural resources and generated carbon emissions at the UK's rate we would need three planets to support us'*.

## Where does London fit in?

Surprisingly, London is not the worst city in the UK in terms of its eco-footprint per person. The WWF's report ranked London 44th out of 60. However, London's total eco-footprint is estimated to extend over an area twice the size of Britain – and equal to the size of Spain (see below). That's how much land Londoners as a whole need to maintain their current standard of living! And if everyone lived as Londoners do, it would require 3.05 Planet Earths to support them.

### Largest five

| City | Planets | Footprint (global hectares) |
|------|---------|------------------------------|
| Winchester | 3.62 | 6.52 |
| St Albans | 3.51 | 6.31 |
| Chichester | 3.49 | 6.28 |
| Brighton and Hove | 3.47 | 6.25 |
| Canterbury | 3.40 | 6.12 |

### Smallest five

| City | Planets | Footprint (global hectares) |
|------|---------|------------------------------|
| Gloucester | 2.81 | 5.06 |
| Stoke on Trent | 2.79 | 5.03 |
| Kingston upon Hull | 2.79 | 5.02 |
| Salisbury | 2.79 | 5.01 |
| Plymouth | 2.78 | 5.01 |
| Newport | 2.78 | 5.01 |

▲ The UK's largest five cities by eco-footprint per person and the

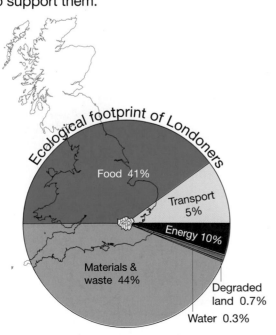

Ecological footprint of Londoners

Food 41%
Transport 5%
Energy 10%
Materials & waste 44%
Degraded land 0.7%
Water 0.3%

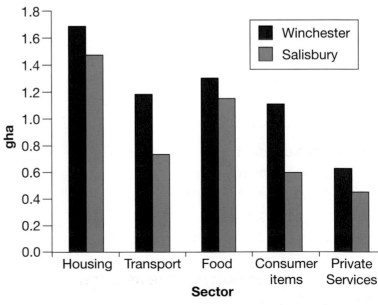

▲ Salisbury and Winchester's eco-footprints compared

## So close and yet so far apart

Salisbury and Winchester are just 80 km apart, have similar populations of around 43 000, and might be expected to have similar eco-footprints per person. Yet Salisbury has the UK's second smallest footprint, while Winchester has the largest. The graph opposite shows that Winchester's impacts are greater in all categories. But why?

- Incomes and house values in Winchester are higher than in Salisbury, so people in Winchester can afford to spend more on everything.
- Both cities have high employment rates.
- More people commute out of Winchester and further to work than in Salisbury. This increases Winchester's overall carbon footprint. Whereas a higher percentage of Salisbury's workers walk, cycle or use public transport to get to work – generating less $CO_2$.

## Why do footprints vary?

The WWF's report showed that consumption is not evenly spread. Internationally, the patterns reveal that the links between wealth and environmental impacts are not clear. Switzerland, Germany and the Netherlands all have higher GDP per capita and longer life expectancies than the UK. But all three have lower eco-footprints.

However, the eco-footprints of developing countries tend to be much smaller than for developed countries, because their lifestyles and consumption patterns are different. The graph shows that countries with a low **HDI** rank make much smaller demands on the Earth's resources. However, as countries develop, their eco-footprints expand.

|  | UK | Switzerland | Germany | Netherlands | India |
|---|---|---|---|---|---|
| **Planets needed to support consumption** | 3.1 | 2.8 | 2.5 | 2.4 | 0.4 |
| **Infant mortality rate (deaths/1000 live births – 2007 estimates)** | 5.01 | 4.28 | 4.08 | 4.88 | 34.61 |
| **Life expectancy at birth (years – 2007 estimates)** | 78.70 | 80.62 | 78.95 | 79.11 | 68.59 |
| **GDP per capita (US$ - 2006 estimates)** | $31 800 | $34 000 | $31 900 | $32 100 | $3800 |

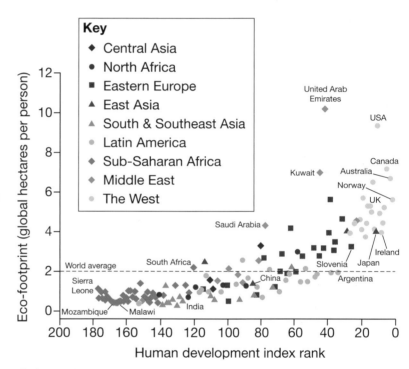

▲ Are eco-footprints and standards of living related?

### your questions

1 Why do you think cities in the UK have different eco-footprints?
2 Compare the countries in the table above.
   a Why does India have such a low footprint?
   b Why does long life expectation increase a country's footprint?
   c What does infant mortality have to do with eco-foot prints?
3 **Exam-style question** Using examples, explain how and why different cities have different eco-footprints. (4 marks)

✚ **In this section you'll learn how large cities like London rely on other places to remain prosperous.**

## Stresses and strains

London's financial core – the City of London – produces 8.8% of the UK's GDP, and Greater London is set to become the fourth largest urban economy in the world. As London's prosperity increases, more people want to live and work there. They will need more energy, food, homes, and transport, putting strains on people and the environment – both inside, and far beyond, the city.

## From local, to global, life support systems

Like most cities, London's early development depended on the ability of the immediate surrounding area to provide all of its needs. Geographers used to talk about the **carrying capacity** of places – where the natural environment tended to limit the number of people who could live there. When transport times were lengthy and costs were high, each city developed its own **sphere of influence** – the neighbouring supply area (see top diagram). However, close ties between cities and their surrounding countryside have now been broken as wealth and technology have increased. London's 7.5 million people now consume items from all over the world, using the carrying capacity of distant places to meet their own needs (see bottom diagram).

*On your planet*

✚ London needs an area 125 times bigger than its own surface area to supply all the resources it consumes.

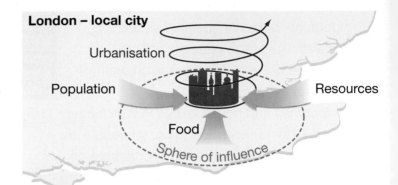

**London – local city**

Urbanisation

Population

Resources

Food

Sphere of influence

London depends on a global network of supply lines:

• 81% of London's food comes from outside the UK.

• 20% of London's water comes from outside the area. Water has to be transferred from Wales to supplement the Rivers Thames and Lea.

• All of London's energy sources are imported. The national grid transmits electricity from power stations across the UK.

The flow diagram (opposite) shows how resources flow into London – the inputs — and what flows out each year.

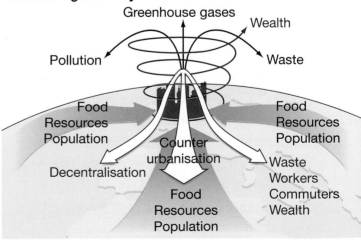

**London – global city**

Greenhouse gases

Wealth

Pollution

Waste

Food
Resources
Population

Food
Resources
Population

Counter urbanisation

Decentralisation

Waste
Workers
Commuters
Wealth

Food
Resources
Population

## London's Life Support System

| Inputs | | Outputs |
|---|---|---|

**FOOD**
6.9 million tonnes per year

**WATER**
866 billion litres per year

94 million litres of bottled water in 2260 tonnes of plastic

**ENERGY**
13.2 million tonnes oil equivalent made up of:
21% electricity
23% liquids
55% gas
<1% renewable

**CONSTRUCTION MATERIALS**
20 million tonnes

**WORKERS**
3 million daily commuters

49 million tonnes of materials consumed

**WASTE**
40.9 million tonnes of $CO_2$

27 million tonnes of food, construction and demolition materials, manufactured goods, chemicals, etc.

Inorganic waste into landfill

Organic waste into rivers

28% of all water is lost through leakage

18% of all energy is wasted

**MANUFACTURED GOODS**
14 million tonnes of manufactured goods, food, building materials, etc.

64 billion passenger km travelled each year – 69% by car

## The dirty secret of where your recycling really goes

Every piece of paper placed in Camden's recycling bins is sent abroad. 10% of newspapers and pamphlets are sent to Malaysia and 90% to Indonesia. A fifth of mixed papers are sent to China, with the remainder ending up in India. Plastic is sent abroad too. Only steel, aluminium and glass are recycled in UK processing plants.

Many recycling experts say that it's no bad thing to send our recycling overseas. The ships that bring China's vast number of imports to the UK would return home empty were it not for the fact that they are stuffed with used British plastics, which in turn makes the next generation of plastic goods.

*Adapted from Camden News Journal*

## The burden of waste

Producing increasing amounts of waste seems to go hand in hand with prosperity. London's waste has traditionally been sent to **landfill** sites. Burying waste was cheap, but difficult in a densely built-up city. Since the 1960s, waste has been taken out to the surrounding Home Counties by road and rail. 70% of London's waste goes outside the city, but this is going to change. The EU has set a limit on the amount of biodegradable waste that can go to landfill, and a £32 per tonne tax imposed by the Government aims to reduce the amount by 66%.

Councils are now encouraging recycling – but that raises other issues as the article on the right shows.

### your questions

1 Briefly explain how cities exceed their natural carrying capacities.

2 **a** Write 100 words describing London's 'life support system' as seen by an environmentalist.

 **b** Now write a further 100 words about the system from the point of view of an economist.

 **c** Complete your writing with 100 words from the point of view of a person living in London.

 **d** Are the views of these three people ever likely to agree? Explain your answer.

3 How sustainable is London as a city? In pairs, discuss how **a** it is **b** it isn't.

✚ In this section you'll find out that people outside London are not always happy to support the capital.

## Mucking out!

One of London's biggest problems is waste – all 27 million tonnes of it! 35 miles east of London is the community of Mucking in Essex. Mucking takes 20% of London's household waste and is home to Europe's largest landfill site. Cory Environmental manages this site and uses barges to transport the waste down the Thames. 680 000 tonnes of rubbish are transported by river each year – saving 400 lorry trips a day. Layers of rubbish up to 30 metres deep cover an area of 200 hectares. A network of underground pipes permit the slow release of methane gas as the waste rots. The old sand and gravel quarry at Mucking was the perfect site for rubbish dumping, but it is almost full now. Time is running out and alternatives are needed.

## Upsetting the locals

Government directives and increasing costs, meant that Mucking landfill was due to close in 2007. Plans to restore the landscape and open a nature park were eagerly anticipated. However, the alternative to Mucking landfill meant the development of an incinerator at Belvedere on the opposite side of the River Thames, near Bexley in Kent. Designed as an 'energy from waste' power station, Belvedere would have provided electricity for 66 000 homes and would have consumed almost all of the annual 680 000 tonnes of waste. But the plan was controversial. In January 2008, the £100 million scheme was postponed and Mucking's life was extended until the end of 2010. People living nearby were not pleased.

| County | % of London's waste |
| --- | --- |
| Essex | 33% |
| Kent | 22% |
| Hertfordshire | 13% |
| Bedfordshire | 8% |
| Buckinghamshire | 8% |
| Berkshire | 8% |
| Oxfordshire | 8% |

◄ Mucking, in Essex, takes 20% of London's waste. Other Home Counties also help out with the capital's waste.

Some arguments for and against landfill and incineration are given below.

### Landfill

| For: | • Makes good use of old quarries<br>• Easily managed and safe<br>• Produces methane that can be used as fuel<br>• Can be safely sealed and landscaped for parkland or building sites |
|------|------|
| **Against:** | • Attracts seagulls and vermin<br>• Loose materials blow around<br>• Gives off smells of methane<br>• Waste materials contaminate groundwater and surrounding soils<br>• Can suffer from subsidence<br>• Generates heavy traffic |

### Incinerator

| For: | • Long life span<br>• Cost effective<br>• Safe disposal of toxic substances<br>• Production of energy from burning waste<br>• The ability to reclaim metals such as aluminium<br>• The residue is useful for road building |
|------|------|
| **Against:** | • Gives off toxic gases (e.g. carcinogenic)<br>• Fine particulates escape into food chains<br>• $CO_2$ (a greenhouse gas) is emitted.<br>• 25% of the original waste remains after burning<br>• If residues are used for path and road surfaces, they seep into the ground as they decay |

## London's taking over ...

London's taking over – and it's not just the waste that's growing. As London grows, more of the surrounding countryside is lost (as the diagram shows) and London's eco-footprint expands. The Thames Gateway project (east of London) is another example of this. This project will regenerate some of the most deprived areas in east London. It will redevelop **brownfield sites**, drain marshlands for new affordable housing, and provide employment opportunities and better public transport links. But in doing so, more land will be used to satisfy London's needs.

The Thames Gateway extends 40 miles along the Thames and will need major new power-generating facilities, including the London Array wind farm. This will be located in the Thames estuary. It will have 270 wind turbines, and be capable of supplying electricity to 25% of London's homes. Two new coal-fired power stations are also planned for North Kent, to replace outdated power stations and support the planned growth.

▼ The urban spread effect

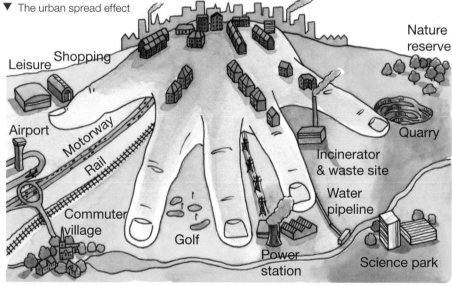

+ **Brownfield site** – an area of land that has been built on before and is suitable for redevelopment.

### your questions

1 Look at the arguments for and against landfill and incineration schemes. Write two letters to a national newspaper arguing for or against the schemes. One should be from someone living in central London, and one from someone living close to Mucking or Bexley.

2 Do you think that the views of people living outside cities are more or less important than the views of those living in cities? Explain your answer.

3 **Exam-style question** Using examples, explain how the footprints of cities often extend way beyond the city boudaries. (6 marks)

**On your planet**

+ Londoners produce enough rubbish to fill Canary Wharf tower every 10 days.

In this section, you'll learn how cities can reduce their eco-footprints by reducing energy consumption and waste generation.

## London – energy and waste

London consumes over 13 million tonnes of fuel each year, with just 1% coming from renewable sources. Nationwide, around 30% of the UK's energy is used in our homes and we waste 60% of it as a result of things like poor insulation.

We're no better off when it comes to rubbish – Londoners produce 3.4 million tonnes of rubbish each year and the problem gets worse the more packaged and disposable goods we buy. At least 80% of our rubbish could be re-used, recycled or composted. The Government has set targets to recycle or compost 30% of our waste by 2010, and 33% by 2015.

## Eco-communities

Plans to create new eco-towns often involve locating them on **greenfield sites** out of town. Better options to create sustainable communities involve building in existing urban areas. The Beddington Zero Energy Development (BedZED) near Croydon, Greater London, is the largest carbon-neutral eco-community in the UK. It is built on reclaimed land and focuses on social and environmental sustainability, while promoting energy conservation. It could be the way forward.

BedZED – a new community on a brownfield site that includes:

- 82 homes (34 for sale, 23 shared ownership, 10 for key workers, 15 affordable rent)
- Commercial buildings
- Children's nursery

▲ BedZED – carbon neutral community of the future – available now.

### What about waste?

The aim is to reduce the amount of waste generated in the first place. BedZED effectively reduces the amount of energy wasted in buildings and transport, but it's difficult to cut down on material waste – packaging comes around most things we buy (see right).

**On your planet**

+ 90% of all products purchased in 2005 became waste within 6 months.

## BedZED – key characteristics

Energy consumption is reduced by:

- making buildings from materials that store heat in warm weather and release it at cooler times
- making buildings from natural, recycled or reclaimed materials
- building the houses to face south to maximize 'passive solar gain'
- backing offices onto homes and facing north with low solar gain and, therefore, reduced need for air-conditioning
- using 300 mm insulation jackets on all buildings
- producing at least as much renewable energy as that consumed

- using heat from cooking and everyday activities for space heating
- using low-energy lighting and appliances throughout
- using energy tracking meters in kitchens
- providing a combined heat and energy power plant, using urban tree waste/off-cuts
- providing homes with roof gardens, rain water harvesting and waste water recycling
- providing a green transport plan – the community layout promotes walking, cycling and public transport with bus, rail and tram links
- the ZEDcars car sharing club and local free electric charging points.

# Future choices

How can we cut down on waste? Turning it into energy is an option – as in the SELCHP (South East London Combined Heat and Power plant), where waste is incinerated and converted into electricity (see the photo).

London Councils are trying two approaches to waste:

- **Direct variable charging** – where recyclables are collected free of charge and 'pay-as-you-throw' policies are applied to everything else. This will involve a shift in consumer attitudes and could encourage 'fly-tipping'.
- **Producer/Polluter pays principle** – where the company that produces the waste items is charged. The costs of products would reflect their life cycle. Items like disposable nappies would be high-cost items.

## your questions

1 In a table, compare the differences between your home area and the BedZED project in terms of energy use and waste produced.
2 Why would BedZED have a much lower eco-footprint?
3 How easy would it be for other urban areas to adapt the ideas used in BedZED?
4 Do an Internet search for 'BedZED' and produce a poster of images of BedZED. Explain how it is different from other urban housing.
5 Design a leaflet to explain to the public why waste reduction is more important than recycling.

In this section you'll learn how cities can reduce their eco-footprints through improved transport strategies.

## Urban transport – avoiding gridlock

We've all grown up in a world of increasing car ownership. Increased personal mobility provides us with the independence to go where we want when we want. As cities become bigger, people tend to live further away from their place of work and have to commute there. Often bus and train routes, or times, are not convenient, so people choose to use their cars to get to work instead.

The last three decades have also seen out-of-town retail, leisure and business parks develop in the rural-urban fringe. As a result, traffic flows into and out of cities, and across them, all the time. Multiple car ownership – families with more than one vehicle – has also added to the volume on the roads, so pollution, congestion and accidents have all increased too. Congestion is a problem – people and goods can't move freely in our cities, so businesses and services tend to move out. This could end up leaving city centres empty.

Traffic congestion leads to costly delays and air pollution.

## London's response

London is a congested and highly polluted city. Road vehicles are responsible for 66% of air pollution, which makes asthma and lung diseases worse.

In 2003, a controversial scheme – the **Congestion Charge** – was put into action in Central London. This meant that all vehicles entering a clearly marked zone had to pay a fee. The zone was extended westwards in 2007. Since 2003, there has been a:

- 21% fall in traffic in the zone
- 45% rise in bus passengers
- 43% increase in the number of cyclists.

In addition, emission-free cars, like the Smart car and G-Wiz, have become a common sight.

In 2008, the Greater London Low Emission Zone (LEZ) was implemented. The most polluting diesel lorries either had to meet emissions standards or pay a daily charge to enter the area. The scheme is being phased in between 2008 and 2012, and will apply to more vehicles over time. The LEZ will help to clean up London's air and meet EU standards by 2010.

## London's Mayor accepts award for transport policy that has inspired the world

'This award recognises London's leading role in taking decisive action to reduce traffic congestion and improve the environment. We will be making the whole of Greater London a Clean Air - Low Emission Zone from 4th February, effectively banning the most polluting lorries from our roads.'

London Mayor Ken Livingstone, January 2008

### your questions

**1 a** Keep a travel diary for a week and include all the journeys each member of your family makes. Note how far they went and what method of transport they used.

**b** Now, draw a series of graphs to show: **i** how frequently each person used a car, public transport, bicycle or walked, and **ii** how far they travelled.

**c** Decide which journeys could have been taken by a more sustainable method. How sustainable are you and your family?

**2** Draw a table of the benefits and problems caused by the Congestion Charge.

**3** Originally both the Congestion Charge and the LEZ schemes were considered controversial. Draw a spider diagram to show: a who you think was against each scheme, and b who was in favour.

**4 Exam-style question** For a city you have studied, explain the benefits and problems caused by transport schemes to reduce its eco-footprint. (6 marks)

## London's sustainable transport policies

How do the schemes work?
- Congestion Charge of £8 pre-paid or £10 on the day. Discount season tickets available.
- Low Emission Zone charges of £100-200 per day for vehicles over 3.5 tonnes in weight.

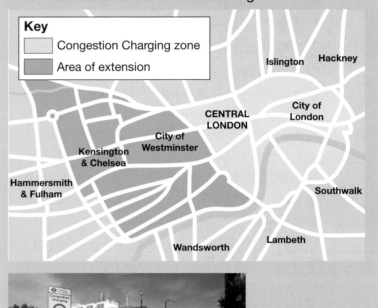

**On your planet**
+ Food miles – the distance food travels from source to plate – increased in the UK by 15% between 1992 and 2002.

+ In this section you'll look at the challenges facing green consumerism in urban areas.

## Bigger shops – less choice

For 30 years, supermarkets have grown and now dominate our shopping patterns. They make shopping in cities convenient and easy – especially if we use a car. In the UK, the five largest retail chains (Tesco, Sainsbury, Asda, Morrison's and Waitrose) account for 80% of the market. Their decisions about what food to sell and where to get it from have a massive effect on producers, retailers and the environment. As supermarkets grow, other shops tend to close – leaving many people with less choice.

▲ Supermarkets end up reducing our choices.

## Year-round food

Competition between supermarkets means that prices are generally held down. To increase profits, supermarkets search for cheaper sources of food, which can often mean importing it from overseas.

- Up to 95% of the fruit and 50% of the vegetables sold in the UK are imported.
- The amount of food flown into the UK doubled in the 1990s and increases every year. Air freight has a bigger environmental impact than road or sea freight.

Items that used to be grown locally, but were only available when in season, are now available all year round. Not only that, but an increase in offers like 'Buy one get one free' means that often we get more than we need and end up wasting food.

**On your planet**
+ Each year, the average UK adult travels about 135 miles by car to shop for food, which usually involves trips to large supermarkets.

## Reasons to choose alternatives

What can we, as consumers, do? We need to know more about what we eat and how it is produced. For example, it can be more environmentally friendly to eat tomatoes imported from Spain than those grown in the UK. Why? The energy needed to heat greenhouses to grow tomatoes in the UK is much more than the energy needed to transport them from Spain, where heating isn't needed because the climate is warmer.

# An ethical approach

Some major retailers, such as the Co-op and Marks and Spencer, are now marketing themselves as more environmentally friendly and ethical. Some supermarkets offer local produce alongside Fair Trade and organic items.

In 2007, Marks and Spencer opened its first 'eco-factory' in Thulhiriya, central Sri Lanka – manufacturing women's underwear on a site designed to be carbon neutral. But the underwear will still have to travel from Sri Lanka to be sold in M&S stores. Will this undermine the company's eco-credentials?

Having an ethical approach to shopping means that we have to think about more than just price. Farmers' markets are now held in many towns in the UK and offer an alternative to the weekly supermarket trip.

## Boost for local produce

Sainsbury's is launching a scheme aimed at getting more small companies in the South West to supply its stores there.

▲ Supermarkets are developing an ethical conscience.

### Advantages

- Good-quality produce; often organic and always seasonal
- Low food miles
- Low packaging
- Supports local economy
- Meet the producer and find out how things are produced
- Certified (by FARMA) as fresh and grown within 30 miles of the market.

► Farmers' markets – a better choice?

### Disadvantages

- Often expensive
- 500 exist in the UK, but they're not close to everyone
- Often only weekly or even monthly
- Unable to provide all we need
- For London, 'Local' is not what it seems, and can mean up to 100 miles away.

## your questions

1 Using information on these pages, explain why people in the UK have so many food miles.

2 a Working in pairs discuss and list ideas about how people could change their shopping behaviour.

  b Now think about other ways that shoppers can reduce food miles and pollution. Go to www.bbc.co.uk and key 'food miles' into the search facility.

3 What is 'ethical shopping'? Give examples and say why some people shop 'ethically'.

4 a Look up farmers' markets in your area and write a paragraph explaining how they help people to reduce their eco-footprints.

  b Write a second paragraph explaining why many people never visit a farmers' market.

✚ In this section you'll look at ways of making urban transport more sustainable

## Doing the right thing

Trying to do the 'right thing' might not be as easy as you'd like. Our view of public transport is often of crowded, late, slow or expensive trains, and the bus doesn't go quite where we want it to. So the car wins, because it is easier and more convenient. But living more sustainable urban lifestyles needs a change in attitude – and that only happens when we see the benefits of change, or when change is made easy for us. Two towns, just a few kilometres apart in East Anglia, have adopted different transport management strategies to influence people's behaviour.

## Colchester v Ipswich

Both Colchester and Ipswich have suffered from serious traffic congestion as workers commute in from the surrounding villages.

- Around 66 000 workers travel into each town every day to work.
- 56% travel by car and 67% travel less than 5 km.

### Ipswich's third Park and Ride site opens

Suffolk County and Ipswich Borough Councils celebrated as their new Park and Ride site won a prestigious design award. The £5 million project places a 550-space car park next to the town's by-pass and eases congestion.

The other two Park and Ride sites have reduced vehicle journeys into Ipswich by 720 000, and this eastern site will provide an opportunity for more sustainable transport in the borough.

The sites have won recognition as 'secure car parks with 24 hr CCTV, fencing and lighting'.

*May 2005 BBC Suffolk*

### Colchester Park and Ride Scheme Dropped

Essex County Council regret to announce that a Park and Ride scheme for the town will not go ahead after Colchester Borough Council failed to support the plans.

County Councillors stated that 'P'n'R would have brought travel and environmental benefits and relieved congestion in the town, but the Borough Councillors do not wish the plans to go ahead.'

Residents in the villages next to the planned sites protested at the potential loss of open spaces and increased noise and traffic pollution close to their homes.

*June 2008 Colchester Gazette*

It makes sense for people to park on the edge of town and travel in by bus to ease congestion in town centres. Colchester's policy of building multi-storey car parks around the town's inner ring roads originally kept traffic away from the town centre. But increased housing and population growth have encouraged more workers to commute into town. A lack of bus lanes (some were allocated but shopkeepers complained about loss of business so they were abandoned) and selling the bus services to private companies has not helped congestion. Plans for 1000-place park and ride sites close to the A12 trunk road were thrown out in 2008.

Ipswich has done things differently. Since the mid 1990s, Ipswich Borough Council has maintained and extended bus services, planned dedicated bus lanes and developed three Park and Ride sites around the outskirts of town. Cheap fares – just £3.20 return journey for the occupants of a car – are much cheaper than the daily return fares of £2.80 each person in Colchester, using the public bus service. The Ipswich bus gyratory system, 'super-routes' with smartcard fare paying systems, and 'real-time' display boards all encourage the use of public transport. Ipswich has a clear transport strategy for the twenty-first century. Is it any wonder that Colchester suffers more from heavy traffic than Ipswich?

# What else can be done?

There are a variety of ways in which transport can be managed, and you are probably aware of the types of schemes in the checklist below.

| Transport options | |
|---|:---:|
| • One way systems – that manage traffic flow | ✔ |
| • Restricted parking – that prevents streets from becoming clogged | ✔ |
| • Traffic calming measures and speed restrictions – that aim to keep traffic moving at a steady pace and reduce accidents | |
| • Red Routes – that restrict roadside stopping | |
| • Park and Ride and strategic multi-storey car parks – to divert traffic to manageable locations and free town and city centres from traffic | |
| • Ring roads – that keep traffic out of the urban centres | |
| • Cycle lanes and footpaths | |
| • Cheap public transport and dedicated bus lanes | |
| • Clean transport – using alternative fuels | |

Each of the schemes in the checklist can be successfully applied on its own. But the aim is to link them together into an **integrated transport policy**, providing a 'door to door' service that rivals the use of the car. However, truly sustainable transport systems involve:

- reducing the number of heavy goods vehicles
- encouraging low emission vehicles (electric cars and LPG buses)
- reducing the need for long journeys.

Land-use planning that puts homes close to places of work, shops and services, or around major public transport hubs, is needed. Many new developments – like the BedZED project (see pages 212-213) have these ideas built in.

## your questions

1 Draw a table to compare the approaches taken by Colchester, Ipswich and your own town to the problem of traffic congestion. Use these headings: The problem; The solution; Its benefits; Its problems.

2 Why do you think different places adopt different approaches?

3 In groups of 2-4 discuss who should make the decisions about developing sustainable transport strategies. Think about the following possible decision makers – national government, local councils, large companies, ordinary people, and the media. Feed back your ideas.

4 Consider everything on the last 4 pages. Design a media campaign (posters/TV and radio commercials/ leaflets) to encourage people to change their shopping and transport habits, under the title 'It's our world!'

5 **Exam-style question** Using examples, explain how attempts are being made to make cities more sustainable. (6 marks)

+ In this section you will learn how rural areas in Malawi face challenges such as isolation, depopulation and economic decline.

## What's Malawi like?

Malawi is one of the world's most rural countries. 85% of its population live in the countryside.

- Its annual GDP per capita (**Gross Domestic Product**, which measures income per person) is only US$800. This is very low, even though the cost of living there is also low.
- 50% of Malawi's population lives below the poverty line.
- It comes 164th out of 177 countries on the UN's Human Development Index, which measures education and health.
- 33% of Malawi's people are under-fed.
- Between 1998 and 2008, when the global economy grew fastest, Malawi's economy hardly grew at all.
-  Malawi owes US$1.8 billion in overseas debt. 25% of its export earnings are soaked up by debt repayments.

In spite of this, Malawi is making progress. As the table shows, it's better off in many ways than it was in 1998.

▼ Collecting water in rural Malawi.

Most families in rural Malawi are trapped in poverty:

- Rural areas have few healthcare facilities, so rural families have worse health than urban families.
- Although primary schooling is free, secondary schooling must be paid for. One child's fee can cost a family most of their year's income.

| Facts | Data for 2008 | Trend since 1998 |
| --- | --- | --- |
| Population | 13.9 million | up 40% |
| Birth rate per 1000 | 42 | stable |
| Death rate per 1000 | 18 | down from 23 |
| Infant mortality per 1000 live births | 90 | down from 132 |
| Fertility rate (Nº births per woman) | 5.7 | stable |
| % population under 15 | 46 | stable |
| % women using contraception | 42 | doubled |
| **Human development indicators** | | |
| Life expectancy in years | 43.5 | up from 36.3 |
| % population with access to safe drinking water | 73 | doubled, but there's a big divide between urban areas (98%) and rural (68%) |
| % of children under 5 malnourished | 22 | down from 48% |
| Maternal mortality per 100 000 births | 807 – but the risk of maternal death over an average mother's 5-6 births is 1 in 18 | slight decrease from 1221 |
| Number of doctors per 100 000 population | 2 | stable |
| % adult literacy rate<br>Male<br>Female | <br>76<br>50 | <br>up from 72%<br>up from 42% |
| **Economy** | | |
| GDP per person | US$800 | up from US$773 |

## Rural isolation in southern Malawi

Rural telecommunications are poor. Service is slow, with congested telephone lines. In 2007, Malawi had 100 000 telephone landlines – 1 for every 139 people. There were 60 000 Internet users – 1 for every 233 people. In each case, most were in the cities. Mobile phone ownership is greater (1 million phones, or 1 in 14 people). But, again, rural coverage is poor.

Malawi's rural economy has hardly grown recently. It has poor **infrastructure**, for example, dirt roads. During the wet season, it can take several hours to travel 20 km. The dirt roads become boggy, which makes rural areas inaccessible for days or weeks at a time.

> ✛ **Infrastructure** refers to the basic services needed for development – water, roads, power supplies.

## Rural poverty

In almost every way, rural areas in Malawi are worse off than urban. The effort needed to stay alive and fed is greater. Each day, rural farmers typically spend:

• 43 minutes collecting firewood
• 48 minutes walking to and from farm plots
• 128 minutes walking to market.

That's nearly 4 hours and doesn't include fetching water! All of these tasks would be shorter in urban areas. The only way to cope is to involve all the family, so many children miss out on education. They have to care for young siblings, the elderly, or weed the crops.

*On your planet*
✛ Every day, some children in Malawi walk 10 miles, each way, to school.

▲ Malawi is about two-thirds the size of England, but with only 30% of its population. Its two largest cities (Blantyre and the capital, Lilongwe) have only 1.25 million people between them.

### your questions

1 In pairs, use the table of data about Malawi to select five ways in which Malawi has improved, 1998-2008. Then find five ways in which it still has a long way to go. Justify each choice.
2 Explain why: **a** roads and **b** telecommunications are important for economic growth.
3 Make a large copy of this table to explain why rural areas are poorer than urban.

| Statement | Reasons for this | Why this makes rural people poorer |
|---|---|---|
| a Health facilities are poor | | |
| b Few children go to secondary school | | |
| c Few rural people have phones | | |
| d It takes a long time to reach towns | | |

# Malawi - who's in charge?

+ **In this section you will learn that what happens in rural Malawi is decided elsewhere.**

## Farming is a roller-coaster

For years, Malawi's rural economy has declined, and farmers have had a rough time as food prices have risen and fallen. The cause lies not in Malawi, but in decisions made elsewhere. How well you survive depends on who you are, because wealth and land aren't distributed evenly. There are three very different types of farmer in Malawi, as outlined here.

### Large estates

In the 19th century, the British colonised Malawi. They took over the best land and developed plantations producing most of Malawi's exports – tea, coffee, tobacco. Many estates remain in British ownership – some by large TNCs (for example, Unilever produces PG Tips tea). Large estates can afford irrigation, fertiliser, storage and transport to get to global markets. They hire local landless labourers, or small farmers seeking extra income. The workers only get paid about 1p per kg of tea leaves or coffee cherries picked, although the estate owners argue that they also get housing, water, firewood and daily lunch.

### Tobacco tenants

Tobacco earns 10% of Malawi's GDP, and 2 million Malawians depend on it for their income. Almost all 900 000 adult growers are labourers working for companies such as British American Tobacco (Malawi), as tobacco tenants. Many workers are children (see right). Malawi has 1.4 million child labourers, the highest rates of child labour in southern Africa. The Catholic organisation, Centre for Social Concern, estimates that 78 000 children are working full- or part-time in tobacco fields. 45% are 10-14 years of age, and 55% are aged 7-9.

▲ A tea estate in Malawi.

### The other face of tobacco

Kirana was 8 when he first worked in the fields. Estate owners took him and his parents 1000 km from their village, with 45 other families, to tobacco fields in northern Malawi. His mother says they left home for a better life. 'Four years later, my whole family is still poor. My son has to work as hard as us if we are to afford necessities. The money that we get from the estate isn't enough.'

Now 12, Kirana has never been to school. His health is failing and he can't work as hard. He's malnourished and gets sick easily. The family often goes without proper meals for 3 days. In the past 2 months, Kirana has had malaria, diarrhoea and pneumonia, and his parents are scared of losing him.

Tobacco tenant farmers are allocated a plot by estate owners to produce a specific amount. The owners lend seed and fertiliser, and deduct the costs from future profits. They are supposed to supply food, but supplies sometimes run out. Many tenants lack medication, proper housing and safe drinking water. These people are among the poorest in Malawi. The Tobacco Control Commission (TCC), a Malawi Government watchdog, estimates that it costs US$1 for workers to grow 1 kg of tobacco, which they then sell for 70c – losing 30c per kilo. The tenants can't make a living, so they send their children to work.

▲ Child workers drying tobacco leaves in the sun in Malawi.

*What do you think?*

+ Can child labour ever be justified?

## Smallholders

Smallholders make up the majority of farmers in most areas. 1.8 million families occupy 1.8 million hectares of land – a quarter of Malawi – and produce 80% of its food. Half have less than half a hectare – enough only for **subsistence** farming. They grow maize, rice and groundnuts to feed themselves. There's little income to invest in fertiliser to improve the soil. The collapse of Malawi's currency has made fertiliser and oil expensive. And transport costs to markets have increased. Until food prices rose in 2007-8, there was little point in travelling to sell crops for an uncertain price.

> + **Subsistence** farming is where farmers grow enough for themselves and their families, with no surplus to sell.

## Rural-urban migration

Faced with such low incomes, many young men move to cities like Lilongwe. Their loss makes rural problems worse because there are fewer people to work. Much land is under-used and there's no money to invest in new methods, or in seed to give higher yields. The only benefit is the income that the men send home from their urban jobs.

### your questions

1 In pairs, think about the three types of farmer. Copy the three large circles as shown.

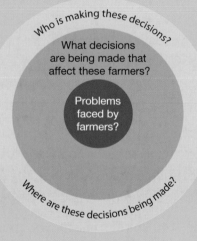

Who is making these decisions?

What decisions are being made that affect these farmers?

Problems faced by farmers?

Where are these decisions being made?

- In the inner circle – write down the **problems** that each farmer faces.
- In the middle circle – **what decisions** are being made that affect these people?
- In the outer circle – **who** is making these decisions, and **where** are they being made?

2 In what ways do these farmers' lives link to the lives of people in the UK?

3 Research child labour in Malawi by logging on to corpwatch.org.

4 Using your research, write a letter to tell shareholders in a large company like British American Tobacco (BAT) how you feel about the use of child labour by their company.

➕ In this section you will learn how new initiatives are bringing growth to rural Malawi.

### The cycle of poverty and AIDS

Since HIV reached Malawi in the early 1980s, 20% of adult Malawians have become infected. Each year tens of thousands die from AIDS-related conditions. The result is increasing poverty among Malawians. HIV/AIDs causes these impacts:

- Those affected become weaker and can't work, so income falls.
- The costs of drugs – needed daily – are high; many can't afford them.
- The death of a wage-earner, plus the cost of a funeral, puts many in poverty.
- Orphans are often left with grandparents, who are less able to work.

The following projects are geared towards families suffering poverty, HIV/AIDS, or both.

▲ Malawi now has over 1.5 million orphans, whose parents have died from AIDS.

### The potential of mushrooms

In rural Malawi, global anti-smoking campaigns have reduced the demand for tobacco exports. Now, the Malawian Government's Sustainable Livelihoods Programme promotes mushrooms as an alternative source of income.

In Ndawambe, 140 km from Lilongwe, mushrooms have already made a difference for the poorest villagers. The project works as follows:

- Ndawambe residents are trained to grow mushrooms.
- They receive micro-loans for the first mushroom compost, which are rapidly repaid from their income.
- The growers use waste from other crops (for example maize stalks) as raw material for the compost.

Villagers in Ndawambe also find other ways of earning income. One family raises chickens; their 900 hens produce 23 trays of eggs a day. They are now building a two-storey home – the village's first ever. Others have started businesses in fruit juice, vegetable oil, honey production, bakeries and fish farming.

*On your planet*
➕ Mushrooms are a delicacy, with high protein value and medicinal properties. Farmers are only just meeting demand from Lilongwe's supermarkets.

# Fish farming

The international charity, World Vision, supports a project to develop fish farming among rural Malawian families affected by HIV/AIDS in 15 locations. The project assists farmers to dig small, rain-fed ponds measuring 20m by 10m. The ponds are designed for species such as tilapia (a common freshwater fish), which feed on aquatic plants and waste. Farm and kitchen waste can be used as food, and tilapia are easy to keep. Children and the elderly can help, making ponds easy to manage.

The benefits are:

- It's cheap. Investment is minimal, because farmers feed the fish using farm and crop waste.
- An income; ponds produce up to 1.5 tonnes of fish per hectare each year!
- Fish is a good source of protein, calcium, vitamin A and other nutrients, critical to those with HIV/AIDS. There has been a 150% increase in fish consumption. World Health Organisation research shows that good nourishment prolongs the survival of HIV/AIDS patients by up to 8 years.
- A reduction in child malnutrition from 45% to 15%.
- Water for crops during droughts. Farms with ponds are 20% more productive than those without.
- Fertiliser produced from pond bottom sediment. Some farmers grow bananas and guava on the pond edges, using water that seeps into the soil.

The project has:

- doubled the income of 1200 households
- increased fish and vegetable consumption among rural communities
- quadrupled the fish farmers in Malawi, 1999-2004
- helped women, who form 30% of those taking part.

▲ Tilapia grow rapidly and taste good!

**On your planet**

+ What feeds on living or dead plant/animal material? Answer – hens, mushrooms and tilapia!

## your questions

1 Explain how HIV/AIDS in the family can have serious effects on the lives of: **a** children, **b** the elderly, **c** the whole family.

2 Make a large copy of this table, and complete it using the details in this section.

| | Mushroom farming | Fish farming |
|---|---|---|
| Initial costs to the farmer | | |
| Economic benefits | | |
| Social benefits | | |
| Environmental benefits | | |
| Spin-off effects (those arising from the projects) | | |

3 Which of the projects do you think is more successful for:
  **a** solving rural poverty
  **b** families affected by HIV/AIDS?
  Explain your reasons.

4 **Exam-style question** Using examples, explain the reasons for rural poverty in developing countries. (6 marks)

➕ **In this section you will think about whether Malawi's future food supplies can be grown sustainably.**

## Growing food sustainably

Malawi's government has big plans. Never again does it wish to import food to feed starving people, as it did in the drought of 2001-2005. Since then, it has spent US$50 million on an aid programme to provide fertiliser for small farmers. By 2008, farmers were reaping bigger harvests than ever before. But is this sustainable? Two new 'big ideas', shown here, could impact on Malawi's future ability to grow food.

## Big Idea 1 – Irrigation

Only 2% of Malawi's land is irrigated. In 2008, the President announced plans for an **irrigation** project to create a 'green belt' of land, using water from Lake Malawi. Lake Malawi is a huge freshwater lake covering one-third of Malawi. His idea is to bring about a 'Green Revolution'. He aims to increase food production to make Malawi a food exporter to other countries.

But the Africa Technology Forum believes that this plan is unsustainable:

- It could reduce the lake's size. Although Lake Malawi is the size of Wales, it's only half the size of the Aral Sea 50 years ago.
- Rapid population growth has increased cultivation near the lake, causing pollution from fertilisers and pesticides. Commercial farming could make this worse.
- Fish stocks are already low because of overfishing. Extracting water could disturb the ecosystem, reducing fish numbers and birds.
- Malawi competes with Mozambique for water from Lake Malawi – not all the water belongs to Malawi (see the map on page 221).

▲ Small-scale irrigation in Malawi. One plan is to increase the amount of water taken from Lake Malawi for irrigation.

> ➕ **Irrigation** is taking water from areas that have it to those that don't.

▼ Lake Malawi – a source of irrigation water for the future?

## Big Idea 2 – GM crops

In 2008, Malawi became the second African country to allow genetically modified (GM) seed into the country. Of all food debates, GM is probably the greatest.

## What the pro-GM argument says

- GM foods are safe and environmentally friendly. Some GM seeds are disease-resistant, reducing the need for pesticides.
- GM crops can solve global hunger – examples are disease-free wheat, and drought-resistant maize.
- GM crops may allow Malawi to grow more food, especially in dry areas, creating jobs and economic growth.
- GM rice and maize are genetically modified to fight disease and malnutrition. GM rice contains Vitamin A/beta carotene. This helps to prevent blindness in poor nations where rice is an important food.

*On your planet*
+ GM scientists have combined an element of anti-freeze into the genes of strawberries. This creates a strawberry that can resist frost.

## What the anti-GM argument says

- GM seed is supplied by American TNCs, so farmers have to pay for new seed instead of using last year's harvest.
- GM pollen could fertilise other plants, producing disease-resistant 'super-weeds'.
- Too little is known about how genes work. GM creates new proteins but will these harm us when eaten? Many GM foods contain genes resistant to antibiotics. Could these be passed on to humans and animals?
- People sometimes won't buy GM food. EU consumers have rejected GM food, although American consumers can't understand the fuss.

**GM – what's that, then?**
**Genetic modification** occurs when scientists take genes from one organism and put them into another. It allows genes to be crossed between organisms that could never breed naturally. A gene from a fish, for example, has been put into a tomato. Now there are no barriers to prevent scientists taking genes from unrelated organisms. The large US companies that created GM seeds have spent huge amounts of money on their research, and want to see returns on their investment.

▲ A GM seeds trial in Malawi. Some see this as the way to solve drought problems, especially with future climate changes.

### your questions

1 In a table, show the advantages and disadvantages to Malawi of using irrigation water.
2 Should the irrigation project go ahead? Explain.
3 Draw a spider diagram to summarise the economic, social and environmental arguments for and against GM foods.
4 In pairs, select the strongest arguments for and against GM crops being introduced into Malawi.
5 **Exam-style question** Using examples, explain how developing countries could develop more sustainably. (6 marks)

# Pressure on the Lake District

**On your planet**

+ The Cairngorms is the largest National Park at 3900 km². The Norfolk Broads is the smallest with 303 km².

✚ **In this section you will learn why the Lake District is under pressure from people outside the area.**

The Lake District, in Cumbria, is one of the UK's most popular National Parks. The third largest of the 15 National Parks of England and Wales, it is home to nearly 45 000 people. The mountain scenery is the highest in England, and its lakes have inspired poets such as Wordsworth. Beatrix Potter's stories have also reached out to generations of children. The spectacular landscape leads to over 15 million visitor days a year (1 visitor spending 1 day = 1 visitor day).

However, the Lake District has a problem. It is popular and accessible to people from all parts of Great Britain and abroad. It is one of the busiest National Parks – and this puts pressure on the landscape.

## What are National Parks?

National Parks are special because of their scenery (e.g. the Pembrokeshire coast), landforms (e.g. the Yorkshire Dales) and ecology (e.g. the New Forest of Hampshire). These varied landscapes lead to many tourist activities. The Cairngorms, Lake District and Snowdonia are hill walking and climbing country. While the South Downs, New Forest and Norfolk Broads offer walks or boating holidays.

Each National Park has its own authority, which makes policies and protects the park from unsuitable development. They also manage different demands on the landscape. But they don't own most of the land. Private landowners own almost all the land, and may not agree with what the authorities say. This leads to conflict.

**Key**

| | |
|---|---|
| ▲ | Main peaks and height |
| — | Roads |
| ⋯ | Railways |
| ▢ | National Park |
| ⟍ | Lakes |
| ╱ | Motorway |
| ● | Main towns and villages |

0       100 km

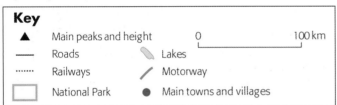

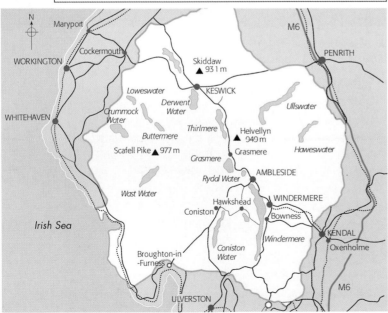

▲ The Lake District and how people get there.

**Key**
1. Brecon Beacons
2. Broads
3. Cairngorms
4. Dartmoor
5. Exmoor
6. Lake District
7. Loch Lomond
8. New Forest
9. Northumberland
10. North York Moors
11. Peak District
12. Pembrokeshire Coast
13. Snowdonia
14. South Downs
15. Yorkshire Dales

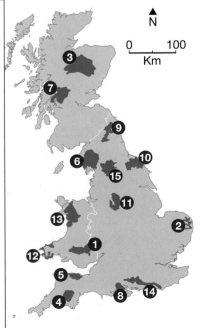

▲ The UK's 15 National Parks. The South Downs is due to become a National Park in 2009.

# What puts pressure on the Lake District?

The photo gives you a clue about why people like the Lake District – its landscape is dramatic and it offers people of all interests a great deal. Visitors go hiking, sail on the lakes, climb the rocky crags – or maybe just sightsee.

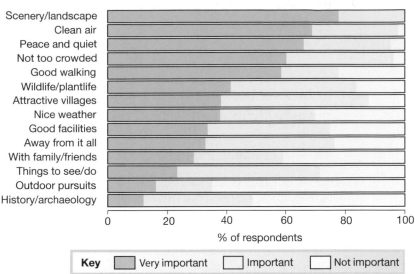

▲ What people like about the Lake District.

▲ Langdale, one of the Lake District's most popular spots.

| Year | Number of cars registered in the UK |
|------|-------------------------------------|
| 1975 | 10 million |
| 1990 | 20.5 million |
| 1996 | 22.8 million |
| 2006 | 27.8 million |

**1** Car ownership in the UK.

| Year | Annual holiday | Average working week |
|------|----------------|----------------------|
| 1960 | 2 weeks | 5.5–6 days, 50 hours |
| 1980 | 3 weeks | 5–5.5 days, 45 hours |
| 2008 | 4 weeks plus | 5 days, 38 hours |

**2** Changing leisure time in the UK.

# Accessibility

The Lake District is under increasing pressure, partly due to the ease of getting there – called its **accessibility**.

- Motorways bring people from Preston, Manchester, Liverpool, Leeds and York. Over 5 million people live within 2 hours' drive.
- UK car ownership has risen. There are now more cars in the UK than households (Table 1).
- Rail access is good and improving. The upgrade of the West Coast Main Line will reduce journey times to the Lake District from London – cut from 4 hours to 3 from 2010.

## your questions

1 Explain what National Parks are, and their purpose.
2 In pairs, research and devise a presentation on one National Park – use nationalparks.gov.uk. For your National Park, find out: **a** where it is, **b** what its landscapes look like, **c** its attractions, **d** what people can do there, and **e** how easy it is to get to.
3 For your park, produce an annotated map to show its location, its nearest cities, and main transport routes.
4 Which parks are: **a** the most accessible, and why, **b** the least accessible and why?
5 Using evidence on this page, describe the major changes to leisure in the UK. Why are these important for National Parks?

✚ **In this section you will learn about the benefits and problems brought by tourism to the Lake District.**

Tourism is vital to the Lake District. In 2007, over 8.3 million people visited. They stayed for more than 15 million visitor days and spent £660 million in hotels and shops. This provided:

- 12 000 jobs directly – working in catering and hotels.
- 12 000 jobs indirectly – providing food from farms or transport services.

Not all tourists are the same!

- Two-thirds are day visitors – contributing 25% of all tourist spending! They arrive without notice, usually by car, and cause congestion on the roads.
- One-third stay overnight – spending 75% of the total. Most stay in hotels. The vast majority (92%) are from the UK, and 8% from overseas. However, overseas visitors spend more and stay longer.

*On your planet*
✚ The biggest charities in the UK are to do with outdoor interests – like the RSPB and the National Trust.

## The benefits of tourism

- Within the Lake District National Park, 33% of jobs occur in hotels, catering and distribution, compared with 6% in the UK as a whole. In Windermere and Keswick, 55% of jobs arise from it, which boosts the local economy because much of the money earnt is spent locally.
- Without tourism, other services – such as local buses – would decline.
- Car park charges generate money for local councils.
- Farmers benefit from selling local foods – milk, meat and dairy produce. Many let out land to campers, or convert old barns into holiday accommodation.
- Tourism brings investment. Farmers are more likely to repair walls, or pub owners improve the quality of their food, and the local council will maintain roads and other infrastructure to encourage more visitors.

| Sector | % Expenditure |
|--------|---------------|
| Accommodation | 40.1 |
| Food and drink | 24.4 |
| Recreation | 8.0 |
| Shopping | 12.4 |
| Transport | 15.1 |

▲ How visitors spend their money.

▲ This field has been turned into a caravan park to bring in extra tourist income for the farmer.

## Problems brought by tourism

- **Traffic congestion** is severe. 85% of tourists arrive by car and public transport brings only 3%. Most local roads are narrow lanes, which become congested. This is especially true at 'honey pot' locations such as Windermere.

- **House prices** are high. Tourism leads to summer letting – so properties are bought as investment holiday lets. Demand pushes prices higher, beyond the reach of local people.

- **Second homes** make matters worse. In Ambleside, 40% of houses are second homes, compared to 15% across the Lake District National Park as a whole. In winter, pubs are quiet and many businesses close. This can destroy communities, even though second homeowners employ local builders and suppliers.

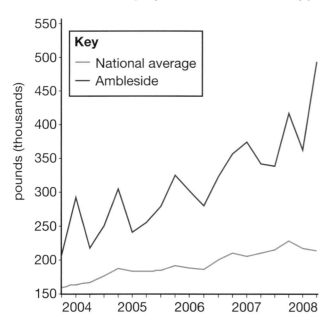

- **Seasonal unemployment** is a problem. Jobs in tourism are seasonal, low paid and part-time. The number of full-time jobs is small and suit students or overseas workers more than people living there full time.

- **Footpath erosion**. Each year, visitors spend 7 million visitor days walking in the hills. Footpaths become trampled. Trodden down plants die, exposing bare soil. Winter rains then erode paths deeply.

▼ Traffic congestion in Ambleside – a bypass is needed for this traffic hotspot. Congestion is serious for local people trying to get to work. And medical staff can't get through when accidents occur.

*On your planet*
+ National Parks in the USA and Australia are state owned – far less hassle than the thousands of landowners in the UK's own National Parks!

◄ Average house prices in Ambleside compared to the UK average (2003-2008).

### your questions

1 Draw up a table with the title 'The benefits and problems of tourism in the Lake District'. List as many as you can find using this section and the results of your own research.
2 Now go through each one, awarding it marks out of 5 to show how strong a benefit or problem it is.
3 Add up a total at the end. Do the benefits outweigh the problems for the Lake District?
4 **Exam-style question** For a rural area you have studied which is under pressure, explain the problems that it faces. (6 marks)

In this section you will learn about attempts to address two of the issues caused by tourism in the Lake District.

## The housing need

Wouldn't we all like a second home if we had the chance? The Lake District National Park has a problem with second homes:

- 18% of houses in the National Park are second or holiday homes.
- In Coniston, 43% of houses are second or holiday homes.
- There are negative impacts on communities as a result of so many second homes. Many local people are priced out of the market.

There is a real need for affordable housing in the Lake District. Those priced out of the market are low-paid workers (hotel workers, farm labourers), young people and the elderly. Social housing has been cut by 50% due to the **Right to Buy** (RTB). Windermere Parish Council carried out a survey in 2008 about housing needs (see the table):

- In just one parish, 200 households needed housing locally for work reasons.
- Only 2 households were able to afford a small one-bedroom property.

| Reasons for housing need | Number of households in need |
|---|---|
| Housing is too expensive | 33 |
| Current house too difficult to manage | 94 |
| Household has people with mobility problem or other special needs | 27 |
| Property needs major repair; household doesn't have the money | 22 |
| Family requires larger accommodation | 60 |

The **Right to Buy** scheme gives people who have been tenants of a council property for more than two years, the right to purchase that property at a discounted rate.

Second home owners don't mix with local people. They don't go to local dances, or to the pub. They turn up with bags of stuff from supermarkets and don't shop locally.

### Solutions to affordable housing?

1 Charge second home owners more than 100% council tax to raise money.
2 Stop the Right to Buy, to preserve social housing stocks.
3 Limit second home ownership.
4 The National Trust and the Forestry Commission could release land to build small-scale affordable housing.
5 Convert disused farm buildings into affordable housing.

## Repairing damaged footpaths

People love walking in the Lake District. But now 150 popular footpaths are in a serious state of erosion. Some methods erode more than others. **Mountain biking** and horse riding can be damaging. **Fell running** also causes erosion. One recent running event brought 300 people to watch, causing damage to footpaths.

Repairing footpaths is expensive, even if the work is done by volunteers and park wardens. It costs £4.2 million over 10 years to repair and maintain eroded paths. Well-maintained paths make for better walking, and protect wildlife habitats from disturbance.

### Solutions to worn footpaths?

1 Leave landowners to repair the paths.
2 Allow walkers but ban fell running, mountain bikes and horses.
3 Central government and local council grants to fix and repair footpaths.
4 Rely on donations from the public.
5 Close down areas where erosion has become serious.
6 Charge tourists to use footpaths.
7 Put up notice boards to educate people about footpath erosion.

*On your planet*
+ A 'people counter' on one path in Wasdale clocked 28 616 walkers in June 2005!

Footpath repair at Coledale Hause (above left and right). Rainwater cascading down the path stripped out hundreds of tonnes of material. This was the worst case of erosion in the Lake District.
The National Park team:
- stopped the water damage by installing a turfed drain to drain water slowly without causing further damage.
- created a good walking surface to stop users trampling vegetation.

### your questions

1 List the problems caused by second homes and holiday lets in the Lake District.
2 In pairs, go through the five solutions proposed. Give the advantages and problems of each and decide which is best.
3 Now do the same for the seven solutions for worn footpaths.

➕ In this section you will learn about attempts to deal with traffic and conflict in the Lake District.

## The problem of traffic congestion

Traffic in the Lake District increases each year. Most visitors arrive by car, and use cars to explore while they're there. Bus services are infrequent and serve main towns only. For most people, their car is the only way to get around. But cars create congestion and put pressure on parking spaces. Long journey times spoil people's holidays, frustrate visitors and make it difficult for local people getting to work.

Local planners have considered ideas for improving access to the Lake District. Seven options are shown on the right.

| Option | How it might reduce parking or congestion |
| --- | --- |
| 1. Build bypasses around key towns, e.g. Ambleside | This has been proposed in the past, but nothing has ever been done. It could improve journey times and increase capacity on the roads. |
| 2. Park-and-ride at key towns, e.g. Windermere | Town centre traffic could be reduced by building park-and-ride car parks, with shuttle buses, on the edge of towns. |
| 3. Encourage hotels to run minibuses | If hotels collected their visitors from railway stations, it might encourage travel to the Lake District by train. |
| 4. Limit car parking | Reduce the number of parking spaces in towns to put visitors off arriving by car. |
| 5. Increase public transport | One company, Stagecoach, provides bus services between the towns, as well as some minor routes. Could it be encouraged to provide more frequent services? |
| 6. Charge vehicles entering the National Park | Like London's congestion charge, cars would pay to enter the Lake District National Park. American and Australian National Parks charge entry fees and have few protests. |
| 7. Expand the coastal rail service | This could improve access to the western Lake District between Barrow and Carlisle. But the line is single track, and needs upgrading to improve speeds – it takes over 90 minutes to get from Barrow to Workington. However, the line could encourage business and be a tourist attraction. |

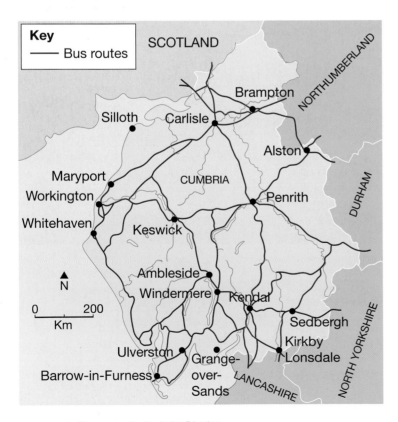

**Key**
— Bus routes

▲ Bus routes in the Lake District.

| Days | Number of return services south of Whitehaven to Barrow | Earliest | Latest |
| --- | --- | --- | --- |
| Monday to Friday | 8 – service roughly every 2 hours | 06.30 approx | 18.36 |
| Saturday | 9 | 06.35 | 18.40 |
| Sunday | No service on Sunday | | |

▲ The frequency of rail services on the Lake District west coast railway, summer 2008.

# Managing conflict in the Lake District

The Lake District is largely owned by private individuals – farmers, hoteliers, homeowners. And each has a view about how it ought to be developed. Some object to development because they believe the Lake District is special. Others feel that development provides jobs for those who live there. The job of the National Park Authority is to manage these conflicting views.

For any proposal, you can show how different people feel using a **conflict matrix**. To make one, follow these steps:

1 List any group with an opinion about a proposal in the left-hand column.
2 Now list the same groups in the right hand box of each row.
3 Each group is then judged by how much they agree or disagree with the others, e.g. tourists may not agree with residents about park-and-ride schemes.
4 The completed table summarises who agrees or disagrees, and by how much.

▲ Cycling in the Lake District – a good way of reducing congestion on the roads?

|  | Hotel owners | National Park Authority | Walkers | Cattle farmers | Local hotel workers |
|---|---|---|---|---|---|
| Hotel owners |  |  |  |  |  |
| National Park Authority |  |  |  |  |  |
| Walkers |  |  |  |  |  |
| Cattle farmers |  | –– |  |  |  |
| Local hotel workers |  |  |  |  |  |

**Key**
++ Strong agreement
–– Strong disagreement
+ General agreement
– General disagreement
0 Neither agree or disagree

**On your planet**
+ By 2035, India will probably have the world's second largest economy.

+ In this section, you'll learn how India's wealth varies between different states.

## Development dilemmas – what's that then?

'Development' means change – usually improvement – for people and the economy. It brings jobs and trade. But sometimes its benefits only go to some people and not to others. Who will benefit most – people in the cities or in rural areas? Will some parts of the country benefit more than others? These questions present problems – or dilemmas. This unit will explore the choices that countries make, and ask whether they are sustainable.

## India – booming for some!

Everything about India is huge – and growing!

- Its population (1.15 billion people in 2008) is the world's second largest, 20 times more than the UK's.
- It has a huge workforce of 500 million people.
- Its economy was the world's fifth largest in 2008, worth US$3 trillion – that's 3 followed by 12 zeros!
- While the world's economy grew by 3% every year from 2000-2008, India's growth averaged 7%.

India's wealth is not evenly shared out. Some Indians are very wealthy, while others live in great poverty. Its wealth varies between:

- states, from the wealthiest (Maharashtra) to the poorest (Bihar)
- cities, which are generally wealthier (e.g. Mumbai), and rural areas – although one state, Punjab, is a wealthy rural state.

▲ Half of Mumbai's population lives on the streets or in poor-quality shacks, while others live in smart homes like these.

**India – Factfile 2008**

Total population: 1.15 billion
Population annual growth rate: 1.6%
Rural population (% of total): 72%
Infant mortality rate (per 1000 live births): 32
Life expectancy at birth (years): 69.3
Population living below the national poverty line 2007 25%
School enrolment, primary, 2001 99.2%
Adult literacy rate (age 15 and above), 2001 65.2%

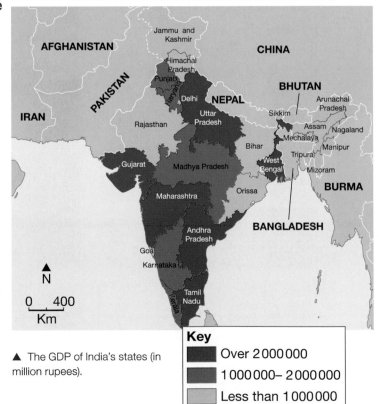

▲ The GDP of India's states (in million rupees).

**Key**

| | |
|---|---|
| ■ | Over 2 000 000 |
| ■ | 1 000 000 – 2 000 000 |
| ▨ | Less than 1 000 000 |

# Maharashtra

Maharashtra is India's richest core region (see right), with the largest **GDP** – both in total and per person. It contains India's largest city, Mumbai, with 13 million people. Maharashtra's economic growth has come from:

- services: e.g. banking, insurance, IT and call centres. Mumbai's universities produce well-educated, English-speaking IT graduates who are employed by large Western companies (e.g. BT), who contract them to provide services – known as **outsourcing**. India's wage rates are low; well-paid call centre workers in Mumbai earn only US$5000 a year.

- manufacturing: half of Mumbai's factory workers work in cotton textiles for export. Other booming industries include food processing, steel, engineering, cement, and computer software. These have led to a construction boom, building factories and offices.

- entertainment: Mumbai has the world's largest film industry – Bollywood – producing the greatest number of feature films each year.

- leisure and business services: e.g. hotels and restaurants.

> **+ GDP (Gross Domestic Product)** is the value of all goods and services produced in one year in a country. It is shown as a **total**, or per person (**per capita**), by dividing the total by the population.

# Core and periphery regions

Much of India's economic growth has taken place around its ports and cities. These are what economists call 'growth poles' – points or places where industries and ports develop. As this happens:

- People move there for jobs.
- As the new workers earn, they spend – creating demand for additional companies.
- They also need services, such as housing, which also creates jobs.

This creates an 'upward spiral' (see below), which economists call a **multiplier effect**. Over time, the multiplier effect gets greater and a whole region develops, called a **core region**. People continue to move there seeking jobs – and the spiral continues.

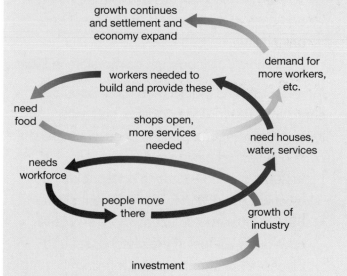

▲ The upward spiral caused by the multiplier effect

Areas which do less well get left behind. These are generally known as the **periphery**. They are explained in the next unit.

## your questions

1 On a sheet of paper, write down these words which are linked to Mumbai's industries: Banks, IT companies, Call centres, Universities, Overseas companies (e.g. BT), Textile factories, Food processing factories, Steelworks, Computer software, Building construction, House building, Entertainment, Hotels and restaurants.

2 Draw links between as many as you can to show how each is connected to the others – e.g. Banks to IT companies. Then write down what the link is.

3 a How does your diagram help to explain how core regions develop?

   b Why do some states in India develop more than others? Use your diagram to explain.

➕ In this section, you'll learn why not everyone shares in India's growth.

## Life in Bihar

Bihar is India's poorest state – and its most rural – 86% of its population live in rural areas, and most work in farming. Farming in India pays poorer wages than urban industries.

- Average incomes in Bihar are only 6000 rupees (£75) per person per year. This is 33% of India's average income, and 20% of Maharashtra's average income.
- 55% of households live below the poverty line, and 80% of people work in low-level jobs paying little.
- 26 of India's 100 poorest districts are in Bihar.

Although Bihar is India's third most populous state, with 100 million people, it gets little investment from companies, because its people can't even afford basic services.

- In 2003, only 58.8% of its households had electricity, and 12% water-flushed toilets.
- 80% of households used wood, and 35% cow-dung, as fuel in their kitchens. Lung diseases are common from inhaling the smoke from these fuels.

## Farming in the rural periphery

Few of India's rural population own land; most land is owned by a wealthy minority. Most farmers rent their land and can only afford small plots. A system called 'sharecropping' is used; farmers pay half of their produce to the landlord.

## Living on India's rural periphery

While Maharashtra has boomed, other states have been left behind, such as Bihar. They attract less investment, often have poor transport, and are a long way from cities. Employers are not attracted there, so there are few jobs. Often, younger people leave to find better jobs. These areas are known as the **periphery** – regions on the outside, away from the **core** – producing less wealth and playing a small part in India's economy.

| Data | Bihar | India |
|---|---|---|
| Urbanisation (%) | 10.8 | 25.7 |
| Literacy (%) | 47.5 | 65.2 |
| Women's literacy (%) | 33.6 | 54.0 |
| Rural literacy (%) | 44.4 | 59.5 |
| Electricity (%) | 58.8 | 75.8 |

**On your planet**
➕ There is no social security system in India – if you are elderly, unable to work, and poor, you beg.

**On your planet**
➕ Educated women in India marry 10 years later, and have half the number of children, compared to those who only complete primary school.

| | |
|---|---|
| Landless households | 54.2% |
| Up to half an acre rented | 27.5% |
| Half to 2.5 acres | 13.3% |
| 2.5 to 5 acres | 4.1% |
| 5 to 25 acres | 0.7% |
| 25 to 50 acres | 0.2% |

▲ The average size of rented family plots in Bihar (2 acres is about the size of a full-sized football pitch).

Few rural families ever have any surplus income and survive on what they grow – called **subsistence farming**. Much of the work is done by hand. Ploughs are traditional wood and iron, pulled by oxen or buffalo. With no surplus crops, farmers cannot invest in machinery or fertiliser to help them grow more. They are trapped in a **cycle of poverty** (see right). Productivity is also low in Bihar, compared to other states in India (see the table).

Landless Indians rely on finding employment with other farmers. In Bihar, 54.2% of households are landless. In bad years, when there is little work, many borrow from moneylenders – often at very high interest rates. With low incomes and debt, malnutrition is common and diets are poor; few households ever have green vegetables, milk or protein.

## Women and poverty

Rural families are larger, compared to those in cities – Bihar's fertility rate is 4.4, compared to Maharashtra's 2.1. This is caused by two factors:

* Rural families need more children to work on the land and to collect wood or water.
* Bihar's women have India's poorest literacy rates. Uneducated women marry early – usually by the age of 17 – and give birth before they are 20. They are poorer than men; they rarely own land, usually working as poorly paid landless labourers, and have few rights.

In Bihar, school attendance is poor.

* In one region, only 35% of children go to primary school; only 8% reach upper primary, and only 2% reach Years 12 and 13.
* Not one woman in 2003 studied for a diploma in engineering or sciences.
* Bihar is a caste-based society, and this affects poverty and literacy. Almost all people in higher castes are literate; almost all in the lowest castes have zero literacy.

| Sugarcane | Maharastra produces twice as much per acre as Bihar |
|---|---|
| Potatos | West Bengal produces 3 times more per acre than Bihar |
| Onions | Gujurat produces 4 times more per acre than Bihar |
| Rice | Punjab produces 2.5 times more per acre than Bihar |
| Wheat | Punjab produces twice as much per acre as Bihar |

▲ How crop production in other states compares to Bihar.

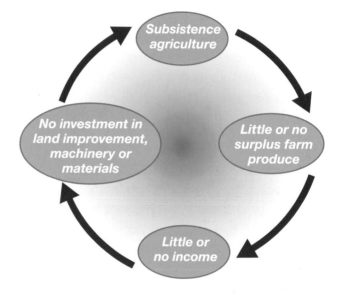

▲ The cycle of poverty.

### your questions

1 Using data, make four statements that show that Bihar is part of India's periphery, and not its core.
2 Make a large copy of the cycle of poverty. Now add arrows and labels to the diagram to show what happens if: **a** a family member becomes ill, **b** a family gets into debt.
3 In pairs, discuss and draw spider diagrams to explain: **a** why land in Bihar produces less than other parts of India, **b** why women are poorer than men, **c** why so few people attend school, **d** why rural families are larger than urban. Label them fully.
4 Write a paragraph to explain why Bihar is much poorer than Maharashtra.
5 **Exam-style question** Using examples, explain why poverty can be seen as a 'cycle'. (6 marks)

➕ In this section, you'll learn about top-down development projects, their advantages and disadvantages.

## Which sort of rice?

Consider the farmer on the right. Like many farmers in Bihar, she grows rice on a small 1-acre plot. What variety should she grow? There are so many to choose from. In a trial, 24 farmers – 12 women and 12 men – sampled 15 rice varieties, scoring them for colour, smell, texture, stickiness, and taste. Women cooked the samples just as they normally would. Between the 24 of them, none could decide which they liked most.

Given this, does it matter which type of rice this farmer chooses to grow? The two 'revolutions' described below show that it matters hugely – both are examples of **top-down development** (see the panel opposite).

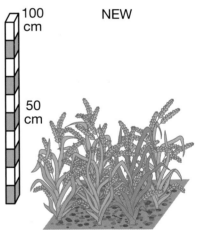

## The 'Green Revolution'

In the 1960s, starvation was common in India. In the 1970s, the **Green Revolution** changed rice growing there by providing farmers with HYV (High Yielding Variety) seeds. These new rice plants are shorter, grow more quickly, and produce more grain than traditional rice (see right). The HYVs were developed by scientists and are sold by multinational companies. However, the results have been mixed:

TRADITIONAL 100 cm 50 cm NEW

0·9 tonnes per hectare

1·6 tonnes per hectare

- Farmers have to buy seeds, instead of using some saved from last year's harvest.
- The new seeds are high-yielding, but need irrigation water, fertiliser and pesticides. Only larger, wealthier farmers can afford this.
- Crop yields are much higher than traditional varieties, so incomes have risen – for the wealthy.
- Large farmers have become much wealthier, especially in Punjab, which has been at the centre of the Green Revolution.

On your planet
➕ Small farmers found the new 'miracle seeds' didn't produce enough straw for their cattle to eat.

India is now a rice exporter. But the environmental impacts of using fertiliser and pesticides have been huge, and the HYVs now have no resistance to some plant diseases.

## The 'Gene Revolution'

A new revolution is now taking place – genetic modification. Scientists take genes from one organism and put them into another, creating new plant varieties. Imagine rice that can resist disease or drought. That's what scientists promise – higher crop yields, and starvation a distant memory.

But genetically modified (GM) crops are controlled by TNCs, e.g. Monsanto. Using patent laws, they 'own' every GM plant grown from their seed. Some force farmers to sign contracts to use only their chemicals, or prevent them from saving seed to use next year. GM engineers can even stop seeds from germinating. Over 1 billion of the world's poorest people rely on saved seed – but, instead, they will now have to spend money every year on new seeds. Again, wealthier farmers will benefit the most. Poorer Indian farmers have been rebelling against this control by TNCs like Monsanto.

▶ Activists dressed as Indian politicians 'force feeding' volunteers with genetically engineered (or modified) aubergines at a protest in New Delhi against GM (or GE) crops.

## Top-down development

Top-down development is where decisions about development are made by governments or by private companies. The decisions are imposed on people because there will be benefits.

Top-down development involves:
- decision-makers looking at a 'big picture' to identify need or opportunity, e.g. for national energy sources, or transport
- experts helping to plan it
- local people being told about it, but with no say in whether it will happen or not.

The argument goes that people gain by a process called 'trickle down', e.g. jobs, wealth and other benefits 'trickle down' to the poor.

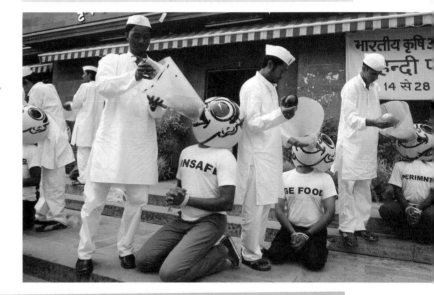

### your questions

1 Divide into four groups:
   - One group is pro-Green Revolution, and one against.
   - One group is pro-GM seeds, and one against.
   In pairs, develop as many points as you can to argue your case. Research by typing 'GM seeds' or 'India Green Revolution' into Google.
2 Present your findings to the class.
3 When presentations are complete, copy and complete the table on the right about the arguments on all sides.

| Type of change | Good points | Bad points | My decision – good or bad? |
|---|---|---|---|
| Green Revolution | | | |
| GM Revolution | | | |

4 Would you advise the farmer opposite to stick with traditional varieties, go for HYVs, or the new GM seeds? Write a brief report explaining in about 300 words.

In this section, you'll learn about the benefits and problems of one top-down development project.

## Top-down – the Government decides!

Over much of India, rainfall is seasonal and unreliable (see right). Parts of north-west India are so dry that semi-desert exists, preventing people from making a decent living. Across the rest of India, between:

- May and September, the Indian monsoon brings huge falls of rain that are difficult to imagine – think of the heaviest rain you have ever seen, and then double it
- November and March, almost no rain falls across large areas of India.

As India's population increases and its economy booms, demand for water is rising. The Government has decided that western India needs super dams to:

- encourage economic development, by providing drinking water and electricity
- open up dry lands for farming by **irrigation**.

Building large dams makes it possible to store monsoon rains to use during the dry season. By 2008, the Government had built over 4500 dams, 14 of which are super dams. Now the Narmada – one of western India's major rivers (see right) – is being tackled with a series of 3000 dams, big and small. The scheme will take 100 years to complete. But how well will it work for people and the environment?

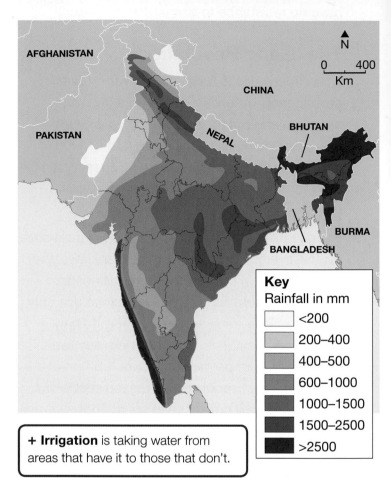

**Key**
Rainfall in mm
- <200
- 200–400
- 400–500
- 600–1000
- 1000–1500
- 1500–2500
- >2500

**+ Irrigation** is taking water from areas that have it to those that don't.

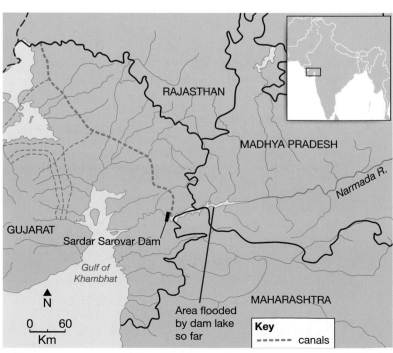

**Key**
- - - - - - canals

# The Sardar Sarovar dam

The Sardar Sarovar dam across the Narmada River is already one of the world's largest dams. When completed, it will provide water all year round to areas of India that suffer drought.

## Benefits of the dam

Originally designed to be 80 metres high, the Government now plans to raise the dam to 136.5 metres – to increase its capacity.

- The dam is multi-purpose – it provides 3.5 billion litres of drinking water a day, and 1450 megawatts of hydroelectric power (HEP), which is more than 750 wind turbines!

- A series of canals distribute water to other states in India. When complete, they will irrigate 1.8 million hectares of farmland in the driest parts of Gujarat, Maharashtra, Rajasthan and Madhya Pradesh. Gujarat and Madhya Pradesh suffer from drought and lose £20 billion in farm production each year.

## Problems caused by the scheme

- 234 villages have been drowned so far, forcing 320 000 people out.
- Religious and historic sites have been flooded.
- Few villages can afford the electricity generated; only the cities benefit.
- Irrigation can increase soil salinity making the soil less usable.
- The silt brought by feeder rivers will collect and reduce the reservoir's capacity.
- Damming the river means that fertile sediment normally deposited on flood plains each year will be lost.
- The area has a history of earthquake activity. Seismologists believe that the weight of large dams can trigger earthquakes.
- Good quality farmland has been submerged.

### What do you think?
✛ An engineering marvel bringing wealth? Or a disaster displacing and ruining the livelihoods of 320 000 people?

### What do you think?
✛ Do you think governments have the right to develop 'top-down' schemes like this if they affect so many people?

## your questions

1 Copy and complete the following table about the economic, social and environmental benefits and problems of the Sardar Sarovar dam.

|  | Benefits | Problems |
| --- | --- | --- |
| Economic |  |  |
| Social |  |  |
| Environmental |  |  |

2 Now highlight in one colour those benefits or problems which are **local**, and in another those which are **further away**.

3 a Which are the greatest benefits – economic, social or environmental? Are they local or further away?
  b Which are the greatest problems?

4 Explain whether you think top-down schemes like this should be built if they cause such problems.

5 Exam-style question For a large development project you have studied, explaiin its benefits and problems. (6 marks)

In this section, you'll learn about bottom-up development projects, their advantages and disadvantages.

## Working from the bottom up

ASTRA (Application of Science and Technology in Rural Areas) has been a major development project in rural India in recent years. It started as a research project at the University of Bangalore, in the state of Karnataka. It is a **bottom-up development** project. Researchers went into villages in rural Karnataka (like the one on the right) to find out what people's lives were like – with no idea about what they might discover. They talked to families, recorded how they spent their time, and listened to problems like those below.

### The problem of time

The ASTRA team found that, for most rural families, the daily routine takes too much time, especially for women and girls. In each family, there is cleaning, fuel collection, preparing and cooking food, fetching water, tending sacred cows, looking after the vegetable patch – all before any paid work is done in the fields!

### The fuelwood crisis

Half of rural India has a fuel crisis. Each family needs 25–30kg of firewood per week – taking up the equivalent of one day a week to collect it. As the population is increasing, fuelwood is becoming more scarce and families have to walk further to find it. Cow dung is increasingly being used as an alternative fuel. Women collect it and shape it into fuel logs. Aside from the health risks involved, the dung could be better used on the land as fertiliser. Burning wood and dung on traditional stoves means that women's health is often poor from the smoke, which causes lung and eye problems.

▲ A village in rural Tumkur district of Karnataka, near Bangalore.

+ **Bottom up development** means
- experts working with local communities to identify their needs
- giving local people control in improving their lives
- experts assisting with progress.

▼ Cooking using a traditional stove (outdoors to reduce the effects of the smoke).

### Lack of education

Traditionally in India, women are responsible for food and the home, as well as one-third of all paid work. Young girls help with domestic chores (such as fuel collection) and raising younger siblings. Therefore, rural girls have little education – few complete primary school. With little formal education, most marry early and maintain a high fertility rate – trapped in a poverty cycle. Even if they go to school, it's dark by 6.30 pm most evenings, which prevents further study.

## Solution – think cow dung!

The answer has turned out to be right under their noses – cow dung! Cow dung has become a highly valued resource; it can be used to provide gas – called **biogas** – for cooking, and for powering electricity generators.

Cow dung is no longer collected and stored in the home for burning instead of wood, but is fed directly into a brick-, clay- or concrete-lined pit that forms part of a biogas plant. The pit is sealed by a metal dome, under which the dung ferments to produce methane. As the pressure builds up, the methane is then piped into homes (see the diagram).

**your questions**

1 Why was it important that the ASTRA team spent time with the villagers in Tumkur first, before thinking up a plan for them?
2 In pairs, decide on a rank order to show which of the three problems – time, fuelwood, and education – was the most serious. Present your findings to the class.
3 Devise a diagram to show how these three problems might be linked for people living in rural areas.
4 Make a copy of the biogas plant diagram and annotate it with these words: dung, where the dung goes in, fermentation, methane/biogas produced, gas pipe to people's houses, slurry produced.
5 How might the interior of an Indian family home change as a result of biogas?

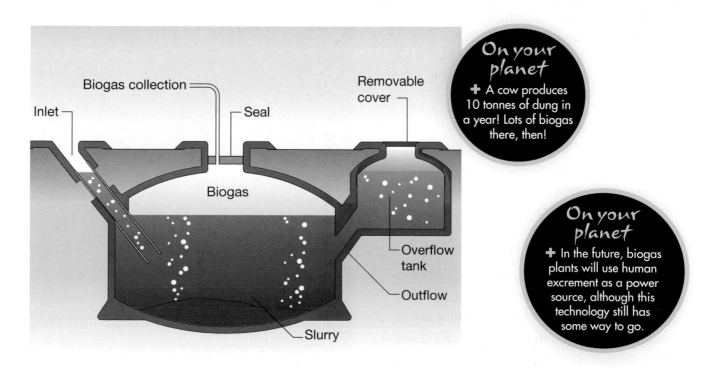

Biogas collection

Inlet

Seal

Removable cover

Biogas

Overflow tank

Outflow

Slurry

*On your planet*

+ A cow produces 10 tonnes of dung in a year! Lots of biogas there, then!

*On your planet*

+ In the future, biogas plants will use human excrement as a power source, although this technology still has some way to go.

➕ In this section, you'll learn about the impacts of biogas.

## Implementing biogas!

Before embarking on the biogas scheme, villagers have decisions to make. Is each cow's dung the property of the cow's owner, or of the village? If the poorest villagers have no cattle, are they entitled to any biogas? Only one-third of households have enough cattle to run family sized biogas plants. But if all the dung resources from the village's cattle are pooled in a communal biogas plant, every family could have enough biogas for their needs.

Most villages in Tumkur chose the community approach, so that everyone gets biogas. The only rule is that the finished slurry belongs to those with cattle, so they get the fertiliser to add to their soil.

## The benefits of biogas

Biogas has brought benefits to villages in Tumkur.

### For people

- It results in a smoke- and ash-free kitchen. Gas stoves reduce eye and lung problems within six months of being installed.
- Women and children are freed from the chore of gathering firewood. They gain two hours a day for other activities. That time allows more children to go to school.
- About 80% of families use this time to earn extra income – some families have increased their income by 25%. Women's self-help groups in Tumkur have started plant nurseries in 24 villages, and 244 women are now earning income.

▲ A village biogas plant with its dried cow dung fuel stacked in the background.

▲ Cooking indoors using a new clean biogas stove.

- Cow dung is no longer stored in the home, but is fed straight into the biogas digester, along with human and kitchen waste.
- Dung fermentation within the pit destroys pathogens (micro-organisms that cause disease). As a result, sanitation and human health have greatly improved.
- Across India, four million cow dung biogas plants have been built and 200 000 permanent jobs created, mostly in rural areas.

## For the environment

- Cattle are now kept in the family compound, to make collecting their dung easier. Previously, cattle would have grazed in the forest – eating saplings and preventing woodland from regenerating.
- The slurry which remains after fermentation is richer in nutrients than raw animal manure. When added to the soil, it guarantees higher yields of vegetables, fruit and cereals. It is also organic, as no pesticides or artificial fertilisers are used in most villages – they are too expensive.
- 277 tonnes of $CO_2$ emissions savings have been achieved from the biogas plants.

▲ Removing slurry from an industrial-sized biogas plant. The slurry has double the nitrogen content as a result of fermentation, making it better fertiliser.

### And there's electricity too!

Many villages have also chosen to use biogas to power an electricity generator, which can be used for:

- pumping domestic water from a borehole. A 500kW generator can pump up 100 000 litres of clean drinking water daily. 53 community boreholes for drinking water and irrigation were drilled in Tumkur villages in 2005.
- pumping irrigation water. Some farmers now get three crops of vegetables a year, because irrigation water can be used during the dry season.
- generating electricity for lighting.

Many decisions about biogas were made by women. Biogas has allowed village women to take part in evening study, literacy classes and other home and community activities – giving them greater identity and feelings of self-worth.

**On your planet**

✚ A 500kW biogas-powered generator can supply enough electricity for a village of 2–3000 people.

## your questions

1 Copy and complete the following table to show the benefits of biogas for Tumkur villages, and whether these are short-, medium- or long-term.

| Impact | Short-term (immediate or in a few months) | Medium-term (over a period up to a year) | Long-term (over a few years) |
|---|---|---|---|
| Social | | | |
| Economic | | | |
| Environmental | | | |

2 Which are the greatest benefits: **a** social, economic or environmental? **b** short-, medium- or long-term?
3 Explain why girls attending school has a long-term impact on **a** the villages in which they live, **b** long-term population growth.

# How sustainable is India's development?

+ In this section, you'll learn what 'sustainability' means, and how to assess whether rural development in India is sustainable.

## How do we know if something is 'sustainable'?

Sustainability is a commonly used word now. Type it into Google, and you get over 32 million hits! What does it mean? How does it work? How do we know if something is 'sustainable' or not?

The term 'sustainable development' came into common use after 1987, when the UN Brundtland Commission published a report on global development called 'Our Common Future'. In it, they claimed that sustainable development *'meets the needs of the present without compromising the ability of future generations to meet their own needs.'* The report showed that:

* the eco-footprint of the USA, even in 1987, was far greater than any other country
* US energy consumption and waste were greater than any other country's
* the vast majority of the world's population had only a small share of its finite resources.

The world doesn't always agree on what sustainability means. Those who believe in Thomas Malthus's ideas (see page 156) believe that the world already has too many people chasing too few resources. Meanwhile Esther Böserup's supporters (see page 159) believe in a 'technological fix' – that the world can be made more sustainable with clean coal, or more efficient car engines. It's unlikely that supporters of these ideas would ever agree.

▲ One of India's coal-fired power stations, generating electricity. How sustainable is burning coal? Can the world really be made more sustainable by cleaning up coal and removing its emissions?

▲ Local groups getting involved in decision-making (Rule 1 opposite) are often effective because they understand each other's needs.

# 10 rules for sustainable development

A number of organisations (such as the UN and Greenpeace) have tried to create ways of defining whether development is sustainable or not. Here are 10 rules that sum up sustainable development and can be applied to different situations.

Sustainable development must:

1. involve local people in decision-making
2. be affordable; it must not put people/ countries into debt that they can't afford, or prevent people from getting basic needs at affordable prices (e.g. housing, energy and water)
3. promote good health – there must be no adverse effects on people
4. protect and encourage native plants and animals, not contaminate plant or animal genes
5. use land that's been developed before where possible (e.g. brownfield sites in urban areas)
6. minimise waste, and encourage re-use and recycling
7. minimise energy use, and use natural energy sources, e.g. solar power
8. minimise water use, and use rainwater or domestic waste water for irrigation
9. minimise pollution; where pollution exists, it should be cleaned up
10. offer benefits to the poor and disadvantaged, as well as to the wealthy.

# Tree planting in Gujarat – how sustainable?

Every year, India's population rises, and farmers are forced to use poorer-quality land for cultivation. Gradually, scrubland and trees are being lost. The National Tree Growers' Cooperative Federation is trying to reclaim India's scrublands sustainably. It organises villagers to reclaim degraded land by planting trees, then finding ways of processing and marketing the wood harvested from those trees by creating village tree growers' cooperatives. The participation of women, landless and marginal farmers, and private landowners is a key focus.

Besides jobs, the project meets villagers' needs for livestock fodder, food, fruit and fuelwood. Fodder from commercial sources is expensive, while that produced by the villagers is cheaper. This, in turn, helps increase milk production. Village industries are being developed using wood, e.g. woodworking, bamboo craft.

On your planet

+ India loses an area of forest nearly the size of Wales each year.

## your questions

1 Why is it difficult to define 'sustainability'?
2 Why wouldn't Malthus and Böserup agree about what sustainability means?
3 In pairs, take the 10 rules for sustainability. Write them out in a list and draw a column. Read about tree planting in Gujarat, and award ticks for each rule that it meets.
4 Now use other case studies in this unit – the Green and Gene revolutions (pages 240-1), Narmada scheme (pages 242-3), and Biogas (pages 244-7). Discuss which examples tick which rules – and how.
5 Which of all the examples is developing **a** most sustainably, **b** least sustainably? Explain your reasons.
6 **Exam-style question** Using the example of tree planting in Gujarat, explain how sustainable this development is.
(6 marks)

✚ In this section, you'll learn how one Indian state has approached the question of development and sustainability.

## Decisions Kerala-style

By now, you have probably realised that top-down and bottom-up development each have their advantages and disadvantages. However, sometimes development has dramatic results. State elections in Kerala elected a Communist government in 1956. One of its first actions was land reform, transferring land to people who previously had only rented small plots. This produced a motivated rural workforce.

## Social welfare in Kerala

What makes Kerala different is its focus on health and education. Its 32 million people average US$293 income per person per year. Kerala is one of India's poorest states, and its GDP per person is 90% lower than the USA's. Yet life expectancy in Kerala is nearly as long as in the USA, and its literacy rate is 91%, compared to 61% for India as a whole.

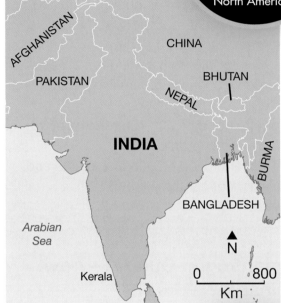

▲ Kerala is about twice the size of Wales. 32 million people live here, making it one of India's most densely populated states.

Kerala's levels of education and healthcare are the highest in India. This results from two things:

• Politically, a decision to invest in education and women's health.
• Economically, Kerala relies less on farming than other states, and has a strong service sector, especially through tourism. It also gets an extra 20% of its GDP from people who work in the Middle East and send their wages home – known as **remittance payments**.

▶ A secondary school in Kerala. The majority of students complete 10 years of schooling.

Kerala's reforms were based on bottom-up approaches to health and education, based around communities. Almost all villages have access to a school and a modern health clinic within 2.5 km. Taxation pays for this, including a tax on remittance payments.

## How Kerala's social welfare compares to India as a whole

- From the late 1970s, Kerala has led India in public services, such as roads, post offices, primary and secondary schools, medical facilities, and banks.
- The caste system was overturned in Kerala, although it survives in some remote villages.
- Rural poverty in Kerala is the lowest in southern India (see the graph).

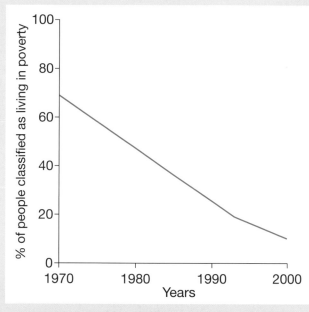

- Kerala has India's fourth lowest industrial output, 11% below the Indian average.
- Kerala's rate of population growth is India's lowest; its 9.4% per decade is less than half the Indian average of 21.3%.

- Indicators of women's health and education are the best in India and the developing world. Food programmes focus on the nutrition of mothers and children, using ration cards and free school lunches.
- Attitudes towards women are enlightened; girls outnumber boys in higher education, and Kerala appointed the first female Chief Justice, Surgeon General, and Chief Engineer in India.
- Women in Kerala marry on average 4 years later, and have their first child 5 years later, than other Indian women. They average only 2 children per woman, and experience very low infant mortality (see the table).
- Over 95% of births are hospital-delivered.
- Kerala has India's lowest birth rate (see the table).
- Kerala's weakness is water-borne diseases, e.g. diarrhoea, dysentery, hepatitis, and typhoid – over 50% of Keralites rely on wells which are contaminated by the lack of sewers.

| Quality of life indicator | Kerala | India | Low-income countries | USA |
|---|---|---|---|---|
| Adult literacy rate (%) | 91 | 61 | 39 | 96 |
| Life expectancy in years (males) | 69 | 67 | 59 | 74 |
| Life expectancy in years (females) | 75 | 72 | n/a | 80 |
| Infant mortality per 1000 | 10 | 33 | 80 | 7 |
| Birth rate per 1000 | 17 | 22 | 40 | 16 |

## your questions

1 Why do you think Kerala's government decided to focus on **a** education, and **b** women's health as their priorities?
2 Explain why improving women's health and education results in **a** later marriage, **b** later childbirth, **c** fewer children and **d** lower infant mortality.
3 Pick out three other differences between Kerala and India as a whole, and explain why they are important.

4 Kerala is the only state in India to have achieved this improvement in social welfare. In pairs, discuss all the possible reasons why this might be. Feed back your ideas.
5 **Exam-style question** Explain how one named development scheme you have studied has had impacts on people. (6 marks)

+ **In this section you will learn what we mean by the 'new economy'.**

The '**new economy**' relates to MEDCs moving towards a global service sector economy. Some people refer to the 'new economy' as the 'network economy'. But what does this mean?

So that we can better understand the 'new economy', we need to take a trip backwards and think about the 'old economy'. This was based on traditional industries, such as ship building and textile manufacture (see the map).

Activities in the 'new economy' are much less fixed to traditional locations, such as coalfields and ports. They rely on being able to secure flows of knowledge and communication. These new industries may be found in science parks, industrial parks or small industrial estates, often near university towns and research centres.

**Globalisation** underpins the 'new economy', since it relies on connections and networks between people and regions. Globalisation takes three forms: economic (including trade and finance), political, cultural.

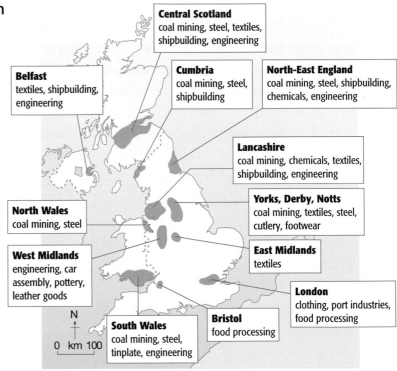

**Central Scotland**
coal mining, steel, textiles, shipbuilding, engineering

**Belfast**
textiles, shipbuilding, engineering

**Cumbria**
coal mining, steel, shipbuilding

**North-East England**
coal mining, steel, shipbuilding, chemicals, engineering

**Lancashire**
coal mining, chemicals, textiles, shipbuilding, engineering

**North Wales**
coal mining, steel

**Yorks, Derby, Notts**
coal mining, textiles, steel, cutlery, footwear

**West Midlands**
engineering, car assembly, pottery, leather goods

**East Midlands**
textiles

**London**
clothing, port industries, food processing

**South Wales**
coal mining, steel, tinplate, engineering

**Bristol**
food processing

N

0  km 100

▲ Traditional industries in the UK (before 1980). These were located close to sources of power, labour and raw materials (such as coal).

+ **Globalisation** refers to growth and the spread of ideas on a global or worldwide scale.

Factors helping globalisation

- International organisations e.g. banks and TNCs
- Large transnational companies
- Better transport
- Computer and Internet technology helps speed up communication
- Expanding markets in developing cities

**Old economy:** Production of manufactured goods. Locally or regionally based. Industry attracted to raw materials, power, cheap land, good transport, etc. Local labour. Mass production. Job-specific skills e.g. engineer or fitter. Examples: iron and steel, textiles – mainly male employment.

**New economy:** Production of knowledge, ideas and services. Globally based and interconnected. Human resources are very important. Risky. Specialised services. Global labour force. Activities attracted to electronic networks (Internet, phone, etc.). Examples: Jobs in ICT, TV production, bio-technology. Equal male and female employment.

## Where are the 'new economies'?

Some geographical areas are more suited to the 'new economy' than others. Global hotspots are the most connected and influential places. Examples are Singapore, Ireland and Switzerland.

But there are also areas where economies still rely on more traditional industries. The least globalised countries include many in Africa, along with Indonesia and Bangladesh. These countries do not engage much in world politics, which keeps them isolated.

Creative industries (the film industry, media, advertising and marketing) are associated with the 'new economy'. In the UK, there is a concentration of these new knowledge-based industries in the south-east of England. The northernmost part of Scotland is also a hotspot (see the map).

## Dot.coms

The 'new economy' is dependent on information, knowledge, networks and technology. Typical of the 'new economy' are the dot.com businesses.

Dot.coms are Internet-based companies. Often they were started with little money but plenty of good ideas. However, many of these new companies failed in the late 1990s and first years of the 21st century. The reasons why include: lack of management experience, poor technological know-how, and keeping costs under control.

Amazon is one of the most famous Internet companies. It is an American electronic commerce (e-commerce) company, based in Seattle. Amazon was one of the first major companies to sell goods through the Internet, and is a good example of a 'new economy' company. It originally started as an online bookstore, before diversifying into electronics and music. Nowadays, it sells a huge range of products. It has over 17 000 employees in the USA, China, India, Japan and across Europe.

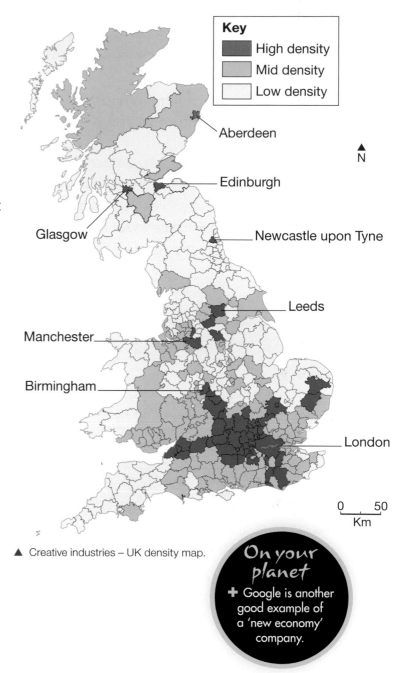

**Key**
High density
Mid density
Low density

Aberdeen
Edinburgh
Glasgow
Newcastle upon Tyne
Leeds
Manchester
Birmingham
London

N

0    50
Km

▲ Creative industries – UK density map.

*On your planet*
+ Google is another good example of a 'new economy' company.

### your questions

1 What are the differences between the old economy and the new economy for **a** people in Britain **b** people in LEDCs **c** technology companies **d** banks **e** skilled jobs **f** unskilled people
2 From this, make a list of the winners and losers of the new economy.
3 In groups of 2-3, study a 'new' company, like Apple. Design a presentation on its locations, products, main markets, and how it keeps its brand name alive.

➕ **In this section you will be looking at the 'new economy' and how it affects people in different parts of the world.**

When we think about the 'new economy', we have to remember that it's all to do with globalisation. There are lots of views about globalisation – some think it's a good thing, while others think that it's extremely bad. Globalisation has different elements – economy, environment, culture, society and politics. When thinking about the 'new economy', we need to bear these in mind (see below).

◀ Globalisation: this mind map shows things that we may associate with the word 'globalisation'.

## Global inequalities?

Some people are anti-globalisation, because they think that it supports the development of an unequal world. However, this is a simple view of a very complex situation. Political and social history, climate and physical geography, differences in trade and flows of money, all interrelate and contribute to the situation.

But the facts of inequality cannot be disputed. In 1960, the richest 20% of people shared 70% of global wealth between them. By 2004, the richest 20% controlled over 90% of the world's wealth. So globalisation has, more than likely, helped to increase the inequality between rich and poor.

## Bangalore 'Boomtown'

Few places have seen the dramatic effects of globalisation more than Bangalore, the Silicon Valley of India. It has experienced an amazing IT boom that has transformed the Indian economy.

But despite this success, there are concerns for the city:

- Poor transport and services, called its **infrastructure**, which have lagged well behind the pace of change.
- Globalisation has brought a change in culture. Some people object on religious grounds to the wild nightlife. Police now enforce a curfew forbidding dancing in bars after 11pm.
- **Rural-urban migration**. This process is causing overcrowding and a lack of affordable housing. It contributes to a rising cost of living.

> ✛ The systems serving a country are known as its **infrastructure**. These include transportation, communication, power plants, healthcare and education services.

## China's growth

China's economy has grown massively in recent years, with its industry largely based in manufacturing (making things). Before long, China's economic focus will probably shift more towards research and development, and technical innovation. The manufacturing industry will become more sophisticated by developing in aviation and aerospace.

But, there has been an environmental cost to the rapid growth that has brought China to its current position:

- 70% of China's rivers and lakes are polluted.
- 100 cities suffer from extreme water shortages.
- 30% of China suffers from acid rain caused by emissions from coal-fired power stations. China emits more $CO_2$ pollution than the USA.
- 85% of trees along the Yangtze River have been cleared. This has led to dust storms and landslides.

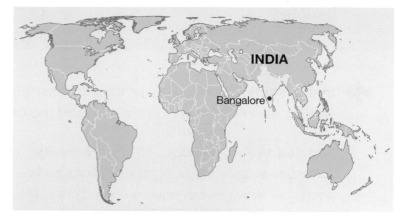

| Occupation | USA | India |
|---|---|---|
| Telephone operator | 14 | 1 |
| Payroll administrator | 18 | 2 |
| Accountant | 25 | 8-15 |
| Financial administrator | 38 | 10-15 |

▲ A highly qualified, but relatively affordable, labour force has allowed India's economy to boom. This table compares hourly wages in 2005 for four jobs in India and the USA (all the figures are in US dollars).

▲ Water pollution in China.

### your questions

1 Study the mind map of globalisation. In pairs, take any 8 of these and draw a table showing 'benefits of globalisation' and 'problems caused by globalisation'. Use examples.

2 Using evidence say whether you think globalisation is good or bad.

3 Are companies which relocate overseas just exploiting cheaper labour? What are the advantages and disadvantages to the employees of a company owned by a foreign transnational company?

✚ In this section you will learn about transnational companies (TNCs) and how they control much of the global economy.

A **global shift** in economic activity has resulted from globalisation. Advances in technology, communications and networking mean that economic activities can now be moved out of MEDCs into LEDCS. The result is a new global economic map.

At first this global shift involved labour-intensive manufacturing, such as making clothing. But, increasingly, it includes all sorts of manufacturing. It also includes services such as call-centres.

General Electric is a giant American transnational company. It was a pioneer when it decided to shift thousands of 'back-office' jobs (such as administration) to India. Now many leading companies in insurance, banking, travel, electricity generation and telecoms are transferring their operations to India. This process is known as **outsourcing** (see sections 16.5 and 16.6).

## Transnationals and multinationals

Transnational companies (TNCs) or multinational companies (MNCs) are big and they are getting bigger. There are also lots of them. What makes them special is that they operate in more than one country. In fact, it's usually over several countries or even several continents. This makes them global in their influence.

### Some amazing company facts

1 Of the 100 largest economies in the world, 51 are global companies; only 49 are countries.
2 Combined sales of the world's Top 200 companies account for more than 25% of the world's economic activity.
3 The Top 200 have almost twice the economic clout of the poorest 80% of humanity.
4 Japanese companies are higher than American in the Top 200 ranking.
5 One-third of world trade is simply transactions between various parts of the same company.

In 1914 there were 3000 multinationals. By 1969 there were over 7000. In 2000 there were 63 000

▼ The growth of multinational companies (1600-2000).

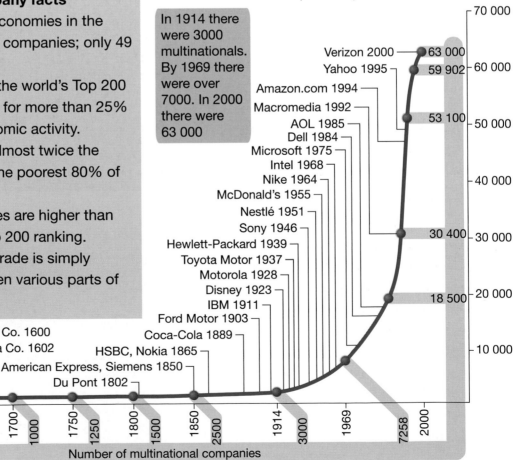

Number of multinational companies

# Proctor and Gamble

In 2005, Procter and Gamble merged with Gillette to form a new global company (P&G), which makes many household products (see below). P&G is typical of the largest TNCs – it has huge global coverage and owns a large number of brands.

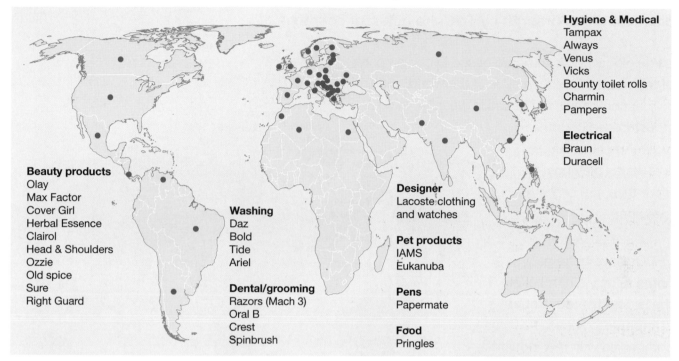

**Hygiene & Medical**
Tampax
Always
Venus
Vicks
Bounty toilet rolls
Charmin
Pampers

**Electrical**
Braun
Duracell

**Beauty products**
Olay
Max Factor
Cover Girl
Herbal Essence
Clairol
Head & Shoulders
Ozzie
Old spice
Sure
Right Guard

**Washing**
Daz
Bold
Tide
Ariel

**Dental/grooming**
Razors (Mach 3)
Oral B
Crest
Spinbrush

**Designer**
Lacoste clothing
and watches

**Pet products**
IAMS
Eukanuba

**Pens**
Papermate

**Food**
Pringles

▲ P&G's global locations and the products they make.

# Dell Computers

In 2001, Dell Computers became the world's largest personal computer (PC) manufacturer. Dell sells 90% of its PCs directly to customers, bypassing shops. This allows customers to 'build' their own customised PC. Ordering is normally done through the Internet. On an average day, Dell sells 150 000 computers worldwide.

Assembling a product from a range of parts is common among computer manufacturers. Dell designs computers in the USA, but assembling is carried out elsewhere (see right), where labour costs are cheaper. Dell is highly competitive in terms of the suppliers it uses and the locations chosen for manufacture.

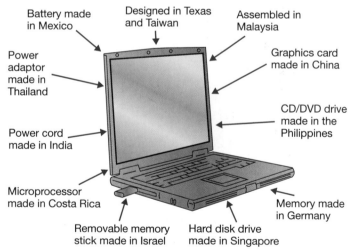

Battery made in Mexico
Designed in Texas and Taiwan
Assembled in Malaysia
Power adaptor made in Thailand
Graphics card made in China
Power cord made in India
CD/DVD drive made in the Philippines
Microprocessor made in Costa Rica
Memory made in Germany
Removable memory stick made in Israel
Hard disk drive made in Singapore

▲ The global origins of a Dell computer.

## your questions

1 What is the 'global shift'?
2 Make a list of the key decision makers that have helped the global shift to happen.
3 Why is the global shift a relatively new process?
4 Classify the origins of computer parts in a Dell computer into 'developed' and 'developing' countries.
5 Work in groups to research one TNC/global brand of your choice. Use the Internet to find out more about its products, profit, employees, etc. Produce a map like the one for P&G.

# TNCs: good or bad?

✚ In this section you will learn about the advantages and disadvantages brought by TNCs to different countries.

Large global companies are based in certain parts of the world. More than 90% of the world's 500 largest companies are based in wealthy industrialised countries. North America, Europe and Japan play host to more than 450 of them (see right).

In the last 20 years, LEDCs have become home to only a handful of the largest **transnational companies**. Those companies with real industrial power are still located in the more developed parts of the world. However, the tide may be turning (see the graph).

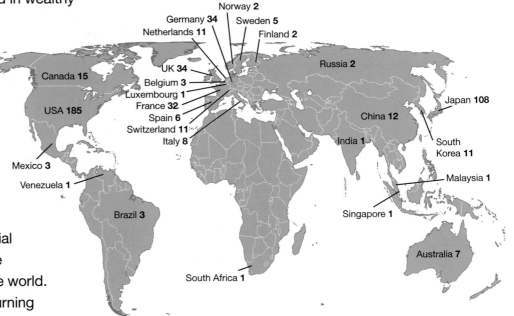

Norway **2**
Germany **34**
Sweden **5**
Netherlands **11**
Finland **2**
Russia **2**
Canada **15**
UK **34**
Belgium **3**
Luxembourg **1**
France **32**
Spain **6**
Switzerland **11**
Italy **8**
USA **185**
Japan **108**
China **12**
India **1**
South Korea **11**
Mexico **3**
Malaysia **1**
Venezuela **1**
Brazil **3**
Singapore **1**
Australia **7**
South Africa **1**

▲ The 'home' countries of the world's 500 largest companies.

## Globalisation 2.0?

According to a 2008 study, TNCs from LEDCs are claiming an increasing share of the global market. This trend is being called 'globalisation 2.0'. Companies from LEDCs are now performing as well as those in MEDCs.

## Winners and losers

The changing economic map resulting from the global shift has created advantages and disadvantages for LEDCs and MEDCs. These are summarised in the table opposite.

► The world's top 1000 companies. For each year, percentage figures are given for those companies based in MEDCs and LEDCs.

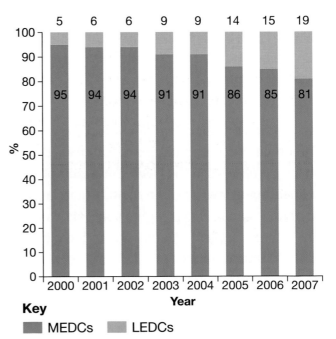

| | 5 | 6 | 6 | 9 | 9 | 14 | 15 | 19 |

| | 95 | 94 | 94 | 91 | 91 | 86 | 85 | 81 |

Year: 2000 2001 2002 2003 2004 2005 2006 2007

**Key**
■ MEDCs  ■ LEDCs

|  | *Advantages* | *Disadvantages* |
|---|---|---|
| **MEDCs** | • Cheaper imports from LEDCs may benefit consumers in MEDCs.<br>• Global shifts allow job opportunities in many different countries.<br>• The growth of economies within LEDCs may lead to a demand for exports from MEDCs.<br>• Greater efficiency caused by outsourcing leads to the development of new technologies in MEDCs. This attracts foreign investment into MEDCs.<br>• The loss of industries to LEDCs can improve the environmental quality in MEDCs. | • The outsourcing of jobs leads to job losses within MEDCs. The job losses are often of unskilled workers.<br>• The job losses are concentrated in certain geographical areas and certain industries. This can lead to deindustrialisation, mass regional unemployment and social problems. |
| **LEDCs (including NICs such as India)** | • Exports rise – this generates income and encourages investment. This can lead to a 'multiplier effect' on the national economy.<br>• Wealth can trickle down to local areas with many new jobs. These jobs are relatively well paid.<br>• New technology leads to people learning new skills.<br>• The growth in employment spreads wealth, and helps to reverse the global inequality between LEDCs and MEDCs. | • New jobs in urban areas promote rural-urban migration. This brings its own problems.<br>• There can be social impacts, e.g. exploitative sweat shops where workers work long hours.<br>• It can lead to over-dependence on a single manufacturing process, e.g. textiles.<br>• It can reduce food supplies as people give up agriculture in favour of these new jobs.<br>• Environmental issues (pollution) are associated with an over-rapid industrialisation. |

Clearly a shift of industry on this scale has pros and cons for the countries themselves and for people living there. It's very easy to spot the disadvantages for MEDCs where the job losses occur. And it's easy to spot the benefits for LEDCs and NICs who gain all these new jobs. In reality, the situation is much more complex.

## The real winners

The real winners are the TNCs. They improve efficiency and increase profits by moving production to more favourable locations where costs are lower. Some people would also argue that, at a global scale, world trade is promoted – thus improving choice and value for the consumer.

▲ Tata (an Indian multinational) now owns UK-manufactured Jaguar cars. In 2008, it paid £1.5 billion for the Jaguar and Land Rover brands.

**your questions**

1 Draw a graph to show the total number of 'largest' companies in a North America b Europe c Asia d the rest of the World. Describe and explain the graph.

2 In pairs discuss whether globalisation brings more benefits to MEDCs or LEDCs.

3 Write a 500 word essay titled 'The winners and losers of the global economy'.

4 **Exam-style question** Using examples, explain how the global economy can bring benefits to developing countries. (6 marks)

✚ **In this section you will learn about the impact of outsourcing.**

**Outsourcing** occurs when a job is given to a company overseas that was formerly done at home. Often this other company is in an LEDC. The decision to outsource is made to reduce costs and make the main company more profitable. There are two types of off-shoring:

- back-office (accounting, IT systems and human resources)
- business processes (call-centres).

China, Malaysia, Singapore, Australia, South Africa and Vietnam are all destinations for off-shoring. Now Ghana, Barbados and Bangladesh are being considered. Parts of your GCSE examination may even get marked online in Australia!

## Call-centres: India

In the mid 1990s, call-centres were shifted to south Asia, especially India. In 2003, when outsourcing reached a peak, 8000 jobs were moved from the UK to India. India now leads the world in outsourcing.

The value of outsourcing to India in 2007 was £24 billion. IT and business process outsourcing services now account for 5.4% of India's Gross Domestic Product (GDP). They have a huge impact on cities such as Bangalore – the centre of the industry.

**On your planet**
✚ In comparison with the global shift of manufacturing, global outsourcing is a relatively recent occurrence – post-1995.

Reduced time for product to get to market

Cost savings and higher profits

Operational expertise

**Reasons why companies might outsource**

Access to a wider range of knowledge and talent pool

Improved quality

Shared risks

Good contracts

**Outsourcing job flows**

In general, outsourcing follows these routes:

- From the UK to countries of the Commonwealth (especially south and south-east Asia and southern Africa).
- From the USA to Latin America and south/south-east Asia.
- From Japan to east/south-east Asia.
- From France to former French colonies, e.g. parts of the Caribbean and West Africa.
- From the EU to eastern Europe.

There are several reasons why India is a particularly attractive destination for outsourcing:

- The Commonwealth link with the UK.
- English language skills are good in India.
- IT is well taught in the Indian education system.
- The low wage costs in India.
- Specialist off-shoring companies are growing in India.
- India produces 3 million graduates a year.
- Low telephone/Internet costs.

Recently, however, some companies are rethinking off-shoring to India. Wages in India are now increasing. And sometimes customers' personal financial details have apparently been sold by unscrupulous employees.

## New Europe

New Europe includes countries such as Poland, Hungary and the Czech Republic. These are attractive low-cost European choices for companies who want to hire skilled people without the difficulty of regularly visiting Asia. Apart from being closer, New Europe countries are attractive because they are part of the EU trading bloc. Many of their workforces also speak a range of languages, such as English and German, which is often useful.

Outsourcing has led to 3000 jobs moving to Poland from the UK. As well as European firms, several American TNCs are also outsourcing – examples are IBM to Poland, the Czech Republic and Slovakia, and Dell and Hewlett-Packard to Slovakia.

*On your planet*

+ The fact that some new destinations are in the same time-zone as Germany and France, and only 1 hour ahead of the UK, is a definite advantage over India, which is 5 hours ahead of GMT.

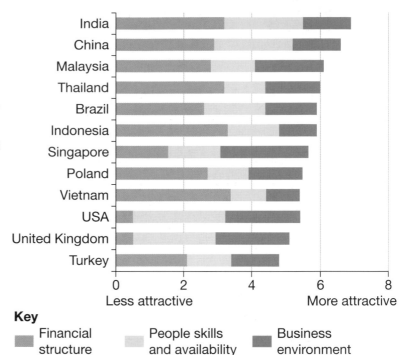

**Key**

| Financial structure | People skills and availability | Business environment |

▲ Off-shore location attractiveness index, 2007, for selected countries

*On your planet*

+ In the UK, there's growing consumer dissatisfaction with telephone calls being answered offshore.

| Company | Original location | Examples of activities |
|---------|-------------------|------------------------|
| Accenture | UK | Financial management, customer support, IT services, parcel tracking, business services |
| DHL | USA | |
| Hewlett-Packard | USA | |
| ExxonMobil | USA | |
| IBM | USA | |
| Siemens | Germany | |
| Tesco | UK | |

▲ Companies outsourcing to the Czech Republic.

### your questions

1 On an A4 blank world map, draw and annotate the main flows that are described in the 'job flows' box opposite. Your flow arrows will be to general regional areas.

2 Using the off-shore location attractiveness index, make up a ranking table to show the top five countries in terms of: financial structure (costs), business environment, people skills and availability.

**In this section you will learn about the benefits and problems brought by outsourcing.**

In the future there may be further outsourcing of services and office functions in skilled professions. For example, global companies are beginning to outsource their research and development – something which has traditionally been done at headquarters. But who are the winners and losers, now and in the future?

## Costs and benefits: LEDCs/NICs

Sometimes it's difficult to assess the costs and benefits of outsourcing.

+ **Westernisation** is converting to the customs and ideas of Western civilisation.

| Costs to LEDCs | Benefits to LEDCs |
|---|---|
| • Westernisation and loss of cultural identity as a result of training and working with English speakers. <br>• Unsocial hours, e.g. a 5-hour time difference between India and the UK means that workers in India don't work from 'nine to five'. <br>• High staff turnover in call-centres – workers may have limited skills when transferring to other jobs. <br>• Abuse from angry customers may result in workers developing anti-Western attitudes. <br>• Some employers call their programmers 'slave coders' – hardly a term to give the workers a feeling of worth. <br>• The economy may get locked into 'low value' activities. <br>• Further relocation will occur as wage costs increase. This may lead to loss of employment in some areas. | • In 2008, outsourcing brought in an estimated $24 billion to the Indian economy. <br>• Outsourced jobs have higher starting salaries compared to the average wage in many LEDCs. <br>• Call-centres and back-office services employed 3-4 million people (2007). <br>• More jobs are created in LEDCs (receiving countries) than are lost in MEDCs. <br>• Young workers have a higher than normal income. <br>• Gender inequalities are being reduced as the call-centre jobs are split 50:50 (male:female). In the UK, the ratio is 20% male to 80% female. |

NO, YOU MAY NOT OUTSOURCE YOUR HOMEWORK TO INDIA.

*On your planet*
+ A range of health, education, planning and architectural functions are now being outsourced to a wide variety of countries.

# Costs and benefits: MEDCs

| Costs to MEDCs | Benefits to MEDCs |
|---|---|
| • Job losses. By 2008, 200000 jobs had been outsourced from the UK. 500000 had been outsourced from the USA.<br>• There is a loss of female jobs in particular.<br>• Many job losses have been from economically vulnerable areas which have suffered deindustrialisation, such as the north-east.<br>• Skilled service jobs are also moving, e.g. Barclays Capital moved 400 jobs from Canary Wharf (London) to Singapore in 2004.<br>• Some companies, such as Dell, are now taking work back because of a lack of quality control in some LEDCs. So outsourcing can involve a loss of control. | • Many companies claim efficiency benefits when they outsource.<br>• Outsourcing means that companies make larger profits which benefit shareholders.<br>• There can be a 40% cost saving in comparison with keeping operations in the home country (such as the UK or USA).<br>• The labour costs of a call-centre worker in the UK are £18-20000 per year, compared with just £2500 per year in India.<br>• Firms often outsource to LEDCs the manufacture of older products. This allows them to focus in the home country on developing new products. |

▲ A call-centre in India: job losses in the home country are balanced by cheaper running costs and wages.

## your questions

1 Look at the cartoon opposite. What are the main messages coming across? Does the person who drew the cartoon think that outsourcing is a good or bad idea?

2 Look at the photo of a call centre in India and think of your own experience in speaking to call centre staff.

Describe in 100 words how you might feel abut their job.

3 **Exam-style question** Using examples, explain how outsourcing produces winners and losers. (6 marks)

✚ **In this section you will think about how work has already changed, and how it might change in the future.**

It's difficult to predict the future. In the 1980s, for example, people writing about the 'important' countries of the future, featured Japan – but not India or China. They also predicted a decrease in the number of working hours. This has happened in some countries. But for many places, such as the UK, there's actually been an increase in the number of hours worked.

There's increasingly a shift towards a **knowledge-based economy**, rather than a product-based and service-based economy. In London, for instance, there are now more people employed in knowledge industries, such as research and development, education and consultancy, rather than the traditional City jobs in banking.

> ✚ A **knowledge-based economy** depends on jobs such as research and development. Jobs in the knowledge-based economy rely on computers.

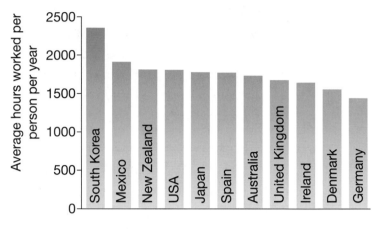

▲ Number of hours per person spent working each year – a comparison

Three major factors are likely to be important in controlling the future world of work:

1 The entry of China and India into the world economy. It's thought that about 40% of the world's population are soon going to be 'new' producers and consumers.

2 Technology – enabling companies to distribute their work around the world and be open for business 24/7.

3 In many MEDCs, the workforce is ageing and getting smaller as birth rates decline.

▲ Computers in the workplace were almost unknown 50 years ago. At first they were vast mainframe machines like this one.

### On your planet

✚ The Intel computer chip company has manufacturing and processing operations round the globe. The organisation expects managers to engage in teleconferencing outside 'normal working hours'. This is due to the 'time zone spread' of the company's workforce.

## Changes in the workplace

Only since the 1970s have computers become the desk tops that we are familiar with today. Computers are now very important to the way that we work. This book couldn't have been written and produced without using a computer! Laptops and WiFi access are already enabling work to take place on the move. And working at home (**teleworking**), often linked to a more rural environment, is also common. Increasingly, locations are being 'wired', so that you can connect to the Internet from a large number of sites.

Flexibility in location means more outsourcing, with fewer city centre office locations. People who can use this technology can now work anywhere and at any time. So the rigid pattern of working particular hours has disappeared for this workforce.

## Changing world of work in China

China is changing the global balance of economic development. As China industrialises, there are plans to encourage the spread of wealth inland from the coastal areas. But who are the winners and losers within China as it becomes more powerful? Look at the table.

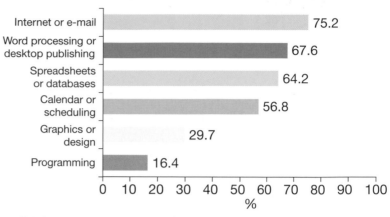

▲ This bar chart shows how we use computers in the workplace (over 75% of workers use computers for the Internet or e-mailing).

▲ WiFi is often available in coffee shops.

| Winners/benefits | Losers/costs |
| --- | --- |
| • Business boosts tourism.<br>• Reduction in the number of people in poverty (490 million in 1981 to 18 million in 2004).<br>• Olympic games 2008<br>• Beijing is now a rising political and financial centre.<br>• The Chinese government is investing in infrastructure and public transport. | • Some regions are suffering from 'overgrowth' and can't house migrant workers.<br>• Growth imbalance – China is a vast country and some regions see very little investment.<br>• Increasing energy demands and worries about use of coal-fired power stations. The Three Gorges Dam has had a big environmental impact.<br>• Few international financial services because of corruption in some cities. |

### your questions

1 Write about how you think the world of work will change for you in the future. Consider **a** where you will work **b** who your employer might be and where, and **c** how you will communicate with your colleagues. What skills do you think are required in the 21st century workplace?

2 What are the advantages and disadvantages of flexible working patterns?

3 How might the number of working hours per week change in the future? Give reasons.

4 In pairs discuss and write up, 'The new economy is much better for people than the old economy'.

➕ **In this section you will think about whether changes in the workplace are sustainable.**

In the future, what will work be like in MEDCs? There will probably be a flexible labour market with trade unions having less power. This goes with fewer benefits, such as healthcare and company pensions. Many workers won't have a 'career for life'. Instead they will have many shorter-term contracts.

For LEDCs, there may be workforce insecurity. Workers in Asian, African and Eastern European countries may experience poverty in old age – in the absence of good pensions or welfare systems. Some countries, especially those in Africa, are very vulnerable to poverty.

## Flexible working

**Flexible working patterns** include flexitime, term-time working, home-working, job-share and part-time working. There are a number of advantages and disadvantages to this (see right).

## MEDCs: the workforce size

Low birth rates in MEDCs are beginning to affect the job market. More people are now retiring from work than are joining the workforce.
- In Western Europe and Japan there's now a declining workforce.
- Australia, New Zealand, Canada and much of the rest of the world will continue to have large workforces fuelled by migration.

**Advantages of flexible working**
- Better health
- Less stress
- Less absenteeism and sickness
- Lower staff turnover, so that people with skills are kept
- Better productivity
- Less traffic congestion and pollution as people don't commute to work

**Disadvantages of flexible working**
- Lower wages (possibly)
- Possible isolation from work colleagues
- It's sometimes difficult to motivate and organise home-workers
- In some industries, flexible working could be an obstacle to promotion

▶ Projections for working-age populations in Japan and Western Europe.

**Key**
- 2000
- 2030
- 2050

*% of total population* (0–100)

Japan — Germany — Italy — UK — France

In MEDCs, migration will become increasingly important as a way of plugging the shortage of workers and skills. Using imported workers (rather than outsourcing jobs) may change the **age profile** of the country. And more people of working age means more tax revenue – this may help to support the elderly. Migrant workers tend to be fairly young. Many migrants will have families which will boost the birth rate.

## Women in the workforce

The increasing proportion of women in the workforce has been a feature of MEDCs over the past 30 years. However, the recruitment of women won't continue at its current rate as more choose to have families – trying to find the right 'work-life' balance. Female work is often in the service sector. In the future, there may be more job sharing and part-time working in MEDCs. This should allow the birth rate to be maintained or to increase.

Women are a fairly constant proportion of the workforce in most countries. The lowest proportions are found in societies where there are cultural constraints, such as the United Arab Emirates (an Islamic country). Countries with the highest proportion of women working are Russia and Tanzania. Here governments have encouraged women to work.

▲ In Africa many women work in agriculture. In the future, more outsourcing to LEDCs may provide a greater range of employment opportunities.

+ The population can be broken down on an age basis. This gives us an **age profile** of a country.

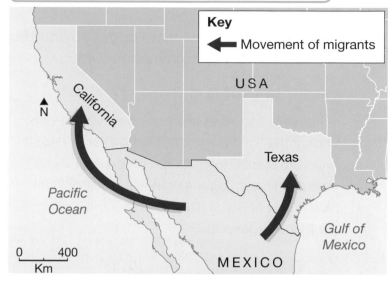

▲ In the USA there are an enormous number of migrants (both legal and illegal) from Mexico. These are now working in low-skilled jobs, especially in California and Texas.

▲ Tourism in Africa: this could increase employment and help to improve the quality of people's lives.

### your questions

1 Draw a spider diagram titled 'The future world of work'. In pairs, discuss ideas and add 'spokes' to show jobs that people might do and where these jobs will be done.
2 Who do you think the world of work in 2050 will be better for? Who will it be worse for? Explain your answers.
3 **Exam-style question** Using examples, explain whether the future world of work will e more or less sustainable than now. (6 marks)

✚ In this section you'll learn about how to prepare for unit 3, the decision-making examination (DME).

## How is unit 3 different?

Chapters 1 to 16 in this book help you to learn about the 'Dynamic planet' (unit 1) and 'People and the planet' (unit 2). You prepare for these exams by learning the work you have done in class and at home. Unit 3 is different, and comes in three stages.

### 1 – The pre-released booklet

In February, before the June exam, you'll be given a resource booklet about an issue from somewhere in the world. It will probably be somewhere unfamiliar. Don't worry – it isn't your knowledge of the place that's being tested, but your ability to understand the issue there. For this exam, you don't have to learn material off by heart – you'll have a fresh copy of the booklet in the exam to refer to. But you **won't** be able to take your notes in.

### 2 – Learning in class

You'll have 4 months to go through the booklet in class (and at home). You can research more if you want – your teacher might even give you extra websites to look at. It might help you to understand the issue if you research using the Internet a bit. But you don't *have* to research – everything you need is in the booklet.

### 3 – Taking the exam

The exam lasts for 1 hour for both Foundation and Higher Tiers, and has 50 marks available. There are three, roughly equal, sections (see right).

**Remember**

✚ It's your ability to understand the issue that's important.

**How do I learn?**
Make notes, practise writing answers, and work in books or files - just as normal.

**Remember**

✚ Everything you need is in the booklet.

| Section A | Short, structured questions for Foundation Tier. For Higher Tier, short questions plus some more open-ended questions. |
| --- | --- |
| Section B | Slightly longer questions, which need detailed evidence, using information from the resource booklet. |
| Section C | Involves decision-making, based on advantages and disadvantages of the options. You'll be asked to write in greater length. |

▲ The sections of your exam.

## What will the topics be?

The topic for the DME will be different each year. It could be a large-scale or a small-scale issue. **It will always be about sustainability**, so you'll need to understand what sustainability means – based on what you've learnt in units 1 and 2. Examples of issues include:

- **small scale** – e.g. Grampound, which is the example chosen for this chapter.
- **global scale** – e.g. Antarctica – look right for a taster of this example.

## How are questions marked?

**Structured short questions** are marked on points – the more you make, the better you do. If there are 4 marks, make 4 points. Include evidence or examples from the resource booklet. In short answers up to 4 marks you can use bullet points, as long as it is clear what you mean.

**Longer answers** of 6 marks or more are marked using levels. This especially applies to section C, where you'll be asked to explain a decision. These questions will test your ability to:

- explain reasons
- use evidence from the booklet to support your reasons
- use details or examples from the course.

There are three levels in the mark scheme – always aim for level 3.

- Level 1 answers – generalised, with few or no examples.
- Level 2 answers – some evidence from the booklet, or what you've learnt, to help explain something fairly clearly.
- Level 3 answers – detailed explanation, always with evidence from the booklet or what you've learnt to support your ideas.

Sections 17.7 and 17.8 give you more information about what the questions in your exam will be like.

▲ The issue is tourism in Antarctica. Should Antarctica be left alone (top photo) or should it be used for human activity, such as tourism

**QWC**

Some longer questions have an asterisk (*). Here, the quality of written communication (QWC) is being assessed, so you need to write in clear sentences with accurate spelling and punctuation. QWC marks are given for:

- legibility of your writing
- accuracy of your spelling and punctuation
- use of geographical terminology (e.g. 'migration' instead of 'people moving')
- organising your answer logically.

In Sections 17.2 to 17.6 you'll learn about the decisions that need to be made about housing in Grampound, a village in Cornwall. This is an example of what you can expect to find in a resource booklet for the DME. Sections 17.7 and 17.8 then show you what sort of questions you would be asked.

## Welcome to Grampound

Grampound is a village in mid-Cornwall, built along the busy A390 road between Truro and St Austell. It was an important historic township, although now its population is only about 640. It is surrounded by a rural, mainly farming, area. The village is a working community, with jobs found either in Truro (the county town) 8 miles west, or in St Austell 6 miles east. There has been some small growth in its population, after a few new houses were built in the 1980s and early 1990s.

Grampound dates back to the 13th century, and there are Roman settlements nearby. Most of the oldest buildings in the village are in the centre, and form a *Conservation Area*, which has many *listed buildings*.

The village has a strong sense of community, and was one of England's top villages – winning the title for West England in the 2007/8 Calor Village of the Year for England competition. In the competition, Calor were looking for well-balanced communities, which gave an excellent quality of life for all residents.

### Grampound is dying!

It is rapidly losing shops (three in recent years), services (the garage has closed) and employment (the tannery closed). The absence of a by-pass has prevented future development, because access to the main road in the village is difficult at peak travel times. There is little wealth in an area where wages are low and house prices are rising rapidly. Few young local families can afford to live here, so the future of the playgroup and the school is threatened.

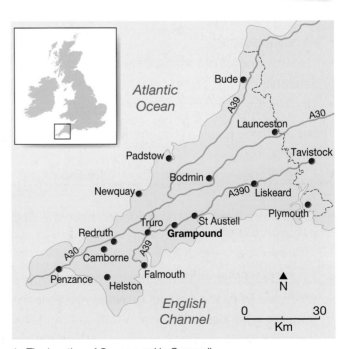
▲ The location of Grampound in Cornwall.

▲ Fore Street, Grampound - the main road leading through the village, and one of the main roads through Cornwall.

# What is the issue in Grampound?

After 30 years of decline, Grampound's population has risen since 1981, following the building of a new housing development on Mill Lane (see the top table). But its population is growing at a slower rate than in Cornwall as a whole, and there is demand for more housing in the village.

In 2002 and 2005, the local parish council carried out surveys to find out Grampound's needs, such as housing. 172 households in the village (about 60%) completed the forms. The surveys found these problems:

- House prices in the village had risen rapidly for 10 years, up to 2005.
- Most new housing was 'executive' style, and was too expensive for many local people to afford.
- More housing is needed for the people who would like to live there.
- Because of conservation areas, the village is short of space where new housing could be built.

The parish council also found that:

- 19 households were looking for alternative housing in the village, most of whom wanted to buy.
- Another 18 people or families wanted to live in Grampound, but could not afford to do so.
- 82% of households said they would approve of new developments in the village, if they included affordable housing.
- 43% of people would accept over 20 new houses by 2015. Nearly a quarter of people would accept over 30 (see the bottom table).

An estimated total of 25 households in the village need affordable housing (assuming the survey is a reliable sample of Grampound's population).

- The main demand is for two and three bedroom affordable houses to buy, although some people do want to rent.
- There is also some demand for smaller flats and bungalows, as well as larger four bedroom houses.

| Year | Population of Grampound (includes outlying farms) |
|------|---------------------------------------------------|
| 1951 | 666 |
| 1961 | 615 |
| 1971 | 604 |
| 1981 | 580 |
| 1991 | 620 |
| 2001 | 638 |

▲ The population of Grampound.

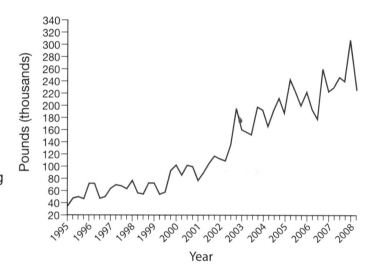

▲ House price rises in Grampound 1995-2008.

| Number of new houses by 2015 | % of residents supporting this |
|------------------------------|--------------------------------|
| no more than 10 | 23% |
| 11 – 20 | 34% |
| 21 – 30 | 19% |
| more than 30 | 24% |

▲ How many houses should be built in Grampound?

## Cornwall's population

Cornwall's population grew from 500 000 in 2001, to about 535 000 in 2008 – making it one of the UK's fastest growing counties. The increase is caused by *inward migrations* of people retiring to live there. People come for the environment, especially the coast.

The population increase puts pressure on housing. Many people sell up in cities like London, where houses are more expensive, and move to Cornwall. They increase the demand for housing, which drives prices up. Local people cannot afford to compete.

## Housing affordability

Cornwall has the lowest wages in England and Wales, making housing difficult to afford for local people. Compared to the rest of the UK:

- unemployment is higher than average, especially for men
- the decline in primary jobs (farming, fishing, china clay and mining) has been severe
- adult weekly wages (£329 in 2005) are 25% below the UK average, and 60% lower than those in central London
- tourism is booming and provides jobs for 25% of Cornwall's population (but jobs in tourism are often seasonal, part-time and minimum wage)
- local buyers have to compete with people buying houses as holiday lets, or second homes. Although Grampound has few of these, a quarter of houses in some villages nearby are holiday homes.

## Historic buildings and countryside

Grampound dates from medieval times. The long, narrow fields near the village survive from that time, and are known as *burgage* plots. They are protected from development. Many houses are small, stone-built cottages, over 200 years old.

▲ Cottages in Grampound.

Any development in Grampound must now suit the landscape. The area is classed as 'very high landscape value' – one category away from 'Area of Outstanding Natural Beauty'. The local council gives guidance on new developments, and says that new housing should be built from natural materials and blend in with local architecture.

## The main road

The A390 road through Grampound is one of Cornwall's busiest main roads, especially in the holiday season. National government statistics show a 22% increase in traffic in Cornwall between 1997 and 2007.

- Pavements in some parts of the village are narrow, or even non-existent.
- The village is on a steep hill, and some vehicles travel at speed. In the past, some have come off the road and collided with houses.
- The local council considered a by-pass for Grampound. A proposal was refused in 1996, and again in 2001, because Grampound's population is not large enough to justify one.

## Shops and services in Grampound

The village has a number of businesses (see the table). The general store contains the post office. It sells local produce and prices are similar to local supermarkets. However, it faces two problems:

- In 2008, the post office was threatened with closure – but survived. However, like many rural post offices, its future is uncertain. Without it, the shop might not survive either.
- It is on the main road through the village and there is nowhere to park. Many people in the village travel to shop elsewhere.

The village primary school (see right) is small, with 70 pupils in two classrooms – one for those in Key Stage 1, the other for Key Stage 2. It is over-subscribed, but has nowhere to expand to.

| Shops | Businesses |
|---|---|
| General store and post office | 2 accountants |
| Atlantis smoked foods | Bags and homeware suppliers |
| Antiques | Bed and breakfast |
| Furniture – kitchens and pine | Car sales service and repairs |
| Pet food and products | Carpentry and joinery |
| **Food and leisure** | Business services |
| Chinese restaurant | Driving tuition |
| The 'Dolphin' pub | Education consultancy |
| **Services** | 2 electricians |
| Doctor's surgery | Fallen stock collection |
| Grampound primary school | Fancy dress hire |
| | Farm beef |
| | 2 furniture makers |
| | Furniture supplier |
| | Handyman services |
| | Hairdressing |
| | 2 plantsmen / landscape gardening |
| | Painter |
| | Picture framing |
| | Plant centre |
| | Potter |
| | Quantity surveyor |
| | Shopfitter |
| | Furniture – upholstery & restoration |
| | Website design |

▲ The school in Grampound, between the village hall and the main road.

### 'It could be anywhere.'

Houses built in Grampound since 1960 have been of mixed quality. Cornwall County Council feels that too many of these:

- have an 'anywhere' feel, ignoring the historic character of the village
- are designed poorly and do not use local building materials.

Now, some builders (e.g. Rosemullion Homes – see section 17.5) design houses that suit the landscape better – but have to use materials that are expensive.

### Protecting the village

Current planning laws limit where housing can be built, in an attempt to protect the landscape. The options are limited to:

- **infilling** spaces between houses
- building on **former gardens** (where there is access)
- developing **brownfield sites**.

These all depend on local landowners agreeing to the redevelopment.

However, some people now feel that Grampound can only survive with more people, and are looking for sites where more houses could be built. This would require an extension of the planning boundary (see the map).

### Brownfield Sites

The local council assessed Grampound in 2007, and found five sites within the conservation area which – they claimed – were under-used or derelict, and which spoilt the village. Some of these are important historic buildings (shown on the map):

1 The Manor Tannery.
2 The site of the former petrol garage.
3 The old weighbridge.
4 The buildings occupied by Fal Valley Pets.
5 Town Mills.

Of these:

- the Manor Tannery is being considered for redevelopment (see page 277)
- neither the site of the former petrol garage, nor the buildings occupied by Fal Valley Pets are for sale
- the old weighbridge is too small for redevelopment
- town Mills is semi-derelict.

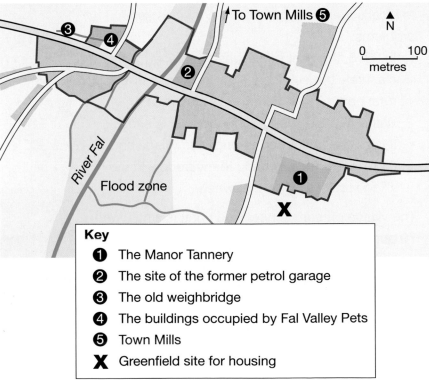

**Key**
- ❶ The Manor Tannery
- ❷ The site of the former petrol garage
- ❸ The old weighbridge
- ❹ The buildings occupied by Fal Valley Pets
- ❺ Town Mills
- X Greenfield site for housing

▲ This map shows the current Conservation Area boundary in Grampound. This should mean that development is only allowed inside the line, and it must be designed sympathetically.

# Grampound on the map

▲ The village general store and post office. Located on a steep hill, there is nowhere to park, so passing trade is limited.

▲ New housing in Mill Lane. Some people feel that new housing ought to be built out of local materials – but this would make it more expensive.

▲ The buildings occupied by Fal Valley Pets. Several buildings here are unused.

▲ The former petrol station in Grampound.

Survey map of Grampound. The scale is 1: 25 000.

▲ The village community hall. Opened in 2003, it was built using Lottery funding and other local sources.

▲ Town Mills.

## Proposal 1 – a greenfield site

### The proposal

In 2004, a local building company, Rosemullion Homes, proposed a new housing development on greenfield land, south of the Manor Tannery. The new development will consist of 49 houses – 18 of which will be affordable (37%). The site is outside the area where new housing is currently allowed. This is shown on the map on page 274.

### What are the issues?

The proposal depends on redeveloping the Manor Tannery. The company claims that it can only afford to provide 18 affordable housing units if it is also allowed to redevelop the Manor Tannery (Proposal 2). This would provide 16 more houses to make a total of 65. If this happened, the 18 affordable houses would only be 27% of the total.

Access to the completed estate would be via Bosillian Lane, to the east of the village. It is a narrow country lane and would need widening slightly.

### What do different people think?

- Grampound Parish Council supports a limited expansion of the existing boundary if it is used for affordable housing and small industrial units or workplaces.
- The UK Environment Agency believes that the new buildings would cause more runoff when it rains – and therefore increase the risk of floods in the village.

▲ What new housing in Grampound might look like. This is new housing built by the same company 8 miles away. It won an award for traditional style and use of materials such as slate.

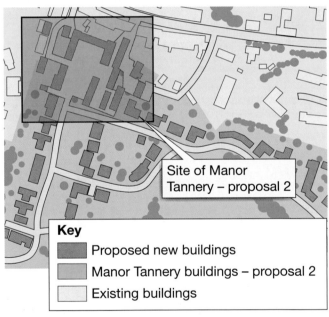

Site of Manor Tannery – proposal 2

**Key**
- ▨ Proposed new buildings
- ▨ Manor Tannery buildings – proposal 2
- ▫ Existing buildings

▲ Plans for housing to the south of the village, outside the area currently allowed for development.

▶ The site of the proposed new housing.

# Proposal 2 – a brownfield site: the Manor Tannery

Rosemullion Homes (see Proposal 1) plan to redevelop the old Manor Tannery. The buildings are semi-derelict, on the site of Grampound's oldest industry. This was a tannery, processing leather for over 300 years until it closed in 2000. The site is within the area where development is allowed.

## The proposal

The proposal is to redevelop the site into:

- 16 residential units plus parking. None of the housing will be affordable. The developers claim that the cost of redevelopment is too high.
- 370 square metres of employment/office space for small, local businesses.

The site is within the conservation area.

## What are the issues?

The site would be expensive to redevelop because:

- tanning leather involves the use of many chemicals, so the site is contaminated and needs cleaning up
- most buildings are listed, and can only be redeveloped in traditional stone and slate
- some parts of the buildings are unsafe and need demolishing first.

## What do different people think?

- English Heritage surveyed the Manor Tannery in 2001 and found that it was 'rare, if not unique' that a complete group of industrial buildings should survive complete with machinery, a water wheel and old water supply system – they are in favour of a redevelopment that would help to preserve the buildings.
- The local council considered making the tannery a visitor attraction but never got any funding. It now thinks that it would be better converted for employment, to preserve its industrial character and fit the character of the village.
- In 2007, the local council's Grampound Development Plan identified the Manor Tannery as a priority for redevelopment.

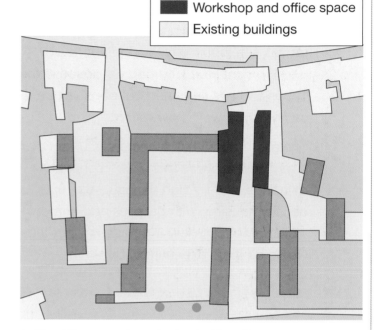

**Key**
- Residential buildings
- Workshop and office space
- Existing buildings

▲ Plan of the proposal to redevelop the Manor Tannery.

▲ The site of the buildings at the Manor Tannery in 2008.

## Proposal 3 – a brownfield site: the old petrol garage

### The proposal

No formal proposal has ever been put forward for this site, but it is within the approved development area. The garage closed in the 1990s and has remained out of use since. The owner, a widow, is now retired.

The site is about half a hectare. Current planning laws could allow as many as 15 cheaper dwellings, at very high density – e.g. flats. Lower density housing (with garages and gardens) would probably allow 3-4 houses – but these would be more expensive.

### What do different people think?

- The Parish Council regards this as a priority for redevelopment.
- The local Borough Council believes that redevelopment of the site will improve the village.

### What are the issues?

- The owner has lived in the house here for over 50 years. She has a strong personal attachment to the site, and is unwilling to sell.
- The site is near the river, although no flood incidents have happened in living memory.

## Proposal 4 – a brownfield site: Town Mills

### The proposal

No formal proposal has been put forward for this site. Town Mills used to be a corn mill. The owners converted it into small workshops for local businesses in the 1980s, but they were destroyed by fire in 2001 and have been derelict ever since. Although it is separate from the main village, redevelopment would be allowed because it would help to improve the look of the buildings. The lane leading to Town Mills has been gradually built along since the 1960s, allowing housing development on an individual basis, with a mix of styles. The site is big enough to allow about 4-6 houses.

### What do different people think?

- Grampound Parish Council would support any re-development which guarantees the future of these historic buildings.

### What are the issues?

- The site has been used for employment workshops in the past, although fire destroyed these in 2001.
- The site is near the river, although no flood incidents have happened in living memory.
- It would be expensive to redevelop because of the historic nature of the buildings.

# What are the options?

| Option 1 | Option 2 | Option 3 | Option 4 | Option 5 |
|---|---|---|---|---|
| Allow only the development of greenfield housing (Proposal 1). | Allow only the redevelopment of Manor Tannery (Proposal 2). | Allow both the greenfield site proposal (Proposal 1) and the Manor Tannery proposal (Proposal 2). | Allow the redevelopment of Manor Tannery (Proposal 2) and redevelop other brownfield sites for housing and employment (e.g. the old petrol garage – Proposal 3), even if it involves **compulsory purchase**. | 'Business as usual' – allow building only when sites become available within areas allowed for development, like those on Mill Lane. |

# What do local people think about the proposals?

+ **Compulsory purchase** is the legal power given to local councils to buy land for development when local landowners are unwilling to sell.

I'm bothered about the increase in traffic in the village – every house seems to have two cars these days.

This village needs more affordable housing. We don't want it becoming just a place for the rich – look what's happened at Padstow. Local people were priced out of there a long time ago.

I'm concerned that the village doctor's surgery won't cope. It's always busy there.

Our school is already over-subscribed. Where would any new children go to school?

We need more people in Grampound to keep the village shop and post office open.

If they build new houses, the priority should be affordable housing for local people.

Most people in Grampound commute elsewhere to work. There aren't enough jobs in Grampound, so more houses will increase the number of cars on the roads.

This village needs more places for small businesses to work from.

New housing could be the start of a lot more development. Look at nearby villages – they've lost their character as more estates get built.

In this section you will look at shorter, longer, and skills questions – of the sort you might meet in your exam.

## Knowing what's expected

Exam success comes from knowing your material, and knowing what to expect. The exam begins with short, structured questions about the background to the issue. These are then followed by longer answers, leading you to make decisions about, for instance, the best ways to provide Grampound's future housing.

The exam tests three things:

- **Knowledge** – e.g. basic things about Grampound, and knowing key words.
- **Understanding** – e.g. understanding whether brownfield or greenfield sites are best.
- **Skills** – interpreting data (e.g. plotting or interpreting a graph) and analysing photographs and maps.

## Short questions

### Knowledge about Grampound

Try all of these questions – do those for whichever tier you are being entered into.

### your questions foundation

1 In which county is Grampound? (1 mark)
2 What kind of settlement is Grampound? (1 mark)
3 Where are the two nearest towns to Grampound? (1 mark)
4 About how many people lived in Grampound in 2008? (1 mark)
5 Describe what is happening to Grampound's population. (2 marks)
6 Give two reasons why affordable housing is a problem in Grampound. (2 marks)

### your questions higher

1 Describe Grampound's location within the UK. (2 marks)
2 Describe Grampound's location within Cornwall. (2 marks)
3 Describe changes to Grampound's population since 1951. (2 marks)
4 Describe why Cornwall's population is increasing. (2 marks)
5 What is the meaning of **a** inward migration **b** conservation area? (2 marks)
6 Give two reasons why affordable housing is a problem in Grampound. (2 marks)

### Key words

### your questions foundation and higher

7 What do the following terms mean – 1 mark each. conservation area, listed buildings, affordable housing, inward migration, brownfield site, greenfield site, semi-derelict, high-density housing, low-density housing.

## Longer questions

For these, the number of marks indicates the number of points you should make:

2 marks = 2 points. To explain strengths and weaknesses, for 4 marks (see question 8, higher) aim for 2 of each. Remember that questions carrying 6 marks or more are marked using levels. (See page 269 for more information on this.)

### Understanding Grampound's issues

**your questions foundation**

8  Describe two advantages of living in Grampound. (2 marks)
9  Describe two disadvantages of living in Grampound. (2 marks)
10  Explain why some people might prefer using brownfield sites for new housing in Grampound, instead of greenfield sites. (4 marks)

**your questions higher**

8  What are the benefits and disadvantages of living in Grampound? (4 marks)
9  Describe and explain Grampound's housing problem. (4 marks)
10  Explain whether brownfield sites or greenfield sites would be better for future housing development in Grampound. * (6 marks)

## Skills questions

**your questions foundation**

11  Plot a graph to show Grampound's population since 1951, using the data on page 271. (2 marks) *In the exam the graph axis would be drawn for you.*
12  Using the OS map extract on page 275, describe the landscape around Grampound. (2 marks)
13  What percentage of people in Grampound support 20 or fewer new houses (see table page 271). (1 mark)

**your questions higher**

11  Describe Grampound's population since 1951. (2 marks)
12  Using the OS map extract on page 275, describe the landscape around Grampound. (2 marks)
13  Using the table on page 271, describe people's thoughts about the amount of new housing that should be built. (3 marks)

*Remember*
+ Questions with this * mark your QWC (page 269) so:
• write legibly (clearly)
• check spelling and punctuation
• use proper terminology
• organise your answer.

In this section you will look at questions which involve you writing at length about which option you prefer (see pages 276-279).

## Analysing the proposed solutions

Before you can decide which option you think is best, you need to analyse each one separately. Two activities will help to do this:

a) An impact assessment table – classifying the impacts of the options into social, economic and environmental.

b) A weighting chart – scoring the options, to determine which is best.

### An impact assessment table

**your questions**

1  **a**  Copy and complete the table below to explain the social, economic and environmental impacts of each option.
   **b**  Underline in blue the impacts that are positive, and in red the impacts that are negative.
   **c**  Now explain:
      **i**  Which is the best option? Why?
      **ii**  Which is the worst option? Why?

|  | Social impacts (e.g. for people and housing) | Economic Impacts (e.g. for jobs and wealth) | Environmental impacts (good or bad for the environment) |
|---|---|---|---|
| **Option 1** Greenfield |  |  |  |
| **Option 2** The Manor Tannery |  |  |  |
| **Option 3** Allow both greenfield and Manor Tannery |  |  |  |
| **Option 4** Brownfield sites |  |  |  |
| **Option 5** Business as usual |  |  |  |

## A weighting chart

This exercise will help you to judge which option is best for improving Grampound's housing situation. Copy and complete the table.

### your questions

2  Some comparison criteria (1 to 4) are shown in the table below, to help you compare the options. Decide about others and complete criteria boxes 5–8 (and more, if you have more ideas).

3  **a**  Write *brief* comments about how well each option fits each criteria.
   **b**  Now score each option out of 5, against each criteria – 1 is poor, 5 is excellent.

4  Add up a score out of 40 for each option.

5  Then explain:
   **a**  Which is the best option, and why?
   **b**  Are brownfield sites best, or greenfield?

| Criteria | Option 1 Greenfield only | Option 2 Manor Tannery only | Option 3 Allow both 1 and 2 | Option 4 Allow only brownfield | Option 5 Business as usual |
|---|---|---|---|---|---|
| 1 Numbers of houses? | | | | | |
| 2 Within the boundary allowed for development? | | | | | |
| 3 Impact on scenery? | | | | | |
| 4 How much affordable housing? | | | | | |
| 5 | | | | | |
| 6 | | | | | |
| 7 | | | | | |
| 8 | | | | | |
| Total assessment score | | | | | |

## And now the big one!

Now try this question, to explain which is the best option for Grampound.
Use the notes you have made in both the impact assessment table and the weighting chart.

### your questions foundation

6  **a**  Select ONE of the five options and explain why it would be the best one to solve Grampound's housing in future. * (6 marks)
   **b**  Explain briefly why you rejected each of the others. * (6 marks)

### your questions higher

6  Select ONE of the five options and explain why you think this would best solve Grampound's housing problems
AND
briefly outline why you rejected the others.
* (12 marks)

✚ In this section you will learn what you have to do for unit 4, the controlled assessment unit.

## What's different about Unit 4?

The other units in this book help you to learn about 'Dynamic Planet' (unit 1), 'People and the Planet' (unit 2), and the 'Decision-Making Exercise' (unit 3). These are each assessed by an exam, which requires preparation and learning, and are each worth 25% of your mark.

Like these, unit 4 *Researching geography* is worth 25% of your GCSE mark, but there is no exam.

- The work is done in school time and you have more control than in an exam. You can plan what to say and do, and go back and correct things. If exams stress you out, you'll probably do better in this unit.
- Although there are controls over how long you can spend (which is why it's called 'Controlled Assessment') you are not as bound by time as in an exam.

The unit is based on fieldwork.

- Your teacher organises a fieldwork investigation for your group.
- You spend class time writing a report of the investigation, which is marked.

## What will the investigation be about?

Your teacher will choose a task from a selection based on these themes:

- coastal environments
- river environments
- rural environments
- town/city environments.

Whichever one your teacher selects, you will not be doing the same investigation as the year above you at school, or the year below.

▲ Your teacher will be able to choose between controlled assessment themes on rivers…

▲ …coasts…

▲ …the countryside…

▲ …or towns.

## The hypothesis

The task will always be a broad statement like this:
The regeneration of the inner city has been successful.

Your challenge is to see whether the statement can be proven right or not. It is more accurate to speak of this statement as a **hypothesis**.

From this broad hypothesis, your teacher will devise a focus for your work, usually in a particular place, for example:

The regeneration of London's Docklands has been successful.

Your fieldwork will then be about proving, or disproving, the hypothesis.

> **+** A **hypothesis** is a statement intended as a prediction for an investigation. After the investigation, it can either be accepted (proved) or rejected (disproved).

---

The regeneration of London's Docklands has been very successful.
Hypothesis proved.
The regeneration of London's Docklands has not been at all successful.
Hypothesis disproved.
The regeneration of London's Docklands has been successful in some ways but not others.
Hypothesis only partly proved.

---

After the fieldwork, you'll spend several weeks putting the results together, presenting them using various methods e.g. graphs, maps and photos. You'll analyse these, and then draw conclusions from them.

You then carry out a final evaluation of your investigation.

This chapter will take you through an investigation so that you

- know what to do at each stage
- get ideas about how to present data
- know what to do when it comes to writing your analysis, conclusions, and evaluation.

## How will I know what to do?

Don't worry – you're not on your own! Your teacher will make sure that you know what to do at all stages. You will have all the fieldwork organised for you and you will be helped closely in the early stages. Until you start doing the presentation of your data, you'll be allowed to work in a group, and to ask your teacher for direct advice.

Later, especially during the write-up, you will have to work alone, but even then your teacher can talk generally about what makes good data presentation or analysis. What you have to do is to make the decision about which styles you'll actually use.

In this section you will learn how to develop an introduction to your study, and how to write your aims.

## Step 1

There are 5 steps to your investigation. Each is covered in the next few pages. Each stage requires a different amount of supervision by your teacher, or different level of 'control'. Take a look at this table to see what's involved in step 1.

| Step 1 | What do I do? | Level of supervision |
|---|---|---|
| Research and planning | Listen to the focus of your task, and make sure you understand what you are investigating, and where. | Low. All you need to do is listen! |
| | Find out about the area where you'll be doing your fieldwork. Research around the issue you are looking at. | Low. You'll probably do some of this in class, led by your teacher, together with some homework research which you do on your own. |
| | Start writing up your aims, with close guidance from your teacher. You'll find that your aims look very similar to everyone else's. | Low. Your teacher is allowed to give you feedback on any of your research to make sure you've been looking in the right places, and to make sure you have written up your aims accurately. |

## The introduction

Your teacher may have pictures or video material of the area you are going to study, and you will probably have been taught a bit about the theme already e.g. about inner cities or about coastal processes.

You need to write up a short introduction (perhaps 300 words) which could answer some of these questions.

- Where is the area located? Locally? Within which part of the UK? Outside of the UK?
- What is the area like? (Try using descriptions and photos, and look at digital maps from sources like Google Earth.)
- What issues are affecting the area, and why?
- What processes are affecting the area, and with what effects?

# The aim and hypotheses

Remember that your teacher can help you in this section. You can also write this part before you do the fieldwork. Writing the aim is very simple. Look at the example hypothesis:

The regeneration of London's Docklands has been successful.

Your aim can be as simple as:

The aim of this investigation is to prove the hypothesis that 'The regeneration of London's Docklands has been successful.'

However, a broad hypothesis like this is usually better when split up. The success of the regeneration being investigated could be measured in different ways.

- Economically – is it creating jobs? Are they well-paid jobs? Does it offer economic opportunity to areas with high unemployment? Is there prosperity e.g. high numbers of car owners, home owners?
- Socially – does everyone gain? Do the poor gain as much as the wealthy? Does regeneration provide better housing, education (improved schools) or leisure facilities (e.g. parks)? Is public transport provided to help people who do not own cars?
- Environmental – is there an improvement in the environment? Are buildings well maintained, or streets free of litter? Is it safe and well lit at night?

You could now write:

I have split my main hypothesis into three smaller ones.

a) The regeneration of London's Docklands has been successful economically. By this I mean...

b) The regeneration of London's Docklands has been successful socially. By this I mean...

c) The regeneration of London's Docklands has been successful environmentally. By this I mean...

▲ Large office blocks overlook the old docks in London.

▲ New homes in east London close to Docklands. These are being built on the sites of old council housing built for low income groups. Many of these will be for private sale. Successful regeneration or not?

## Add some colour

You could add photographs with captions designed to make the reader think. If you use other people's photos – e.g. from the internet – make sure you **always** give the source. Never try to pass something off as your own if it isn't.

➕ In this section you will learn some basic Ordnance Survey map skills, and ideas for using GIS in your Controlled Assessment.

## Using OS Maps

A good way of demonstrating geographical skill in your Controlled Assessment is to make use of an Ordnance Survey (OS) map to locate your area of study, to show a route, or to identify key features. Combining this with the use of GIS (global information systems) could be a strong feature of your planning and research.

### Finding places

Every OS map is divided into a grid, and each gridline has a number. You can use these numbers, the grid reference, to give the location of a particular place. To locate a grid square, use a 4-figure grid reference.

- Find the eastings gridlines first (the numbers along the top and bottom) e.g. Tower Bridge is between eastings 33 and 34. Take the lower value (33).
- Next, find the northings (the numbers running top to bottom) – Tower Bridge is between northings 80 and 81. Again take the lower value (80).
- So Tower Bridge lies in grid square 3380.

To pinpoint more specific locations, use 6-figure grid references.

- Again take the eastings first. Look at the square containing Tower Bridge (33 to 34). Imagine the square is divided into tenths. The bridge (not just its name!) lies about seven tenths across. So write down 337.
- Next, look at the northings and imagine the square to be in tenths between 80 and 81. The bridge is two tenths across. So write down 802.
- So the full 6-figure grid reference for Tower Bridge is **337802**.

### Measuring Distance

The scale of this map is 1:50 000 so 2 cm equals 1 km. Measure the distance between two places on the map with your ruler. Divide this number by 2, and that is the distance in km. Use one decimal place for part-kilometres. The distance on the map between Tower Bridge and Liverpool Street station is about 3 cm. Divide this by 2, and we know the real distance is approximately 1.5 km, in a straight line.

### Using Directions

On OS maps, north is always at the top of the map. To find the direction of Liverpool Street station from Tower Bridge, work from Tower Bridge. Move an imaginary line (or your ruler) like the hand of a clock around Tower Bridge, until it meets the station. Then look at the angle of this line compared to the compass. Liverpool Street station is NNW of Tower Bridge.

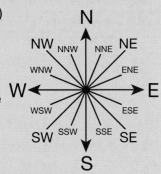

### Geographical Information Systems (GIS)

Several GIS are available on the web. You have almost certainly used Google Maps or Google Earth. You should try to use GIS in your Controlled Assessment.

- Match an OS extract like this. Photos bring places alive.
- Update recent developments. GIS can be ahead of printed maps e.g. in showing new roads or housing.
- Provide screenshots of areas of your study to give more detail. An OS map can't show age or density of housing – but a digital image can.

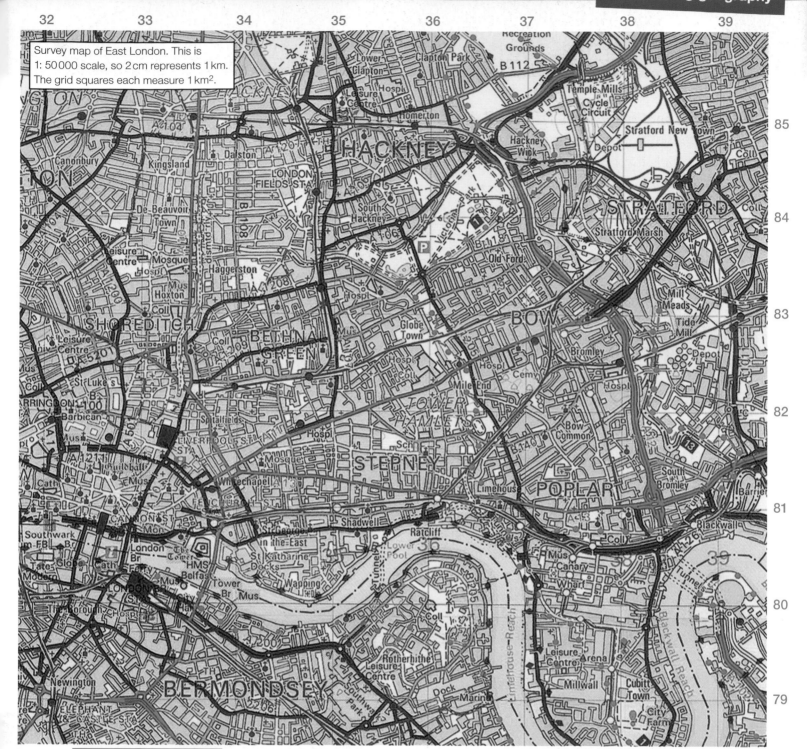

Survey map of East London. This is 1:50 000 scale, so 2 cm represents 1 km. The grid squares each measure 1 km².

## your questions

1 Give 4-figure grid references for Canary Wharf, and Liverpool Street station.

2 Give 6-figure grid references for Canary Wharf, and Liverpool Street station.

3 Measure the straight line distance **a** from Canary Wharf to Liverpool Street station **b** from Canary Wharf to Tower Bridge.

4 Give the direction **a** from Canary Wharf to Liverpool Street station **b** from Tower Bridge to Canary Wharf.

5 Use Google Maps to find the area shown in this OS map. Take a screenshot, and paste it into a Word document.

6 Identify five things on the photo that can't be identified on the OS map.

7 Take screen shots of close-ups of **a** Canary Wharf **b** Shadwell **c** Victoria Park.

8 Describe what you can see about land use from these three areas.

➕ **In this section you will find out about the fieldwork activity, what you have to do, and how to get the best out of it.**

## Step 2

This is the second step in your investigation. Take a look at this table to understand what level of control is required by your teacher.

| Step 2 | What do I do? | Level of supervision |
|---|---|---|
| Fieldwork activity | Take a full part in the field trip! Go out and collect data. Follow carefully all the instructions about what to do. | Low. You do this in groups. Your teacher does not have to be watching you all the time. |
| | Remember you are essential to successful data collection as everyone else is depending on you, and you on them. | Low. You can share your results with the class to get a better set of results. |
| | Your teacher will decide most of the fieldwork data collection methods e.g. land use maps or questionnaires. | Low. Your teacher can help you to make sure you have collected data in the right ways. |
| | You can show initiative by collecting some of your own data. For example, take photos and collect information that may be freely available e.g. from tourist offices, or estate agents. | Low. Your teacher will encourage you to do this anyway. But make sure any data you collect is relevant – don't just collect anything – and ask your teacher if you are unsure. |

## Fieldwork tips

Fieldwork methods depend upon the focus. You'll have very different tasks to do if you're studying a river compared to an inner city area. Whatever your focus, you need to make sure that you

- collect a range of data
- make sure you are clear about how every piece of data collected fits into what you are trying to investigate
- get experience in collecting all data
- remember that you're part of a team – the data that you collect could be important to everyone in your year group.

▲ Students collecting data in groups. Like these students, you're allowed to use each other's data and work together.

## Collecting data

Look again at the earlier hypothesis: The regeneration of London's Docklands has been successful. The table below shows some possible data collection methods. You could use these ideas to suggest methods for your own investigation.

| Method | Purpose |
|---|---|
| Land use map for your study area | Show what kinds of land use exist now, and how they may have changed – can you find a historic map from before the regeneration? |
| Age-of-buildings map for your study area | Identify the most recent buildings which have resulted from the regeneration. |
| Environmental Quality Survey (EQS) at different places | Compare environmental quality (building quality, noise, open space) in areas that have been regenerated with areas that haven't. |
| Questionnaire to find out people's ideas about the success of regeneration | Find out whether people think they have benefited from regeneration or not, and to see whether their ideas vary with age or gender. |
| Field sketches | Give your own impressions of the areas you visit – the annotations are more important than artistic quality! |
| Photographs | Take photographs of different areas to be used alongside the EQS, to give a visual reference. |

▼ An Environmental Quality Survey (EQS).

| Qualities being assessed | | Good +2 | Fairly good +1 | Av. 0 | Fairly poor -1 | Poor -2 | |
|---|---|---|---|---|---|---|---|
| Building quality | 1. Well designed / pleasing to the eye | | | | | | Poorly designed / ugly |
| | 2. In good condition | | | | | | In poor condition |
| | 3. Evidence of maintenance / improvement | | | | | | Poorly maintained / no improvement |
| | 4. Outside – land, gardens or open space are in good condition | | | | | | Outside – no gardens, or land /open space in poor condition |
| | 5. No vandalism evident | | | | | | Extensive vandalism |

## Collecting secondary data

Your own data may not be the only source that will help you. You can use secondary data – that is, data which has been collected by other people or organisations.

- Census data – although it may be out of date by a few years, it is still a valid, complete sample of everyone living in the area that you are studying.
- Data on house prices – either from estate agent publicity, or from upmystreet.co.uk, which will give you an idea of how much house prices have changed.

✚ **In this section you will find out about some data collection methods you could use and how to write about them.**

## Step 3

This is the third step in your investigation. Take a look at this table to understand what you have to do. You'll see that the level of control at this stage is still low.

| Step 3 | What do I do? | Level of supervision |
|---|---|---|
| Methodology | Describe, explain, and justify the methods used in your fieldwork to collect your data.<br><br>Give some indication of how you sampled different places or people to survey. | Low. You can find out about different methods from your teacher to check that you understand them. |

## Writing the methodology

In your methodology, you need to do the following.

- Show on a map the places that you actually visited – or where you sampled people or features (e.g. the points along a river that you took readings from).
- Describe how and why you selected the places in which you recorded your data. Use the sampling table below to help you.

| Method | What this means | When you might use it |
|---|---|---|
| Random | Places or people chosen at random – just like pulling numbers out of a hat. | • For a river study, when deciding how far apart to sample sediment. |
| Stratified | You base a survey on what you know to be there already e.g. if you know that 20% of the population is over 65, then 20% of the people sampled for your questionnaire should be over 65. | • Questionnaires.<br>• EQS surveys of different land uses, making sure each land use is surveyed.<br>• A river course with straight, winding and meandering sections, where you ensure that you include an example of each. |
| Systematic | You decide on a fixed interval between points, even if it means that you miss out certain interesting features. It might be places 50 metres apart, or every 4th person to walk past you. | • A river study where you sample places every 200 metres or every quarter of a mile, etc.<br>• An inner city where you carry out an EQS survey every 100 metres. |
| Opportunistic | You take whatever comes your way, either because the sample size might be so small, or because something looks particularly interesting. | • Rural communities, where the number of people is so small that you interview everyone you see! |

▲ Different sampling methods.

# Writing up your method

You will need to describe and justify the methods you have chosen to use. You might find the following table useful as a format – make sure you write something in every box. The example here is for a study of inner city regeneration, but you could easily use the same idea for a river, coastal or rural study.

| Data collected | Reasons why this data was needed | How you collected data | Problems you came across and how you tried to overcome them |
|---|---|---|---|
| Land use map | | | |
| Age of buildings map | | | |
| Environmental Quality Survey | | | |
| Questionnaire about people's views of the regeneration | | | |
| Field sketches | | | |
| Photos | | | |
| Property values | | | |

▲ Using a table to describe and explain methodology.

# Sharing your results

The important thing is to make sure that you have all the data from your group or class. Your teacher might be able to place them on the school website or VLE, which means you can access them from any point in or out of school.

✚ **In this section you will find out about deciding on the best way to present your data.**

## Step 4

This is the fourth step in your investigation. Read this table so you understand what kind of help is allowed by your teacher.

| Step 4 | What do I do? | Level of supervision |
|---|---|---|
| Data presentation | Present your results in ways that make them easy for you and the reader to interpret.<br><br>This will take place over several lessons. Data presentation is very time-consuming!<br><br>Use varied methods of presenting data – the greater the variety, the better. Use maps, graphs, charts, and annotated photographs. | Low. Although you are expected to work on your own, you can be assisted by a teacher and can talk to friends to discuss ideas. You complete your presentation of results over a few lessons and can give work in for checking. Your teacher will tell you how long you have to complete this. |

Follow these hints for productive lessons:
- Discuss with your teacher (and look through this book!) to see what the possibilities are for graphs, maps etc. Then you simply make the choice.
- If you don't know how to draw a particular kind of graph, you can ask people for help.
- You don't have to hand draw any graphs, though you can if you prefer – everything can be done on the computer.
- Go to each lesson with a plan or list of what you want to complete, so you make the best use of your time.

## Presenting data

Choose varied ways of presenting your data.
- If you are dealing with statistics, produce graphs.
- If you are dealing with statistics used to compare places, then place graphs around a map, with lines to link your graphs to the relevant places.
- Find ways of linking different data together. For example, you could link together your EQS results on the same page as photographs that illustrate environmental quality (see page 297).
- Think creatively. Use a folded piece of A3 paper, if A4 paper is too small.

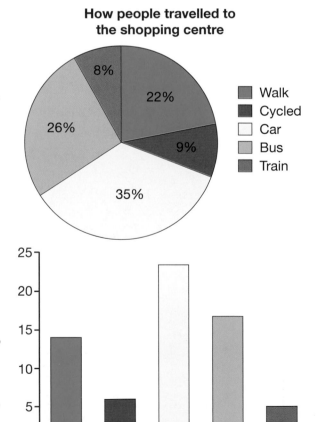

**How people travelled to the shopping centre**

8%
22%
26%
9%
35%

- Walk
- Cycled
- Car
- Bus
- Train

▲ Two different graphs showing the same results – but which is the best way? You decide!

## Graphs

Choose the right kind of graph for the data you want to display. You can use a computer to create them, or draw them by hand. These are just some examples, but there are many more you could use.

## Maps

Choose the right kind of map for the information you want to display. You can use GIS software to produce these, or other graphics software, but you can also draw them by hand.

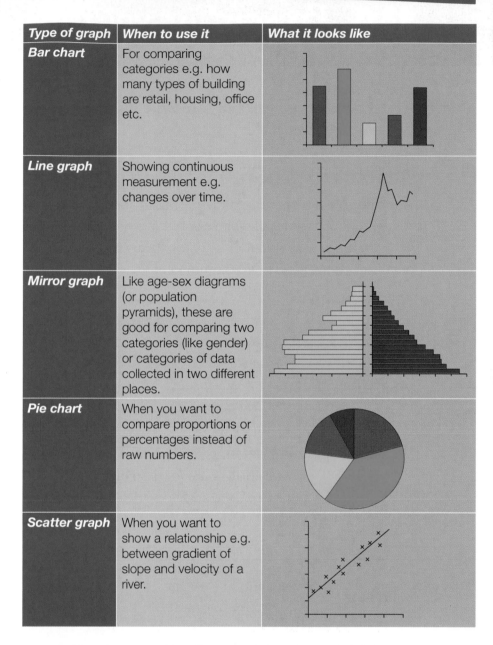

| Type of graph | When to use it | What it looks like |
|---|---|---|
| **Bar chart** | For comparing categories e.g. how many types of building are retail, housing, office etc. | |
| **Line graph** | Showing continuous measurement e.g. changes over time. | |
| **Mirror graph** | Like age-sex diagrams (or population pyramids), these are good for comparing two categories (like gender) or categories of data collected in two different places. | |
| **Pie chart** | When you want to compare proportions or percentages instead of raw numbers. | |
| **Scatter graph** | When you want to show a relationship e.g. between gradient of slope and velocity of a river. | |

| Type of map | When to use it | What it looks like |
|---|---|---|
| Choropleth map | • Where there are categories e.g. building heights grouped by number of storeys. | Shaded areas with a key. Uses categories which range from light shades of a colour for low values to dark shades of the same colour for high values. |
| Dot map | • To show distribution e.g. particular kinds of shop in a city or town. | Dots displayed on a map to indicate an occurrence. The dots might be coloured e.g. a red dot represents a café, a blue one shows an antiques shop. |
| Isoline | • Rainfall map with areas between isolines coloured in.<br>• Building heights within a city centre.<br>• Pedestrian counts. | A map showing lines which join values of the same amount e.g. rainfall, or temperature. Often, the areas in between each line are shaded in like choropleth maps. |

In this section you will see some examples of well-presented data.

## Annotating photos, maps and graphs

**Annotations** are specific to the particular item they are linked to. Therefore, for each of your graphs, maps or photos you should try to annotate what you can actually see on them. Try the following as a guide, and look at the annotated photo and chart on the right.

a) On a set of data
  - describe what you see e.g. Place X has a high percentage of offices…
  - compare what you see e.g. Place Y has the highest percentage of shops but the lowest percentage of offices…

b) On a photo
  - describe what you see e.g. Place X has a high percentage of offices…
  - combine evidence from the photo with information you have collected elsewhere e.g. in an EQS survey. See the photo below.

> **+ An annotation** is a piece of information that you add to some text, or an illustration, to describe or explain a particular aspect of it.

The buildings are quite well designed and are in good condition. There is evidence of good maintenance.

There is no litter. The precinct is well maintained and close to public transport, shops and amenities.

Although there is a lot of traffic noise from the nearby main road, this precinct is traffic-free and safe for people.

Although there is no green space or parkland, this is used by young people for biking and leisure. Very few trees and shrubs are visible.

▲ A well-annotated photo from Shadwell in east London.

# Combining your results

Sometimes presenting results in combination with each other can work really well. In this excellent example, a student has annotated the photos by linking them to the EQS results from the same place (Padstow, in Cornwall). Text in green type shows positive comments, and red shows negative.

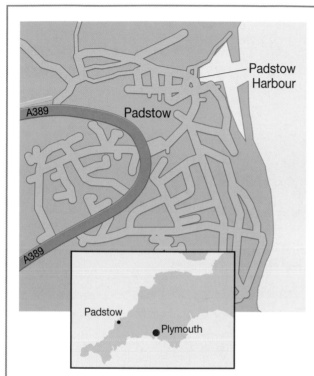

▲ Padstow Harbour.

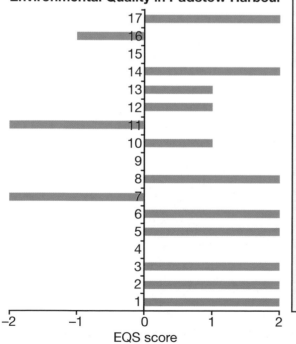

**Environmental Quality in Padstow Harbour**

EQS score

**EQS key**

1 – Design
2 – Condition
3 – Maintenance
4 – Open space
5 – Vandalism
6 – Traffic congestion
7 – Parking
8 – Traffic noise
9 – Safe for people
10 – Pollution smell
11 – Large gardens
12 – Trees and shrubs
13 – Public parks
14 – Litter
15 – Road maintenance
16 – Public transport
17 – Amenities and services

## The Harbour, Padstow

- Got a total score of 14.
- Received high scores on most criteria.

### Good points include:

- The old buildings are attractive, and well designed.
- Doors opening straight onto the street.
- There is little traffic congestion in this area, and there is good access to parking.

### Negative points include:

- There is almost no open space in or near this area, no trees or plants and no access to a public park.
- Few houses have any gardens.

In this section you will learn about the analysis – what it is, and how to do it well.

## Step 5

This is a vital step in your investigation. It's at this stage that you have to work on your own and cannot ask for help. The analysis is worth a high percentage of your marks and so should be given plenty of thought.

## What is analysis?

Analysis is about using the data you've collected to see whether your hypotheses are correct. In your analysis you need to cover a number of things.

- **Describe** results that you have not already covered in your annotations e.g. place X has high environmental quality.
- **Compare** places for the highest and lowest e.g. place C has the highest buildings of all our sample points in Chichester city centre.
- **Explain likely** reasons for what you see e.g. place H has the highest pedestrian count because this is where all the large stores are such as Debenhams and M & S.
- **Explain possible** reasons for what you see e.g. place Q probably has a higher number of people working part-time because ….
- **Link** different graphs or photos e.g. place Z has the lowest environmental quality score (Graph 2c) which is due to the amount of derelict land (Photo 1A).

| Step 5 | What do I do? | Level of supervision |
|--------|---------------|----------------------|
| Analysis | Look at your results and interpret what they mean, describe what they show, and explain the reasons for what you observe. | High. This section must be done on your own, supervised by a teacher. This will be done over a few lessons, and you must hand your work in at the end of each one.<br><br>Your teacher will tell you how long you have to do each section but they cannot help you or give you feedback. You can't talk to your friends. |

## Where do I start?

**1** Start with general points.
Generally, the inner city areas I studied prove my hypothesis correct, because environmental quality is poorer in places which have not been regenerated than in those which have been.

**2** Then go for more specific points to show differences between places. Always illustrate with data.
Canary Wharf has the highest percentage of people in professional and managerial jobs (78%) whereas Shadwell has the lowest (16%). However, Shadwell has the greatest percentage of people in work (only 5% unemployed) so the pattern is not definite.

**3** Then compare data within places.
Shadwell may have the lowest unemployment, but it also has the highest proportion of those in part-time work (21%).

# Good analysis

Here is an extract of some analysis that a student wrote.

Disposable income and home ownership are also key indicators. We see in Fig. 9 that the majority of people in all three areas own a car, over 85% in both Tintagel and Padstow, and that most families in the settlements own more than one car, with every family in St. Teath owning two. We can see from Figure 4.6, as we saw in the social analysis, that the majority of people in all three areas own their own homes, with very few people (less than 30% in each area) who rent their homes, suggesting available housing, and wealth in the areas. This is emphasised by Figures 4.7 and 4.8, where we see that Cornwall's housing prices are some of the lowest in the UK, especially compared to London and the UK average. We can also see that people in the three areas have a large amount of disposable income available to them, with most people (around 30% for St. Teath and Padstow, and over 40% for Tintagel) being able to afford 7 out of the 10 possessions. This would suggest economic welfare is not dependent on settlement.

This is well structured analysis because the student has
- referred to data and figure numbers in the presentation (yellow)
- used a range of geographical terminology (green)
- quoted data from their results to support their ideas (blue)
- used words to show links, strengths, weaknesses, suggestions and assumptions (pink).

# Know the mark scheme!

Level 3 quality is the highest on the mark scheme. Marks are awarded at this level if you
- illustrate your work with data
- compare more than one set of data e.g. compare environmental quality scores with home ownership
- refer to other sources such as your own interviews or research e.g. People I interviewed in Canary Wharf showed that… .

➕ **In this section you will find out how to write your conclusion and evaluation.**

## Step 6

This is the final step in your investigation. Like the analysis, your conclusion and evaluation is worth doing well as it is worth a high percentage of the marks.

## Writing the conclusion

In your analysis, you look at your separate hypotheses (the examples here were on the economic, social, and environmental effects of regeneration). In your conclusion you need to return to your main hypothesis – our example is: The regeneration of London's Docklands has been successful.

Your conclusion must take this statement and try to decide whether it is true or not. Consider the possible outcomes.

- The statement in your aim might be correct, partly correct or incorrect. You have to decide which it is.
- Whichever you decide must be on the basis of evidence – not just from general impressions you have in your mind.

To help you do this, try summarising everything in a table (see right) before the lesson so that your mind is clear when you go into class to write it up.

If you are not sure about whether the statement is correct or not, it is alright to say so. But you must give plenty of examples to support each side of the argument, to show why you cannot decide.

At the end, list the main findings of your work. What firm conclusions can you come to?

| Step 6 | What do I do? | Level of supervision |
|---|---|---|
| Conclusion | Here you draw out the answer to your main aim e.g. How successful has regeneration in the inner city been? | High. Both sections must be done on your own in lesson time, supervised by a teacher. You can't take the work home. Your teacher will tell you how long you have to do each section. They can't help or give you feedback. You can't talk to your friends. |
| Evaluation | Here you look back at the whole task, and comment on the strengths and weaknesses of your investigation. | |

| Evidence that regeneration has been successful | Evidence that regeneration has not been successful |
|---|---|
| 1. The large number of office blocks, shops and restaurants shows that employment ought to have increased. | 1. Most of the people we talked to and questioned were low paid, and told us many of the offices were empty. |
| 2. | 2. |

# Writing the evaluation

In your evaluation, you assess the **validity** of what you have done. Is what you have found out reliable? Can your data be trusted? Did the study go well? You need to be able to assess the success of the whole investigation. Again, before your lessons when you will be writing this up, think about the following questions.

In paragraph 1 consider your methods. Don't use the excuse: if we had more time, we could have done more. Think about these points instead.

- Did your data collection methods work well? Did some work better than others?
- Would any results be different at another time of the day? Of the week? Of the year? For instance, would environmental quality data be different in summer compared to winter?
- Were the results accurate? If so, why? If not, why not? If you went back at different times, would you get similar conclusions?
- Do you have any **anomalies** (strange results) that cannot really be explained?

▲ Cornwall. How would this place look in a different season or at a different time of day?

> **+** An **anomaly** is an individual result that does not seem to fit into the general trend being displayed.

In paragraph 2 consider the reliability of your conclusions.

- Even though you took a small sample of places or people to be studied, are your conclusions valid?
- If you went to other similar places, would you find out similar things, or is your study area unique?
- Would you go about the investigation in the same way again?

In paragraph 3, think about the relevance of the investigation.

- How is your investigation useful to others?
- Who might be interested in your investigation, and why? What might they do with the information?

Finally, think about whether the investigation could be extended.

- Could your investigation be taken any further?
- Which parts might be worth developing, and how?

✚ In this section you will find out how to present your report. Remember to leave plenty of time – if you complete all the points in the checklist, you'll need a few days! Don't leave it until the last hour before handing it in.

## Final Checklist

Follow this checklist of things to do before you hand in your work.

### 1. Section headings

Are the following clearly identifiable by name?

a Introduction and aims

b Methodology

c Presentation of results   ⎫ You can combine these if you wish but

d Analysis                  ⎭ results **must** be clearly shown.

e Conclusions

f Evaluation

### 2. Presentation of results

- Give all your maps, diagrams, photos and graphs figure numbers with captions e.g. Figure 1 Map showing the location of the River Skerne.
- Any photo should be given a caption. Try and connect it to some other results.
- Refer to figure numbers in your writing.
- Don't take figure numbering to extremes! If you have, for example, an EQS graph with a photo next to it, the whole thing can be Figure 3.6.

### 3. Numbering and index

- Number your pages. All pages need to be numbered in sequence. Do this last!
- Create a contents page.
- Include an index of all your figures.

## 4. General presentation

- All material must be handed in as A4 size. Anything that is A3 size must be trimmed down, hole-punched, and be accessible without having to undo clips or ring binders.
- Stick down any loose material (e.g. acetate overlays) so they don't fall out and get lost.
- All secondary sources (including photos taken by other student or data found on a website) must be given in a bibliography at the end, quoting the source and what you used the source for e.g. www.bbc.co.uk/news for inner city projects in Manchester.
- Your coursework must be handed in as a loose-leaved project in a card or plastic wallet file. Every page must be in the correct order. You are not allowed to hand in ring binders or individual plastic sleeves.

## Who am I writing this for?

Your report can be written either in the first person (e.g. I carried out a questionnaire in...) or third person (e.g. a questionnaire was carried out in...). Write for an intelligent adult, but one who does not know anything about the area you are investigating. Therefore, make all your explanations really clear.

## Does it have to be a written report?

Your report does not have to be continuous writing. You can choose to create short movies, PowerPoint presentations, web pages, or perhaps a GIS map.

But if you want to do this **you must talk to your teacher** so that you can gain maximum marks from your work. The important thing is that a movie has to be just as well scripted as a piece of written analysis. In the same way, a PowerPoint must have notes which make crystal clear what it is you are saying about each slide. Remember – the information must be clear to someone who doesn't know you and who does not know your study area.

## And one other thing

Check through the mark scheme. Make sure you understand it properly, so that you know what is needed to hit the highest levels.

> *Remember*
> ✚ Make a plan for yourself ahead of each lesson so that you make the most of the time available.

> *Remember*
> ✚ All secondary sources must be properly acknowledged.

> *Remember*
> ✚ Everything you include must be clear to somone who has no previous knowledge of your study.

> *Remember*
> ✚ Always keep a photocopy or electronic copy of your whole project.

## Exams – how to be successful

If you want to be successful in the exams then you need to know how you will be examined, what kinds of questions you will come up against in the exam, how to use what you know, and what you will get marks for. That's where this chapter can help.

### What is Edexcel Geography specification B like?

Edexcel's specification B is a brand new specification. It includes new ideas e.g. 'Living Spaces' and offers a new take on some familiar subjects e.g. 'Coastal change and conflict'. It is so broad that you're bound to find several interesting topics to give you a good understanding of the world.

Your course consists of four units.

- The taught part of the GCSE specification consists of two units – Dynamic Planet (chapters 1 to 8) and People and the Planet (chapters 9 to 16). Each unit consists of 8 topics, but you only need to study 6 of them. Make sure that you know which options you have to study.
- Unit 3 is a decision-making exercise. It draws on what you have been taught in units 1 and 2. Chapter 17 tells you all about this unit.
- Unit 4 is called Controlled assessment – what teachers and students used to call coursework. Look at chapter 18 for lots of information on this.

## How will you be assessed?

Each unit counts for 25% of your final grade. Units 1, 2 and 3 are assessed by an exam. Unit 4 is done over a period of about one term. You create a report on your fieldwork, which is then marked.

The tables on the next two pages show what the exams for units 1, 2 and 3 will be like. Look through them carefully so that you know what to expect.

# Units 1 and 2

| Foundation tier | Higher tier |
|---|---|

**For each tier in each paper:**

- the exam is one hour long
- it is worth 50 marks in total
- it counts 25% towards your final grade
- it has three sections – A, B and C.

Any resources you need are included in with the questions – there won't be a separate booklet.

You must answer all parts of questions 1-4 in section A (each question is worth 8 marks, making a total of 32 marks) plus

- **one** question from section B (worth 9 marks)
- **one** question from section C (worth 9 marks).

## Section A

Questions 1 to 4 are broken up into short questions, rarely worth more than 2 marks.

- There are resource materials (data, photos, cartoons etc.) on which you'll be asked questions.
- You'll be expected to know what these resources are getting at from what you've learnt in class.
- Detailed case study knowledge is NOT needed in this section.
- Section A is point marked.

## Sections B and C

Questions 5-6 in section B and 7-8 in section C allow you to write more about what you've learnt. You will need to have learnt examples and case studies to complete the longer answers.

- The longest answer you will have to write is for 6 marks.
- 6 mark answers are marked by levels – level 1 (lowest) to level 3 (highest).

## Section A

This has several short questions, worth from 1 to 4 marks.

- There are resource materials (data, photos, cartoons etc.) on which you'll be asked questions.
- You'll be expected to know what these resources are getting at from what you've learnt in class.
- Detailed case study knowledge is NOT needed in this section, but you will be expected to give detailed answers for 4 marks, based on what you've learnt in class.
- Section A is point marked.

## Sections B and C

Questions 5-6 in section B and 7-8 in section C allow you to write more about what you've learnt. You will need to have learnt examples and case studies to complete the longer answers.

- The longest answer you will have to write is for 6 marks.
- 6 mark answers are marked by levels – level 1 (lowest) to level 3 (highest).

# Unit 3

| Foundation tier | Higher tier |
|---|---|

For each tier in each paper:

- the exam is one hour long
- it is worth 50 marks in total
- it counts 25% towards your final grade
- there is no choice of questions – you have to answer every one
- it has three sections – A, B and C. Normally these will be about the resources and follow a sequence
- each section is worth roughly the same number of marks – between about 15 and 18 per section.

The resources are in a separate booklet. This booklet is given to you in February before you do the exam in June. Although you can write and make notes on the copy you have been given, you can't take it or your notes into the exam with you – you'll be given a fresh copy.

You must answer all questions in the paper.

## Section A

Section A normally looks at the background to the issue in the resource booklet.

Questions are generally short – between 1 and 4 marks.

- Some questions will ask you to interpret the resource materials (data, photos, cartoons etc.) so you must look carefully to find the answers.
- You'll be expected to know what these resources are getting at from what you've learnt in class before the exam.
- Section A is point marked.

## Section B

Section B normally looks at the problems and issues in more depth. Like section A, questions are based on the resources, and whatever you have learnt in class before the exam.

- Questions vary between 1 and 4 marks
- Section B is point marked.

## Section C

Section C involves you in making a decision about which is the best way forward. This is usually the longest piece of writing you'll have to do in exam conditions.

Usually, you will be presented with options to choose between. These will normally be a choice of schemes that will be set out clearly in the resource booklet.

- Questions vary between 3 and 8 marks.
- Questions of 6 marks or more are marked by levels – level 1 (lowest) to level 3 (highest).

## Section A

Section A normally looks at the background to the issue in the resource booklet.

Questions are generally short – between 3 and 6 marks.

- Some questions will ask you to interpret the resource materials (data, photos, cartoons etc.) so you must look carefully to find the answers.
- You'll be expected to know what these resources are getting at from what you've learnt in class before the exam.
- Questions of 5 marks or under are point marked. Those of 6 marks or more are marked by levels – level 1 (lowest) to level 3 (highest).

## Section B

Section B normally looks at the problems and issues in more depth. Like section A, questions are based on the resources, and whatever you have learnt in class before the exam.

- Questions vary between 3 and 6 marks
- Questions of 5 marks or under are point marked. Those of 6 marks or more are marked by levels – level 1 (lowest) to level 3 (highest).

## Section C

Section C involves you in making a decision about which is the best way forward. This is usually the longest piece of writing you'll have to do in exam conditions.

Usually, you will be presented with options to choose between. These will normally be a choice of schemes that will be set out clearly in the resource booklet.

- Questions vary between 6 and 10 marks.
- Questions of 6 marks or more are marked by levels – level 1 (lowest) to level 3 (highest).

# How are exam papers marked?

Examiners have clear guidance about how to mark. They must mark fairly, so that the first candidate's exam paper in a pile is marked in exactly the same way as the last. You will be rewarded for what you know and can do; you won't lose marks for what you leave out. If your answer matches the best qualities in the examiner's mark scheme then you will get full marks.

The tables on the previous two pages tell you which parts of the exam are point marked and which parts are level marked. Are you clear about what that means?

## Point marking

Look at this question.

> Forests are ecosystems. Suggest two ways in which humans could protect ecosystems. (2)

There are two marks for this question and you have to suggest two ways of protecting ecosystems. So, you get one mark (or point) for each way that you give. The mark scheme tells examiners which methods they can accept as correct.

## Level marking

Look at this question.

> Using examples, describe the key physical characteristics of either polar or hot arid areas. (6)

There are six marks for this question and examiners award these using the following mark scheme.

| Level | Mark | Descriptor |
|---|---|---|
| Level 1 | 1-2 | Describes a few physical characteristics with no detail; does not move beyond 'very cold' or similar. No link to chosen location. Lacks focus and organisation. Basic use of geographical terminology, spelling, punctuation and grammar. |
| Level 2 | 3-4 | Some structure. Describes a range of characteristics with some detail for some areas, and with some use of appropriate examples. Some detailed description of chosen examples although this is variable. Examples are appropriate, with a range generally, or one or two areas more specifically. Clearly written, but with limited use of geographical terminology, spelling, punctuation and grammar. |
| Level 3 | 5-6 | Structured answer. Describes a good range of characteristics in some depth with appropriate details/data. Uses examples to illustrate descriptions. Accurate and clear link to chosen extreme environment. Well written with good use of geographical terminology, spelling, punctuation and grammar. |

So to gain the most marks your answer needs to meet the criteria for level 3.

# Choosing and answering the questions

### Command words

When you look at an exam question, check out the command word – that is, the word that tells you what the examiner wants you to do. This table gives you some of the most commonly used command words.

| Command word | What it means | Example |
|---|---|---|
| Account for | Explain the reasons for – you get marks for explanation rather than description. | Account for the earthquakes along destructive plate margins. |
| Compare | Identify similarities and differences between two or more things. | Compare flood management methods in Sheffield with those in Darlington. |
| Define | Give a clear meaning. | Define 'erosion.' |
| Describe | Say what something is like; identify trends. | Describe the trends shown in the graph. |
| Explain | Give reasons why something happens. | Explain, using examples, why the floods in Sheffield affected some parts of the city more than others. |
| How far? | You need to put both sides of an argument. | How far were the Sheffield floods due to human causes? |
| Justify | Give evidence to support your statements. | How could Grampound best provide new housing for people who want to live there? Justify your suggestions. |
| List | Just state the factors; nothing else is needed. | List the different ways of protecting coasts affected by erosion. |
| Outline | You need to describe and explain, but more description than explanation. | Using the graph, outline the trend in population. |
| To what extent? | The same as 'How far?' | 'Coastal protection – throwing good money after bad.' To what extent is this true? |

## Interpreting the questions

You have probably already been told by your teacher to read the exam questions carefully, and answer the question set, not the one you think it might be. That means you need to interpret the question to work out exactly what it is asking. So, look at the key words. These include:

**Command words** – these have distinct meanings which are listed above.

**Theme or topic** – this is what the question is about. The examiner who wrote the question will have tried to narrow the theme down, and you need to spot how so that you do not write everything you know about the theme.

**Focus** – this shows how the theme has been narrowed down.

**Case studies** – look to see if you are asked for specific examples.

Here is an example of a question that has been interpreted using the key words opposite.

Using named examples, explain how rapid erosion on coastlines can be managed using traditional and more modern strategies. (6)

### Command words
'explain how' You must give reasons.
'Using named examples' You are being asked to include specific examples you have studied.

### Theme and focus
This question is for Coastal Change and Conflict, looking specifically at how coastlines can be managed.

### Case studies
You can name examples from one or more places you've studied – the examples (plural) are for strategies, not for places you've studied. So three different strategies in one place is fine.

## Choose the right question for yourself

Do not be seduced by an attractive cartoon or photo – how difficult is the whole question?

If you choose the wrong question, have the guts to change in the first 5-6 minutes. If you find yourself running out of time, bullet points are better than no points at all.

### Choose the right question

Always choose questions from the part of the specification you have studied – the other questions only look easy because you haven't studied them.

Read all the questions carefully and then choose. At this stage, you might want to sketch out a few points in pencil, to make sure that you include examples you think you might forget.

## Using case studies

Look again at the question on the previous page. There is one case study in this book that you could use to answer this question – Christchurch Bay. You might have studied another example – perhaps seen a video of Holderness. You may also have studied coastlines for your controlled assessment – perhaps somewhere like Walton-on-the-Naze or Norfolk.

Throughout your course you will look at lots of different case studies, and will need a good way to learn and remember them. Try this method – draw a spider diagram with all the key information on it. An example is shown below, but it's only a start – you would need to fill in more details.

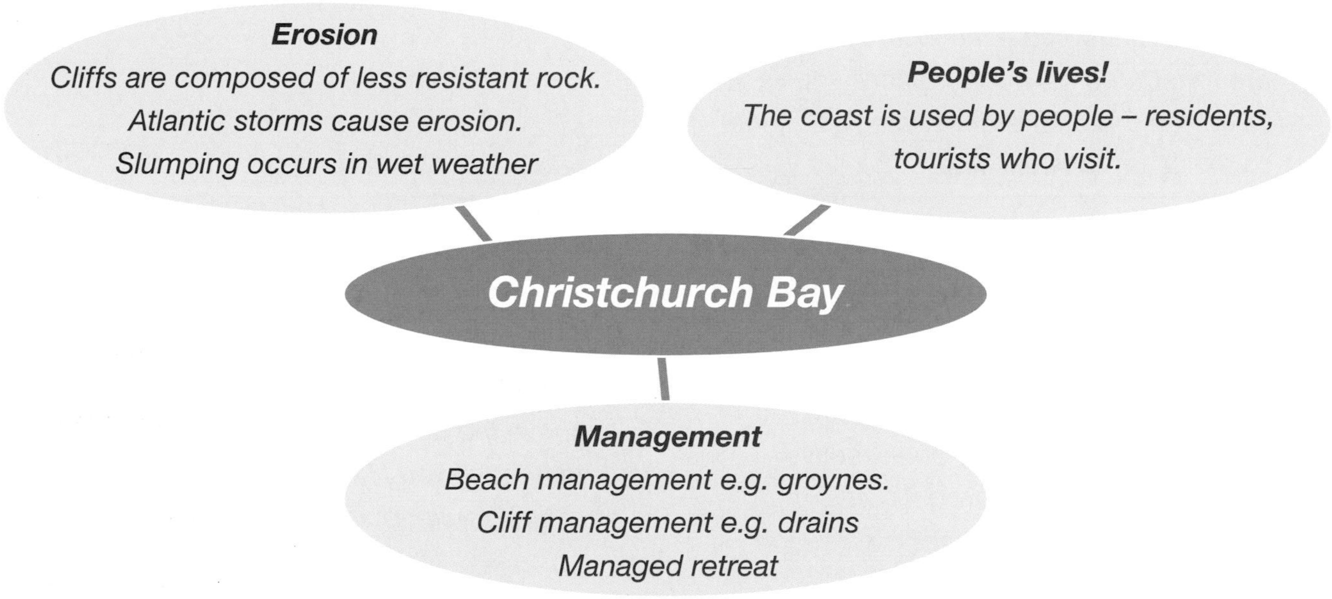

*Erosion*
*Cliffs are composed of less resistant rock.*
*Atlantic storms cause erosion.*
*Slumping occurs in wet weather*

*People's lives!*
*The coast is used by people – residents, tourists who visit.*

*Christchurch Bay*

*Management*
*Beach management e.g. groynes.*
*Cliff management e.g. drains*
*Managed retreat*

Once you have revised and learned your case study you can use it to answer exam questions. Look again at the example question. It asks you how rapid erosion on coastlines can be managed. It asks you to explain how the coastline can be managed using traditional and more modern strategies.

The ways are:
• hard engineering
• soft engineering
• holistic management.

# Ordnance Survey symbols

## ROADS AND PATHS

 M I or A 6(M) — Motorway

A 35 — Dual carriageway

A 31(T) or A 35 — Trunk or main road

B 3074 — Secondary road

Narrow road with passing places

Road under construction

Road generally more than 4 m wide

Road generally less than 4 m wide

Other road, drive or track, fenced and unfenced

Gradient: steeper than 1 in 5; 1 in 7 to 1 in 5

Ferry — Ferry; Ferry P – passenger only

Path

## PUBLIC RIGHTS OF WAY

(Not applicable to Scotland)

| 1:25 000 | 1:50 000 | |
|---|---|---|
| - - - - - - - | · · · · · · · | Footpath |
| + + + + + + | - - - - - - | Road used as a public footpath |
| -+-+-+-+- | -+-+-+- | Bridleway |

## RAILWAYS

Multiple track

Single track

Narrow gauge/Light rapid transit system

Road over; road under; level crossing

Cutting; tunnel; embankment

Station, open to passengers; siding

## BOUNDARIES

+ — + — + — National

· — · — · — District

County, Unitary Authority, Metropolitan District or London Borough

## HEIGHTS/ROCK FEATURES

50 — Contour lines

· 144 — Spot height to the nearest metre above sea level

outcrop    cliff    scree

## ABBREVIATIONS

| P | Post office | PC | Public convenience (rural areas) |
|---|---|---|---|
| PH | Public house | TH | Town Hall, Guildhall or equivalent |
| MS | Milestone | Sch | School |
| MP | Milepost | Coll | College |

## ANTIQUITIES

 VILLA Roman    ⋏ ✝ ✤ ▼ ▼ ● ✳ ✳ ✳ ● ✳

 Castle Non-Roman    ☆

## LAND FEATURES

 ruin — Buildings

 — Public building

 — Bus or coach station

Place of Worship — with tower / with spire, minaret or dome / without such additions

∘ Chimney or tower

Glass structure

Ⓗ Heliport

△ Triangulation pillar

Mast

Wind pump / wind generator

Windmill

Graticule intersection

Cutting, embankment

Quarry

Spoil heap, refuse tip or dump

Coniferous wood

Non-coniferous wood

Mixed wood

Orchard

Park or ornamental ground

Forestry Commission access land

National Trust – always open

National Trust, limited access, observe local signs

National Trust for Scotland

## WATER FEATURES

Marsh or salting    Slopes    Cliff    High water mark
Towpath    Lock    Flat rock    Low water mark
Aqueduct    Canal    Ford    Lighthouse (in use)
Weir    Normal tidal limit    Sand    Lighthouse (disused)    Beacon
Lake    Bridge    Dunes    Mud    Shingle
Footbridge
Canal (dry)

## TOURIST INFORMATION

P Parking

P&R Park & Ride

V Visitor centre

i Information centre

📞 Telephone

⋏ 🚐 Camp site/

Golf course or links

Viewpoint

PC Public convenience

✕ Picnic site

Pub/s

Museum

Castle/fort

Building of historic

Steam railway

English Heritage

Garden

Nature reserve

Water activities

Fishing

☆ Other tourist feature

# Map of the world (political)

GREENLAND
(Den.)

Arctic Circle

Alaska
(U.S.A.)

Nuuk
(Godth b)

Reyk

C A N A D A

Ottawa

U.  S.  A.

Washington
D.C.

Azores
(Port.)

Bermuda
(U.K.)

Tropic of Cancer

Hawaiian Is.
(U.S.A.)

M xico

Havana

Nassau

THE
BAHAMAS

CUBA

JAMAICA   HAITI

DOMINICAN
REPUBLIC

ST. KITTS-NEVIS

ANTIGUA &
BARBUDA

CAPE
VERDE IS.

M E X I C O

Belmopan
BELIZE

Kingston

Puerto
Rico
(U.S.A.)

DOMINICA
ST. LUCIA

GUATEMALA

Guatemala
San Salvador
EL SALVADOR

HONDURAS

Tegucigalpa

ST. VINCENT &
THE GRENADINES

BARBADOS
GRENADA

NICARAGUA

Managua

TRINIDAD AND
TOBAGO

San Jos

Panam

Caracas

COSTA RICA

PANAMA

VENEZUELA

GUYANA

Georgetown

Paramaribo

SURINAME

Cayenne

FRENCH GUIANA (Fr.)

Bogot

COLOMBIA

Equator

Galapagos
Is. (Ec.)

Quito

ECUADOR

B R A Z I L

Tokelau Is.
(N.Z.)

P E R U

American Samoa
(U.S.A.)

SAMOA

Lima

Bras lia

French Polynesia
(Fr.)

La Paz

BOLIVIA

Niue   Cook Is.
(N.Z.)   (N.Z.)

Tropic of Capricorn

TONGA

PARAGUAY

Pitcairn Is.
(U.K.)

Asunci n

C H I L E

A R G E N T I N A

URUGUAY

Santiago

Buenos
Aires

Montevideo

Falkland Is. (U.K.)

Stanley

South Georgia
(U.K.)

# The continents and oceans

North
America

Europe

Asia

PACIFIC

OCEAN

NORTH

ATLANTIC

OCEAN

PACIFIC

OCEAN

Africa

INDIAN

OCEAN

South
America

SOUTH

ATLANTIC

OCEAN

Oceania

SOUTHERN OCEAN

Antarctica

# Glossary

*cross reference

## A

**abrasion** the scratching and scraping of a river bed and banks by the stones and sand in the river

**alluvium** all deposits laid down by rivers, especially in times of flood

**antecedent rainfall** the amount of moisture already in the ground before a rainstorm

**artesian water** water that rises out of the ground out of natural pressure

**asthenosphere** Part of the Earth's *mantle. It is a hot, semi-molten layer that lies beneath the *tectonic plates. It is between 20 and 70km thick

**attrition** the wearing away of particles of debris by the action of other particles such as river or beach pebbles

## B

**bankful** this is the discharge or contents of the river which is just contained within its banks. This is when the speed, or velocity, of the river is at its greatest

**bar** an accumulation of sediment below the water which may be exposed at low tide. It is caused by heavy wave action along a gently sloping coastline. The crest of a bar normally runs parallel to the coast and can extend across a bay or estuary

**basalt** a dark coloured volcanic rock. Molten basalt spreads rapidly and is widespread. About 70% of the Earth's surface is covered in basalt *lava flows

**base flow** the usual reliable level of a river

**biodiversity** the varied range of plants and animals (flora and fauna) found in an area

**biofuels** these are any kind of fuel made from living things, or from the waste they produce

**biogas** a gas produced by the breakdown of organic matter, such as manure or sewage, in the absence of oxygen. It can be used as a biofuel

**biome** a very large ecosystem. The rainforests are one biome. Hot deserts are another

**bioproductive area** the area of land, sea and air required to provide the goods and services we consume. Our *eco-footprint shows how much of this area we are using

**biosphere** the zone where life is found. It extends 3m below ground to about 30m above ground and up to 200m deep in the oceans

**bleaching** the whitening or fading of the bright coral colours, due to the loss of algae, from the coral system. Bleaching is a sign that the coral reef is undergoing stress

**blue water** water that is stored in dams and then used for domestic purposes

**borehole** a deep hole drilled in the ground, especially to find water or oil

## C

**carbon footprint** a measurement of all greenhouse gases we individually produce

**carrying capacity** the maximum number of people (or plants or animals) who can be supported in a given area

**closed system** where there are clear boundaries with no movement of energy across them. The *hydrological system is a closed system. The water goes round and round

**community forests** areas in England where there is a programme of environmental improvement which involves planting and protecting trees and other natural habitats. The idea is to improve the quality of life of people living there

**compact community** a community where the best use of space is made

**conflict matrix** this is a technique for identifying possible areas of disagreement by putting different interest groups into a table or matrix. It is also known as a 'clash table'

**congestion charge** a fee for motorists travelling within a city. The main aims are to reduce traffic congestion, and to raise funds for investment in the city's transport system. London's congestion charge was one of the first to be introduced

**conservation farming** these farming methods aim to conserve soil and water while at the same time providing a sustainable livelihood for the farmer

**constructive plate boundaries** where two tectonic plates are moving apart and new crust is constructed. Also known as 'plate margins'

**continental crust** the part of the Earth's crust that makes the continents; it's between 25 km and 100 km thick

**convection currents** these transfer heat from one part of a liquid or gas to another. In the Earth's mantle the currents which rise from the earth's core are strong enough to move the tectonic plates on the earth's surface. Convection currents also occur in the atmosphere

**convergence** there are two meanings. a) the coming together of *tectonic plates and b) when air streams flow to meet each other

**cost benefit analysis** looking at all the costs of a project, social and environmental as well as financial, and deciding whether it is worth going ahead

**counter-urbanisation** when people leave towns and cities to live in the countryside

**cycle of poverty** a vicious spiral of poverty and deprivation passing from one generation to the next

## D

**deprived areas** an area, usually in a developed nation, where there is high unemployment and crime and poor health and education services and housing

**development** the use of resources to improve the standard of living of a nation

**diguette** a line of stones which are laid along the contours of gently sloping farmland to catch rain water and reduce soil erosion

**discordant coast** a coast which alternates between bands of hard rocks and soft rocks. Discordant coastlines will have alternating headlands and bays

## E

**eco-footprint** your eco-footprint is the area of land and sea that supplies all of the 'stuff' that you need to live – land for your home, your food, the energy you use plus all the materials you buy. Your eco-footprint is about 6 football pitches. If everyone lived like us we would need several Earths to sustain us!

**ecosystem** a unit made up of living things and their non-living environment. For example a pond, a forest, a desert

**El Niño** this occurs in the Pacific Ocean every three to seven years. Unusually warm ocean conditions off the western coasts of Ecuador and Peru cause climatic disturbances. El Nino can affect climates around the world for more than a year

**energy** a source of power

**environmental refugee** a displaced person caused by environmental disasters as a result of climate change

**environmental sustainability** what we need to consume in order to protect the Earth's environment. We need to reduce our demands on the planet to a level where future generations will not suffer

**epicentre** the point on the ground directly above the focus (centre) of an earthquake

**eutrophication** the process by which ecosystems, usually lakes, become more fertile as fertilizers and sewage flow in. The resultant loss of oxygen in the water kills off all species that need oxygen to survive, such as fish

**evacuate** when people move from a place of danger to a safer place

**evaporation** the changing of a liquid into a vapour or gas. Some rainfall is evaporated into water vapour by the heat of the sun

## F

**fault** a fractured surface in the Earth's crust along which rocks have travelled relative to each other

**fell running** the sport of running and racing, off road, over upland country with steep climbs and descents. It is called after the fells of northern Britain especially those in the Lake District

**flood plain** flat land around a river that gets flooded when the river overflows

**focus** the point of origin of an earthquake

**food chain** a chain of names linked by arrows, showing what species feed on. It always starts with a plant

**friction** the force which resists the movement of one surface over another

## G

**genetic modification** any alteration of genetic material (DNA or RNA) of an organism by means that could not occur naturally. It is also called 'genetic engineering'

**glacial** a long period of time during which the Earth's glaciers expanded widely

**global warming** the way temperatures around the world are rising. Scientists think we have made this happen by burning too much fossil fuel

**goods** things that are of value to us

**greenhouse gases** gases like carbon dioxide and methane that trap heat around the Earth, leading to global warming

**Gross Domestic Product (GDP)** the total value of goods and services in a nation measured over a year

## H

**HDI** a standard means of measuring human development

**helical flow** a continuous corkscrew motion of water as it flows along a river channel

**holistic management** in coastal management this means looking at the requirements of a long stretch of coast when planning, rather than just a single beach or short stretch of coastline. This is also known as 'integrated coastal zone management' (ICZM)

**hot spot** there are two meanings a) where volcanoes occur away from plate margins; probably due to strong upward currents in the mantle b) areas with rich animal and plant life

**hunting and gathering** a form of society with no settled agriculture or domestication of animals, which has little impact on the environment

**hydraulic action** the force of the water within a stream or river

**hydrological cycle** the movement of water between its different forms; gas (water vapour) , liquid and solid (ice) forms. It is also known as the water cycle

**hydrosphere** all the water on, or close to, the surface of the earth. 97% is in the seas and oceans

## I

**impermeable** doesn't let water through

**industrial** societies where the main industries are in the manufacturing sector

**industrialization** where a mainly agricultural society develops and begins to depend on manufacturing industries

**infiltration** the soaking of rainwater into the ground

**integrated transport policy** a government policy aimed at improving and integrating public transport systems, and of making cars and lorries more environmentally acceptable and more efficient

**interception** the capture of rainwater by leaves. Some *evaporates again and the rest trickles to the ground

**interglacials** a long period of warmer conditions between *glacials

**interlocking spurs** hills that stick out on alternate sides of a V-shaped valley like the teeth of a zip

**internal migration** the movement of people within a country, in search of seasonal or permanent work or for social reasons

## L

**lagoon** a bay totally or partially enclosed by a *spit or reef running across its entrance

**landfill** an area of ground where large amounts of waste material are buried under the earth

**landslide** a rapid *mass movement of rock fragments and soil under the influence of gravity

**lava flows** *lava flows at different speeds depending on what it is made of. Lava flows are normally very slow and not hazardous but, when mixed with water, lava can flow very fast and be dangerous

**lava** melted rock that erupts from a volcano

**longshore drift** the movement of sand and shingle along the coast

**low income** the World Bank's main criterion for classifying economies is gross national income (GNI) per capita. Based on its GNI per capita which is adjusted every year, every economy is classified as low income, middle income (subdivided into lower middle and upper middle), or high income

**M**

**magma** melted rock below the Earth's surface. When it reaches the surface it is called lava

**magnetosphere** this is a highly magnetized region (a 'force field') around and possessed by an astronomical object. The Earth is surrounded by a magnetosphere as are some other planets in the solar system like Mercury, Jupiter and Neptune

**magnitude** of an earthquake, an expression of the total energy released

**mantle** the middle layer of the earth. It lies between the *crust and the *core and is about 2900km thick. Its outer layer is the *asthenosphere. Below the asthenosphere it consists mainly of solid rock

**marine hotspots** areas rich in marine species. These are small areas and highly vulnerable to extinction

**marram grass** grasses that are found on coastal sand dunes. They have extensive systems of creeping underground stems which allow them to survive under conditions of shifting sands and high winds

**mass movement** the movement downslope of rock fragments and soil under the influence of gravity. A *landslide is a rapid mass movement

**meander** a bend in a river

**mega-city** a many centred, multi-city urban area of more than 10 million people. A mega-city is sometimes formed from several cities merging together

**middle course** the journey of a river from its source in hills or mountains to mouth is sometimes called the course of the river. The course of a river can be divided into three main sections a) *upper course b) middle course and c) lower course

**Milankovitch cycles** the three long-term cycles in the Earth's orbit around the sun. Milankovitch's theory is that *glacials happen when the three cycles match up in a certain way

**multi-cropping** the practice of producing two or more crops at the same time on the same parcel of land during a 12-month period

**N**

**national park** an area that is protected from human exploitation and occupation. Most developed countries have national parks

**natural increase** the birth rate minus the death rate for a place. It is normally given as a % of the total population

**Newly Industrialised Countries (NICs)** countries which were recently less developed but where *industrialization has happened quickly

**NGO – non-governmental organisation** NGOs work to make life better, especially for the poor. Oxfam, the Red Cross and Greenpeace are all NGOs

**nomads** people who move with their animals from place to place in search of pasture

**O**

**oceanic crust** the part of the Earth's crust which is under the oceans; it's made of basalt and is between 5 km and 10 km thick

**outsourcing** where part or all of a project, such as the design or the manufacturing, is handed over to a third-party company. This is normally done to save time or costs or both

**overhang** the part of a cliff above the reach of waves which is unaffected by wave action

**ox-bow lake** a lake formed when a loop in a river is cut off by floods

**P**

**Pangea** a supercontinent consisting of the whole land area of the globe before it was split up by continental drift

**percolate** to move gradually through a surface that has very small holes or spaces in it

**permeable** letting water pass through

**plate boundaries** where *tectonic plates meet. There are three kinds of boundary a) *constructive – when two plates move apart b) destructive – when two plates collide c) conservative – when two plates slide past one another

**plumes** upwellings of molten rock through the *asthenosphere to the lithosphere, forming a *hot spot

**point bar** the accumulation of river sediment on the inside of a meander

**population balance** where births and deaths are almost equal

**population decline** where the number of live births is less than the number of deaths

**population increase** where the number of live births exceeds deaths

**post-industrial** societies where service and high technology industries are dominant and where heavy manufacturing industries are less important

**predict** saying that something will happen in the future. A scientific prediction is based on statistical evidence

**pre-industrial** the situation before the industrial revolution. It can be used to describe poor countries which are mainly agricultural

**primary industry** where people extract raw materials from the land or sea. For example farming, fishing and mining

**Q**

**quaternary industry** where people are employed in industries providing information and expert help. For example IT consultants and researchers

**Quaternary** the last 2.6 million years, during which there have been many *glacials

**R**

**Ramsar sites** wetlands of international importance. They are named after The Convention on Wetlands, signed in Ramsar, Iran, in 1971

**recurved** hooked

**remittance payments** money sent home to their families by people working overseas. Remittance payments are an important source of income for some less developed countries

**re-urbanisation** when people who used to live in the city and then moved out to the country or to a suburb, move back to live in the city

**river basin system** a river basin system is the area of land drained by a river and its tributaries. It is also known as a drainage basin

**rock outcrop** a large mass of rock that stands above the surface of the ground

**rural idyll** when people move to the countryside because they think it will offer them a better quality of life and lower crime, particularly when they have children

**rural-urban migration** the movement of people from the countryside to the cities, normally to escape from poverty and to search for work

**S**

**salt marsh** salt tolerant vegetation growing on mud flats in bays or estuaries. These plants trap sediments which gradually raise the height of the marsh. Eventually it becomes part of the coast land

**sand dune** a hill or ridge of sand accumulated and sorted by wind action

**saturated** soil is saturated when the water table has come to the surface. The water then flows overland

**scree** shattered rock fragments which gather below free rock faces and summits. It is also used to refer to the slope below a rock face which is made up of these fragments

**secondary industry** where people make, or manufacture, things. For example turning iron ore into steel, making cars and building houses

**seismometer** a machine for recording and measuring an earthquake

**services** things that satisfy our needs

**Shoreline Management Plan (SMP)** this is an approach which builds on knowledge of the coastal environment and takes account of the wide range of public interest to avoid piecemeal attempts to protect one area at the expense of another

**siltation** to become filled with silt

**sink** a container, a pool or a pit into which waste goes

**Smart car** an urban car designed to be small enough to park 'sideways on'.. An electric, rechargeable version is being released in the UK

**socio-economic** a term used to classify people depending on what they do for a living

**soil creep** the slow (sometimes very slow) gradual movement downslope of soil, *scree or glacier ice

**source** a place, person or thing that you get something from

**spatial** relating to space and the position, size and shape of things in it

**sphere of influence** area around a settlement (or shop, or other service) where its effect is felt. London has a very large sphere of influence

**spit** a ridge of sand running away from the coast, usually with a curved seaward end

**stratovolcano** a cone shaped volcano formed from layers of different kinds of lava

**sub-aerial** processes occurring on land, at the Earth's surface, as opposed to underwater or underground

**subduction** the transformation into *magma of a denser *tectonic plate as it dives under another less dense plate

**subsistence farming** where farmers grow food to feed their families, rather than to sell

**suburbanisation** when people leave cities and towns to live in suburbs

**surface run-off** rainwater that runs across the surface of the ground and drains into the river

**sustainable management** meeting the needs of people now and in the future, and limiting harm to the environment

**T**

**tectonic plate** the Earth's surface is broken into large pieces, like a cracked eggshell. The pieces are called tectonic plates, or just plates

**tertiary industry** where people are employed in providing a service. For example the health service (doctors, nurses, dentists) and education (teachers)

**thalweg** the line of the fastest flow along the course of a river

**thermal expansion** expansion as a result of heating. When sea water warms up, it expands

**through-flow** the flow of rainwater sideways through the soil, towards the river

**tombolo** a *spit, resulting from *longshore drift which joins an offshore island to the mainland

**transnational companies** these are companies which operate across more than one country. They are similar to multinational companies but may operate in only two countries

**transpire** when plants lose water vapour, mainly through pores in their leaves

**U**

**unsustainable use** ways of catching wild fish that threaten the fish stock itself by overfishing, or threaten the environment the fish need to thrive

**upper course** the journey of a river from its source in hills or mountains to mouth is sometimes called the course of the river. The course of a river can be divided into three main sections a) upper course b)*middle course c) lower course. In the upper course the water flows slowly, winding around *interlocking spurs

**urban heat island** where temperatures in cities are much higher (up to 4°C) than the surrounding countryside because of heat released from buildings including factories and offices and from air pollution

**W**

**water table** the level below which the ground is *saturated

**water wars** conflicts between countries for access to and control over water. Global warming and population increase increases the chances of this happening

**wave- cut notch** an indentation cut into a sea cliff at water level by wave action

**wave-cut platform** the flat rocky area left behind when waves erode a cliff away

**whole ecosystem approach** an approach that aims at protecting the whole marine ecosystem and that acts first before the damage is done

**windpump** a type of windmill used for pumping water from a well, or for draining land

# Index